Autodesk Fusion 360:
A Tutorial Approach

CADCIM Technologies
525 St. Andrews Drive
Schererville, IN 46375, USA
(www.cadcim.com)

Contributing Author
Sham Tickoo

Professor
Purdue University Northwest
Department of Mechanical Engineering Technology
Hammond, Indiana, USA

CADCIM Technologies

Autodesk Fusion 360: A Tutorial Approach
Sham Tickoo

CADCIM Technologies
525 St Andrews Drive
Schererville, Indiana 46375, USA
www.cadcim.com

ISBN 978-1-936646-66-1

NOTICE TO THE READER

Publisher does not warrant or guarantee any of the products described in the text or perform any independent analysis in connection with any of the product information contained in the text. Publisher does not assume, and expressly disclaims,any obligation to obtain and include information other than that provided to it by the manufacturer.

The reader is expressly warned to consider and adopt all safety precautions that might be indicated by the activities herein and to avoid all potential hazards. By following the instructions contained herein, the reader willingly assumes all risks in connection with such instructions.

The Publisher makes no representation or warranties of any kind, including but not limited to, the warranties of fitness for particular purpose or merchantability, nor are any such representations implied with respect to the material set forth herein, and the publisher takes no responsibility with respect to such material. The publisher shall not be liable for any special, consequential, or exemplary damages resulting, in whole or part, from the reader's use of, or reliance upon, this material.

www.cadcim.com

DEDICATION

To teachers, who make it possible to disseminate knowledge
to enlighten the young and curious minds
of our future generations

To students, who are dedicated to learning new technologies
and making the world a better place to live in

THANKS

To the faculty and students of the MET department of
Purdue University Northwest for their cooperation

To employees of CADCIM Technologies for their valuable help

Online Training Program Offered by CADCIM Technologies

CADCIM Technologies provides effective and affordable virtual online training on various software packages including Computer Aided Design, Manufacturing, and Engineering (CAD/CAM/CAE), computer programming languages, animation, architecture, and GIS. The training is delivered 'live' via Internet at any time, any place, and at any pace to individuals as well as the students of colleges, universities, and CAD/CAM training centers. The main features of this program are:

Training for Students and Companies in a Classroom Setting

Highly experienced instructors and qualified engineers at CADCIM Technologies conduct the classes under the guidance of Prof. Sham Tickoo of Purdue University Northwest, USA. This team has authored several textbooks that are rated "one of the best" in their categories and are used in various colleges, universities, and training centers in North America, Europe, and in other parts of the world.

Training for Individuals

CADCIM Technologies with its cost effective and time saving initiative strives to deliver the training in the comfort of your home or work place, thereby relieving you from the hassles of traveling to training centers.

Training Offered on Software Packages

CADCIM provides basic and advanced training on the following software packages:

CAD/CAM/CAE: *CATIA, Pro/ENGINEER Wildfire, Creo Parametric, Creo Direct, SOLIDWORKS, Autodesk Inventor, Solid Edge, NX, AutoCAD, AutoCAD LT, AutoCAD Plant 3D, Customizing AutoCAD, EdgeCAM, and ANSYS*

Architecture and GIS: *Autodesk Revit (Architecture/Structure/MEP), AutoCAD Civil 3D, AutoCAD Map 3D, Navisworks, Primavera Project Planner, and Bentley STAAD Pro*

Animation and Styling: *Autodesk 3ds Max, Autodesk 3ds Max Design, Autodesk Maya, Autodesk Alias, Foundry NukeX, and MAXON CINEMA 4D*

Computer Programming: *C++, VB.NET, Oracle, AJAX, and Java*

For more information, please visit the following link: ***http://www.cadcim.com***

Note
If you are a faculty member, you can register by clicking on the following link to access the teaching resources: ***http://www.cadcim.com/Registration.aspx***. The student resources are available at ***http://www.cadcim.com***. We also provide **Live Virtual Online Training** on various software packages. For more information, write us at ***sales@cadcim.com***.

Table of Contents

This page is intentionally left blank

Preface

Autodesk Fusion 360

Autodesk Fusion 360, developed by Autodesk Inc., is one of the world's fastest growing cloud based software. It is a parametric feature-based solid modeling tool that not only unites the 3D parametric features with 2D tools but also addresses every design-through-manufacturing process. The adaptive technology of this solid modeling tool allows you to handle extremely large assemblies with tremendous ease.

This solid modeling tool allows you to easily import the AutoCAD, AutoCAD Mechanical, and other related CAD files with an amazing compatibility.

The drawing views that can be generated using this tool include orthographic view, isometric view, section view, detailed view, and so on. You can use predefined drawing standard files for generating the drawing views. Fusion 360 drawing views remain associative to the designs they are created from, implying that the changes made in the design in the model workspace of Autodesk Fusion 360 automatically reflect in drawing as well.

Autodesk Fusion 360: A Tutorial Approach textbook is written with the intention of helping the readers effectively use the Autodesk Fusion 360 solid modeling tool. The mechanical engineering industry examples and tutorials are used in this textbook to ensure that the users can relate the knowledge of this book with the actual mechanical industry designs. The salient features of this textbook are as follows:

- **Tutorial Approach**

  The author has adopted the tutorial point-of-view and the learn-by-doing approach throughout the textbook. This approach guides the users through the process of creating the models in the tutorials.

- **Real-World Projects as Tutorials**

  The author has used about 40 real-world mechanical engineering projects as tutorials in this book. This enables the readers to relate the tutorials to the models in the mechanical engineering industry. In addition, there are about 20 exercises that are also based on the real-world mechanical engineering projects.

- **Tips and Notes**

  The additional information related to various topics is provided to the users in the form of tips and notes.

- **Learning Objectives**

  The first page of every chapter summarizes the topics that are covered in that chapter.

- **Self-Evaluation Test, Review Questions, and Exercises**

  Every chapter ends with a Self-Evaluation Test so that the users can assess their knowledge of the chapter. The answers to the Self-Evaluation test are given at the end of the chapter. Also, the Review Questions and Exercises are given at the end of each chapter and they can be used by the Instructors as test questions and exercises.

- **Software Version**

  As a cloud-based platform, updates are frequently available for the Autodesk Fusion 360 software.This text book has been using software version 2.0.3803. If you are using a version of the software later then version 2.0.3803, you might some variances between images and workflows in this student guide and software that you are using.

Symbols Used in the Textbook

Note

The author has provided additional information to the users about the topic being discussed in the form of notes.

Tip

Special information and techniques are provided in the form of tips that helps in increasing the efficiency of the users.

Formatting Conventions Used in the Textbook

Please refer to the following list for the formatting conventions used in this textbook.

- Names of tools, buttons, options, panels, and Ribbon are written in boldface.

  Example: The **Extrude** tool, the **OK** button, the **Modify** panel, and so on.

- Names of dialog boxes, drop-down lists, list boxes, areas, edit boxes, and check boxes, are written in boldface.

  Example: The **REVOLVE** dialog box, the Pattern Type drop-down list in the **RECTANGULAR** dialog box, the **Distance** edit box of the **EXTRUDE** dialog box, the **Tangent Chain** check box in the **SHELL** dialog box, and so on.

- Values entered in edit boxes are written in boldface.

  Example: Enter **5** in the **Distance** edit box.

- The methods of invoking a tool/option from the **Ribbon**, **Quick Access Toolbar**, **Application Menu** are enclosed in a shaded box.

  Ribbon: SKETCH Panel > Line
  Marking Menu: Sketch > Line

Naming Conventions Used in the Textbook

Tool

If you click on an item in a toolbar or a panel of the **Ribbon** and a command is invoked to create/edit an object or perform some action, then that item is termed as **tool**, refer to refer to Figure 1.

For example:
To Create: **Line** tool, **Dimension** tool, **Extrude** tool
To Edit: **Fillet** tool, **Draft** tool, **Chamfer** tool

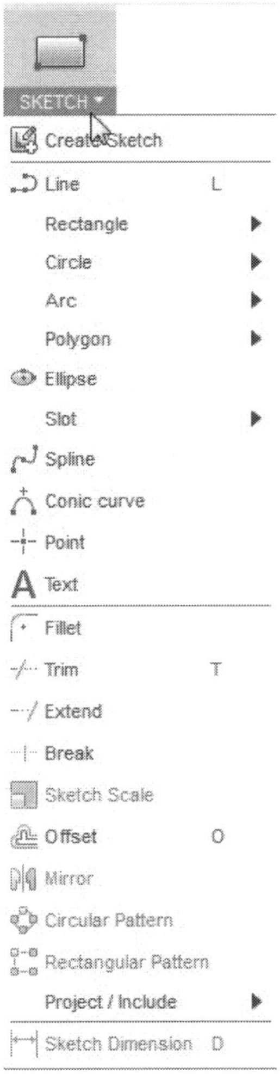

Figure 1 *Various tools in the **SKETCH** panel of the Ribbon*

Button

The item in a dialog box that has a 3d shape like a button is termed as **Button**. For example, **OK** button, **Cancel** button, **Apply** button, and so on.

Dialog Box

In this textbook, different terms are used for referring to the components of a dialog box. Refer to Figure 2 for the terminology used.

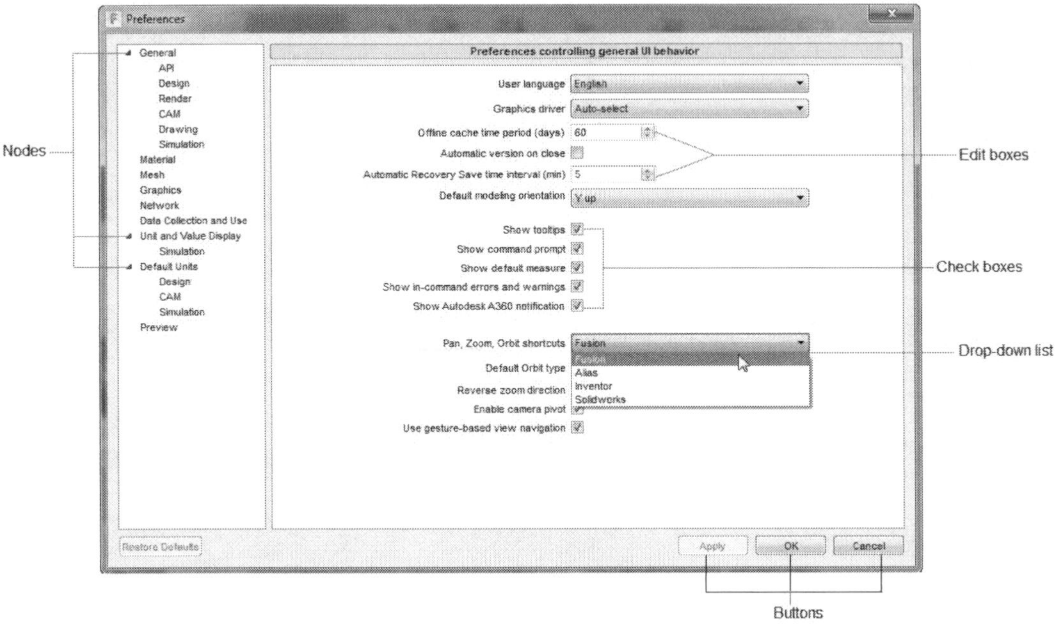

Figure 2 The components in a dialog box

Drop-down List

A drop-down list is the one in which a set of options are grouped together. You can set various parameters using these options. You can identify a drop-down list with a down arrow on it. For example, **Type** drop-down list, **Orientation** drop-down list, and so on; refer to Figure 3.

*Figure 3 Partial view of the **Type** and **Operation** drop-down lists*

Options

Options are the items that are available in shortcut menu, Marking Menu, drop-down list, dialog boxes, and so on. For example, choose the **Edit View** option from the Marking Menu displayed on right-clicking in the drawing area; select the **Distance** option from the **Extent** drop-down list; choose the **New Body** option from the **Orientation** area, refer to Figure 5.

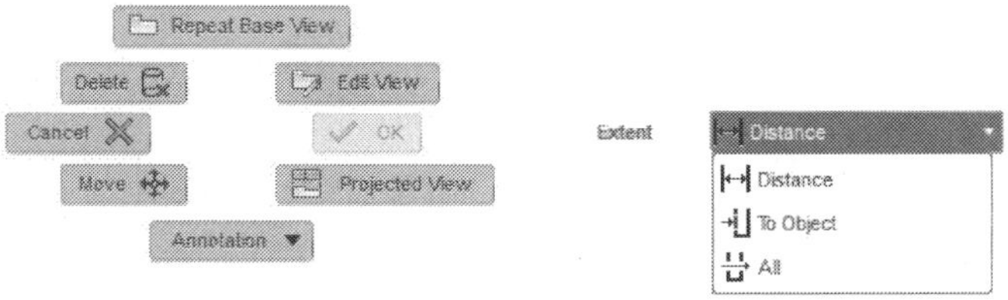

***Figure 4** Options in the Marking menu and the **Extent** drop-down list*

Free Companion Website

It has been our constant endeavor to provide you the best textbooks and services at affordable prices. In this endeavor, we have come out with a Free Companion website that will facilitate the process of teaching and learning of Autodesk Fusion 360. If you purchase this textbook, you will get access to the files on the Companion website.

The following resources are available for the faculty and students in this website:

Faculty Resources

- **Technical Support**
  You can get online technical support by contacting ***techsupport@cadcim.com***.

- **Instructor Guide**
  Solutions to all review questions in the textbook are provided in this guide to help the faculty members test the skills of the students.

- **Part Files**
  The part files used in illustrations, tutorials, and exercises are available for free download.

Student Resources

- **Technical Support**
  You can get online technical support by contacting ***techsupport@cadcim.com***.

- **Part Files**
  The part files used in illustrations and tutorials are available for free download.

If you face any problem in accessing these files, please contact the publisher at ***sales@cadcim.com*** or the author at ***stickoo@pnw.edu*** or ***tickoo525@gmail.com***.

Stay Connected

You can now stay connected with us through Facebook and Twitter to get the latest information about our textbooks, videos, and teaching/learning resources. To stay informed of such updates, follow us on Facebook (***www.facebook.com/cadcim***) and Twitter (**@cadcimtech**). You can also subscribe to our YouTube channel (***www.youtube.com/cadcimtech***) to get the information about our latest video tutorials.

Chapter 1

Introduction

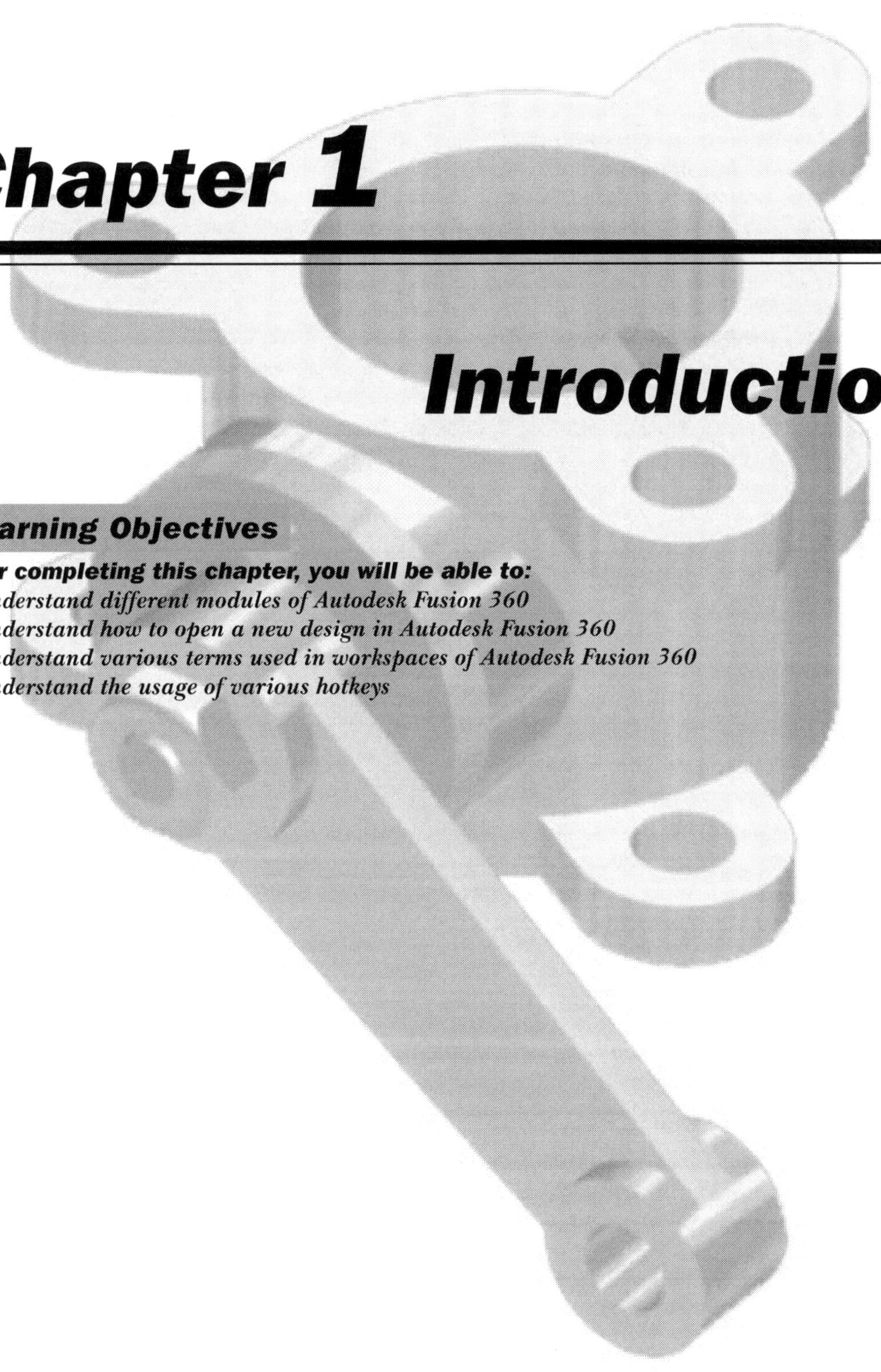

Learning Objectives

After completing this chapter, you will be able to:

- *Understand different modules of Autodesk Fusion 360*
- *Understand how to open a new design in Autodesk Fusion 360*
- *Understand various terms used in workspaces of Autodesk Fusion 360*
- *Understand the usage of various hotkeys*

INTRODUCTION TO Autodesk Fusion 360

Autodesk Fusion 360 is a cloud-based CAD/CAM/CAE tool which is designed for a collaborative product development. It allows you to make HYBRID model that is manufacturable. You can communicate your designs with renderings and drawings. In addition, you can use its CAM capabilities to generate a toolpath for CNC machining and also you can send your design to a 3D printer for rapid prototyping. Your data is kept safe and secure on the cloud, with unlimited storage and access. It connects your entire product development process in a single cloud-based platform that works on both Mac and PC, with unlimited installs, and users invited by you to your Fusion 360 projects.

Autodesk Fusion 360 is a parametric and feature-based solid modeling tool. It allows you to convert basic two-dimensional (2D) sketch into a solid model using very simple but highly effective modeling options. This solid modeling tool does not restrict its capabilities to the 3D solid output but also extends them to the bidirectional associative drafting. This means that you only need to create the solid model. Its documentation, in the form of the drawing views, is easily done by this software. You just need to specify the required view. This solid modeling tool can be specially used at places where the concept of **"collaborative engineering"** is brought into use. Collaborative engineering is a concept that allows more than one user to work on the same design at the same time. This solid modeling package allows more than one user to work simultaneously on the same design.

As a product of Autodesk, this software package allows you to directly open the drawings of the other Autodesk software like AutoCAD, Mechanical Desktop, AutoCAD LT, Inventor, and so on. This interface is not bounded to the Autodesk software only. You can easily import and export the drawings from this software package to any other software package and vice versa.

The Autodesk Fusion 360 software combines the related tools and functions into groups called workspace. This control the commands that are available and the type of data that is created. There are multiple workspace that are available in Autodesk Fusion 360 namely Model, Sheet Metal, Patch, Render, Animation, Simulation, CAM, Drawing.

SYSTEM REQUIREMENTS

The system requirements to ensure the smooth functioning of Autodesk Fusion 360 on your system are as follows:

- Microsoft Windows 10, Windows 8.1 or Windows 8 (64 bit only) or Windows 7 (SP1 required).
- Intel or AMD Processor with SSE2 support.
- 2 GB RAM minimum (4 GB recommended).
- Hard disk space 5 GB minimum (10 GB recommended).
- A certified graphics card and driver.
- Microsoft Office 2007 or later.
- Adobe Acrobat higher than 8.0.7.
- DVD drive and Mouse or any other compatible pointing device.
- Internet Explorer version 8 or higher.
- Good Internet connection

GETTING STARTED WITH Autodesk Fusion 360

On installing Autodesk Fusion 360 on your system; a shortcut icon of Autodesk Fusion 360 will automatically be created on the desktop. Double-click on this icon to start Autodesk Fusion 360.

When Autodesk Fusion 360 is started for the first time, the system prepares itself by loading all the required files and the initial interface of Autodesk Fusion 360 will be displayed, as shown in Figure 1-1.

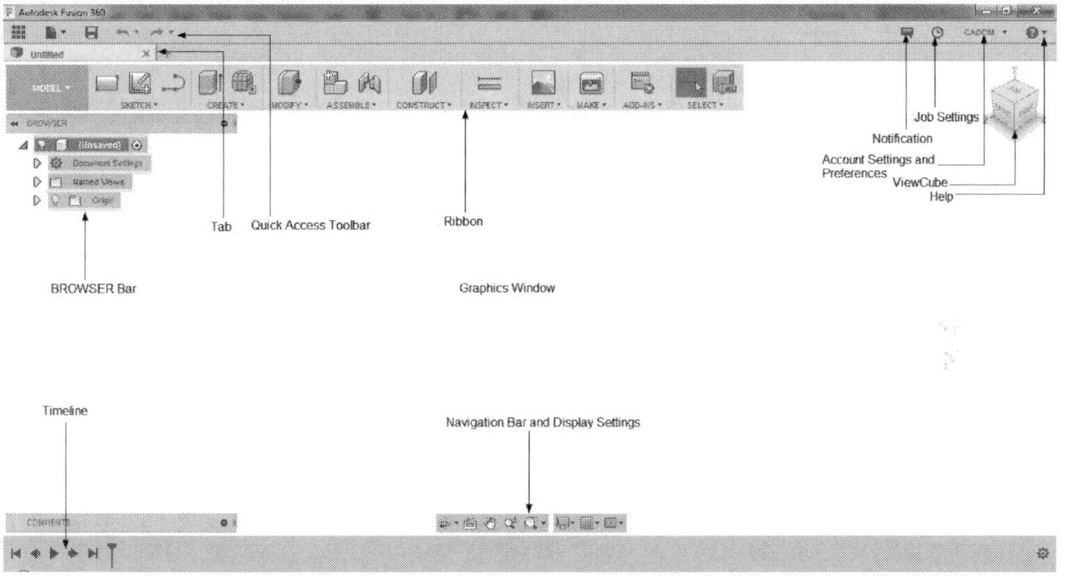

Figure 1-1 *Initial interface of the Autodesk Fusion 360*

Here, you will learn the basics of Sketching, Part modeling, Assembly, Animation, Sculpting, and Drawing through related tutorials.

By using the tools available in the Autodesk Fusion 360, you can view the information related to starting a Autodesk Fusion 360, new design, opening an existing design, setting a project, and so on. The **Help** tool at the extreme right corner of the window enables you to access resources such as Community Forum, Gallery, Product Roadmap. You can also clear user cache data, submit an idea, learn about Fusion 360, and so on.

To start a new design, choose the **New Design** option from the **File** menu in the **Quick Access** toolbar; the new tab gets added to the right side of the previously created tab. This tab is used to create a new design of Autodesk Fusion 360.

It is evident from Figure 1-1 that the interface of Autodesk Fusion 360 is quite user-friendly. Apart from the options shown in Figure 1-1, you are also provided with various shortcut menus which are displayed when you right-click in the graphics window. The type of the shortcut menu and its options depend on where or when you invoke the menu. For example, when a command is active, the options displayed in the shortcut menu will be different from the options displayed when the command is exited.

Quick Access Toolbar

This toolbar is common to all design workspaces of Autodesk Fusion 360 and is located at the top of the Fusion 360 window, as shown in Figure 1-2. The tools available in this toolbar are discussed next.

***Figure 1-2** The Quick Access Toolbar*

Select Panel

While creating a sketch, you often need to select some elements. The tools that are required to make these selection are available in the **Select** panel, as shown in Figure 1-3. Some of these tools are **Window Selection**, **Freeform Selection**, **Paint Selection**, and so on. By default, the **Window Selection** tool is chosen. This tool is used to select entities by creating a selection trap. A trap is a rectangular box drawn by clicking and dragging the mouse to define the diagonally opposite corners. All the objects that lie completely inside this selection trap gets selected. The **Freeform Selection** tool is used to select the entities by clicking and dragging the mouse to draw a freeform shape around them. The **Paint Selection** tool is used to select the target by dragging the mouse over the target object. The tools in the **Selection Tools** flyout are used to select the model on the basis of name, boundary and size. You can also set the selection priority by selecting the tools from the **Selection Priority** flyout. The **Select Body Priority** tool is chosen to set the selection priority for bodies. The **Select Edges Priority** tool is chosen to set the priority for edges and the Select Face Priority for faces. You can also filter the selection of check boxes with the help of the **Selection Filter** flyout.

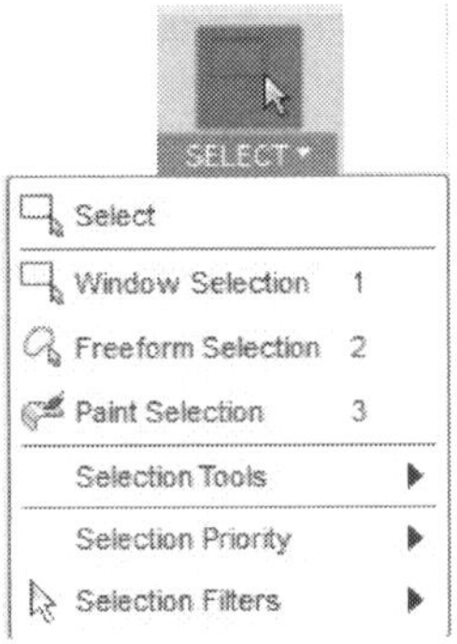

***Figure 1-3 The** SELECT panel*

Data Panel

The Data Panel located on the left side of the window. The **Show/Hide Data Panel** button is used to display or hide the data panel which is displayed on the left of the interface. This panel will be displayed on the left side and display the project folders, refer to Figure 1-4. All files are securely stored in the cloud by default. The Data Panel represents a new way of working with your documents, saving them, uploading the files and collaborate with other users on your designs. Also, you can add members to a project and import the data from other sources. The options available in the Data Panel are discussed next.

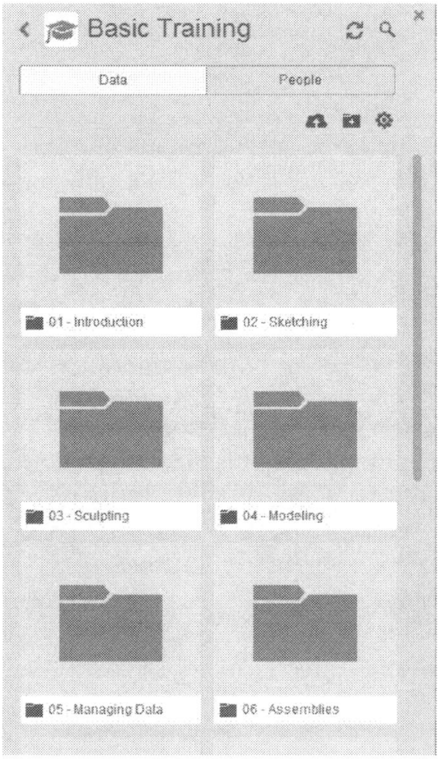

Figure 1-4 *The Data Panel*

Refresh: By clicking on the **Refresh** button, you can refresh the data in the Data Panel.

Search: This option is used to search the desired documents.

Data: This option displays the list of folders and documents, if activated.

People: This option when activated enables the users to invite enables to invite people to a project, and displays a list of people working on a project.

Upload: This option uploads the selected file or files to an active project.

New Folder: Click on this option to create a new folder.

Options: This option is used to choose how to sort and list data.

File Menu

File menu contains options relating to the handling of files, such as open, save, print, export, and so on.

Save

Choose this option to save the design.

Ribbon

There is no command prompt in Autodesk Fusion 360 and the complete designing process is carried out by invoking the tools from the Ribbon. The Ribbon is a long bar available below the **Quick Access** toolbar, as shown in Figure 1-5.

Figure 1-5 *The Ribbon*

Account Settings and Preference

In Autodesk Fusion 360, you can control your profile settings as well as set preferences for default settings. To do so, select the **Preferences** option from the User's profile drop-down list in the upper right corner; the **Preferences** dialog box will be displayed, as shown in Figure 1-6. The options available in this dialog box are discussed next.

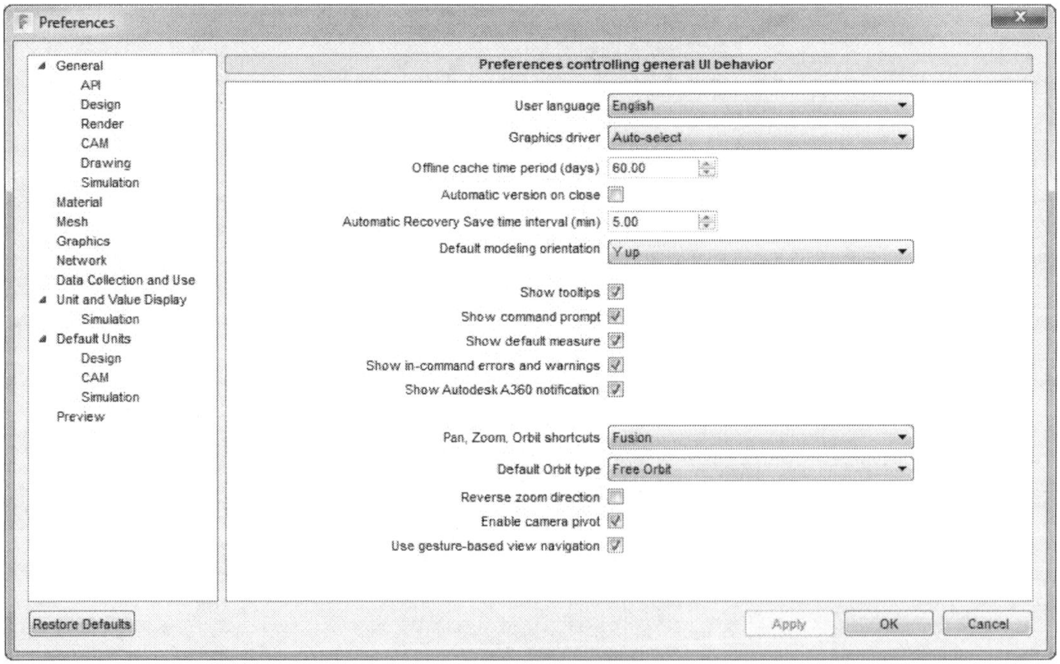

Figure 1-6 *The **Preferences** dialog box*

General

The options in this node are discussed next.

API: The options in this sub-node are used for setting preferences for general UI behavior.

Design: The options available in this sub-bode are used to set preferences for Model, Sculpt, and Patch workspaces.

Render: The options in this sub-node are used to specify default render settings for newly created designs.

CAM: The options in this sub-node are used to configure the settings for the CAM workspace.

Drawing: Use this option in this sub-node to set the preferences of newly created design.

Simulation: The options in this sub-node are used to set the preferences for the simulation of an assembly.

Material

This options in this node are used to set the default physical material and appearance.

Mesh

The options in this node are used to set preferences that control mesh.

Graphics

The options in this node are used to control the graphics display.

Unit and Value Display

The options in this node are used to set the precision and display of units.

WORKSPACE

Autodesk Fusion 360 serves the basic design tasks by providing different workspaces. This software has different workspaces that are unified into one design package. When you are working on a design, switching workspaces updates the toolbar to show the tools relevant to that workspace, but retains the model as it is shown in the design environment. A workspace is defined as a specified environment consisting of a set of tools that allows the user to perform specific design tasks. The basic workspaces in Autodek Fusion 360 are Model, Patch, Render, Animation, Simulation, CAM, Drawing, and Sculpt. These workspaces are discussed next.

MODEL Workspace

In this workspace you can draw the sketch using the tools in the SKETCH panel. While drawing the sketch, some constraints are automatically applied to it. You can also apply additional constraints and dimensions manually. After drawing the sketch, choose the **STOP SKETCH** option and convert the sketch into a sketch-based feature using the tools from **CREATE** panel. This workspace also provides tools in the MODIFY panel, such as fillets, chamfers, and so on.

Use construction planes to create a planar area when a face is not available. These features are called dress-up features. In this workspace, you can make different construction planes, axes, and points. You can make construction planes such as offset plane, tangent plane, mid plane and so on by choosing the respective option from the **CONSTRUCT** panel of the **MODEL** workspace. You can also assemble the components with the help of different mates and joints available in the ASSEMBLE panel. Additionally, you can insert the images, measure the components, and set the selection priority with the help of others panel available in this workspace. Some of the panels are discussed next. The MODEL workspace with associated panel is shown in Figure 1-7.

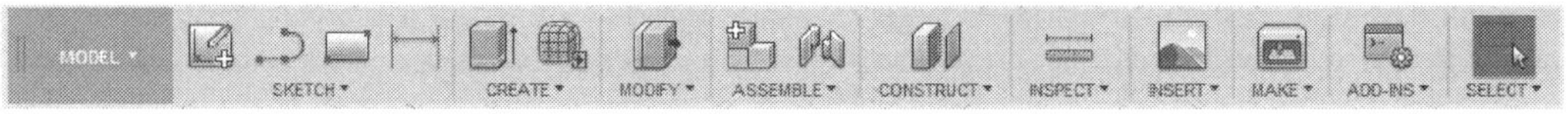

Figure 1-7 *The MODEL workspace*

SKETCH Panel

This is one of the most important panels in the **Ribbon** as the sketches are the foundation for creating any three dimensional object or model. All tools for creating the sketches of the parts are available in this tab. The SKETCH panel is shown in Figure 1-8.

CREATE Panel

Once the sketch is completed, you need to convert it into a feature using the modeling tool available in this panel. This panel provides all the modeling tools that can be used to convert a sketch into a feature. The tools in the CREATE panel are shown in Figure 1-9.

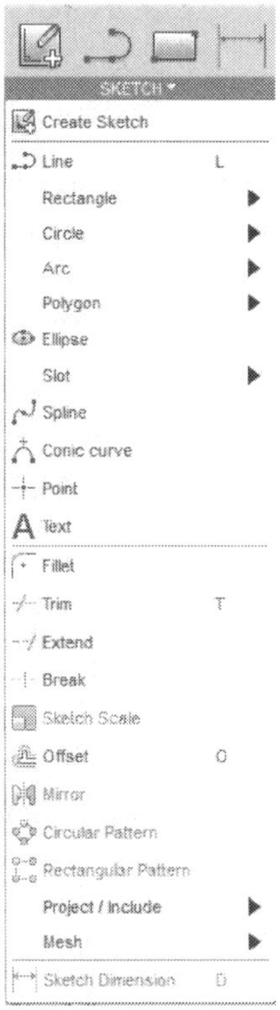

Figure 1-8 *The SKETCH panel*

Figure 1-9 *The CREATE panel*

MODIFY Panel

This panel provides a set of tools to make modification in solids, surfaces and sculpt bodies. The tools in the MODIFY panel are shown in Figure 1-10.

ASSEMBLE Panel

This panel is used to insert a component and apply various types of mates and joints to assemble them together. Mates are the constraints that can be applied to components to restrict their degrees of freedom. You can also move and rotate a component in the assembly, change the hidden states of the assembly and individual components, edit the component of an assembly, and so on. The ASSEMBLE panel is shown in Figure 1-11.

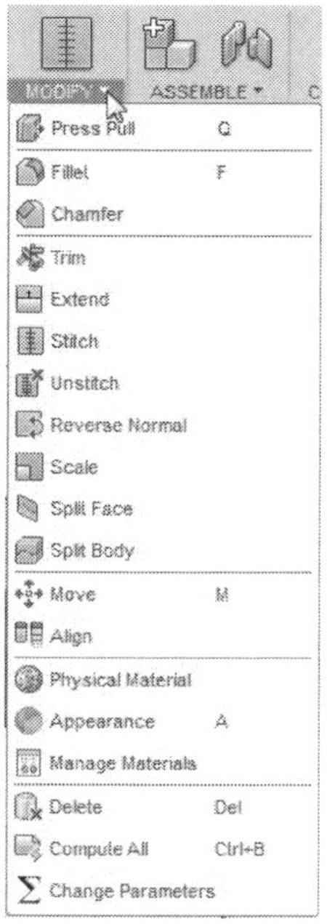

Figure 1-10 *The **MODIFY** panel*   ***Figure 1-11*** *The **ASSEMBLY** panel*

PATCH Workspace

Autodesk Fusion 360 helps you to work with surfaces and surface bodies. A surface is geometry within the software that has no mass properties or thickness, but still allows users to form and create the shape and style of a model. In Autodesk Fusion 360, a surface or surface body is considered a patch. The Patch workspace is used to create and repair these surface bodies for the modeling purpose. The associated panel with this workspace are same as that of the Model workspace, as shown in Figure 1-12.

Figure 1-12 *The **PATCH** Workspace*

RENDER Workspace

This workspace is used to create photo-realistic renderings. The associated panels for the RENDER workspace are shown in Figure 1-13.

*Figure 1-13 The **RENDER** workspace*

SHEET METAL Workspace

This workspace can be invoked by choosing the **SHEET METAL** workspace from the MODEL workspace drop-down list. This workspace provides the tools that are used to create sheet metal parts. The associated panel for the SHEET METAL workspace is shown in Figure 1-14.

*Figure 1-14 The **SHEET METAL** workspace*

ANIMATION Workspace

A major drawback of most solid modeling tools is their limitation in displaying the working of an assembly. You cannot show the working of an assembly in most of the software because they do not have proper tools to display an assembly in motion. As a result, the designers cannot show the working of the assemblies to their clients or they have to take the help of some other software packages such as 3D Studio MAX, 3D Studio VIZ. However, this software package provides a module called the ANIMATION workspace using which you can animate the assemblies created in the Assembly design and view their working. The assemblies can be animated using easy steps. The associated panel for the ANIMATION workspace is shown in Figure 1-15.

*Figure 1-15 The **ANIMATION** workspace*

SIMULATION Workspace

Simulation perform various types of analyses to determine how loads lead to deformation and failure, helping you understand if and how the part will fail. The associated panel for the SIMULATION workspace is shown in Figure 1-16.

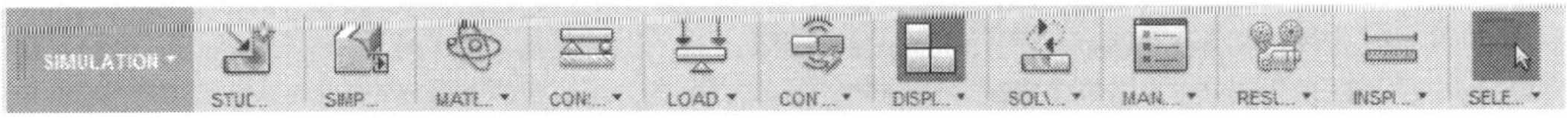

*Figure 1-16 The **SIMULATION** Workspace*

CAM Workspace

This workspace helps you to combine the CNC programming, simulation, and design, with real-time collaboration, online project and data management. In this workspace you will be able to work with almost all the CAD formats. Fusion 360 breaks down the barriers of traditional CAD to CAM workflows by giving you access to professional CNC programming tools. You can easily import files in various formats and apply the CAM capabilities of this workspace on the imported files. This workspace enables you to quickly generate toolpaths for the design to be manufactured to reduce machine and tool wear, and produce the highest quality finished parts. The associated panels for the CAM workspace is shown in Figure 1-17.

Figure 1-17 The CAM workspace

DRAWING Workspace

This workspace is used for the documentation of the parts or assemblies in the form of drawing views. You can also create drawing views of the animation created in the Animation workspace. All parametric dimensions added to the components in the model during the creation of the parts are displayed in the drawing views in this module. The associated panel for the DRAWING workspace is shown in Figure 1-18.

Figure 1-18 The DRAWING workspace

Navigation Bar and Display Settings

The Navigation Bar and Display Settings is located at the bottom of the graphics window and contains tools that are used to navigate the model in order to make the designing process easier and quicker, as shown in Figure 1-19. The navigation tools also help you to control the view and orientation of the components in the graphics window. The display settings control the appearance of the interface.

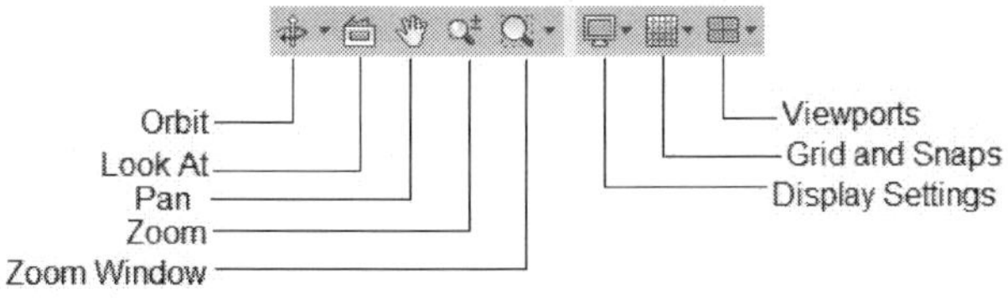

Figure 1-19 The Navigation Bar and Display Settings

BROWSER Bar

The BROWSER bar is available below the Ribbon, on the left in the graphics window. It displays all the operations performed during the designing process in a sequence. All these operations are displayed in the form of a tree view. You can undock the BROWSER bar by dragging it. The contents of the BROWSER bar are different for different workspaces of Autodesk Fusion 360. For example, in the MODEL workspace, it displays various operations that were used in creating the part. Similarly, in the ANIMATION workspace, it displays all the tools that can be used to animate them.

ViewCube

Autodesk Fusion 360 provides you with an option to change the view of a solid model freely in 3D space using the ViewCube. A ViewCube is a 3D navigation tool, which allows you to switch between the standard and isometric views in a single click. By default, ViewCube remains in the inactive state. When you move the cursor closer to the ViewCube, it gets activated.

UNITS FOR DIMENSIONS

In Autodesk Fusion 360, you can set units at any time by using the **CHANGE ACTIVE UNITS** dialog box. You can invoke this dialog box by expanding the **Document Settings** node in the **BROWSER** Bar and choosing the **Change Active Units** button which appears on hovering the cursor on the **Units:mm** subnode. After invoking this dialog box, select the required unit from the **Unit Type** drop-down list. Next, choose the **OK** button to apply the specified settings.

IMPORTANT TERMS AND THEIR DEFINITIONS

Before you proceed in Autodesk Fusion 360, it is very important for you to understand the following terms widely used in this book.

Feature-based Modeling

A feature is defined as the smallest building block of a design that can be modified individually. In Autodesk Fusion 360, solid models are created by integrating these blocks. Therefore, the models in Autodesk Fusion 360 are a combination of a number of individual features. These features understand their fit and function properly. As a result, these can be modified whenever required. Generally, these features automatically adjust their values if there is any change in their surroundings.

Parametric Modeling

The parametric nature of a software package is defined as its ability to use the standard properties or parameters in defining the shape and size of a geometry. The main function of this property is to drive the selected geometry to a new size or shape without considering its original dimensions. You can change or modify the shape and size of any feature at any stage of the design process. This property makes the designing process very easy.

For example, consider the design of the body of a pipe housing shown in Figure 1-20. In order to change the design by modifying the diameter of the holes and the number of holes on the front, top, and bottom faces, you need to select the feature and change the diameter and the number of instances in the pattern. The modified design is shown in Figure 1-21.

Figure 1-20 Body of pipe housing

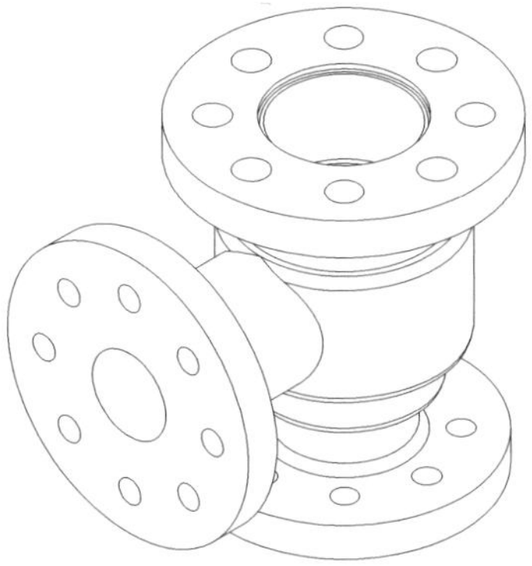

Figure 1-21 Design after modifications

Bidirectional Associativity

As mentioned earlier, this solid modeling tool does not restrict its capabilities to the 3D solid output. It is also capable of designing and drafting highly effective assembly design. There exists a bidirectional associativity between all environments of Autodesk Fusion 360. This link ensures that if any modification is made in the model in any of the workspaces, it will automatically reflect in the other workspace as well.

MARKING MENU

Marking menu is a type of menu that consists of tools and options which are commonly used in Autodesk Fusion 360. Marking menu replaces the conventional right-click context menu. The Marking menu consists of different tools in different workspace. For example, in the Model workspace, the Marking menu consists of commonly used tools such as **Sketch**, **Move/Copy**, **Hole**, **Press Pull**, and so on. In the Drawing Workspace, it consists of tools and options such as **Projected View**, **Move**, **Annotation**, and so on.

You can invoke a tool from Marking menu by right-click anywhere in the graphics window. On doing so, all the menu items surrounding the cursor will be displayed. After invoking the Marking menu, you can choose the desired tool or option from it. To do so, move the cursor toward the desired tool; the tool is highlighted along with a marker ray. Next, choose the highlighted tool to invoke it.

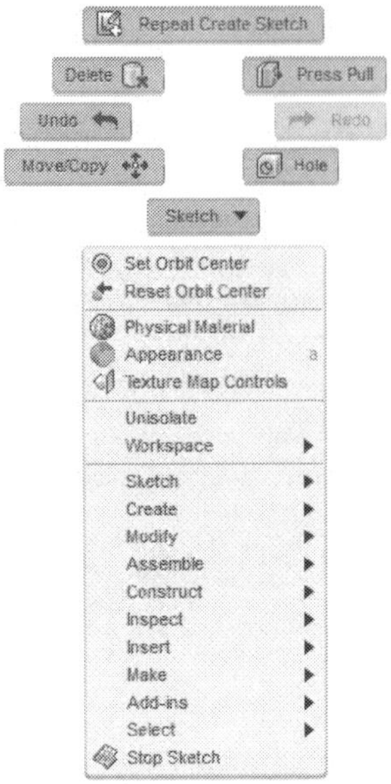

Figure 1-22 Marking menu available in the Model workspace

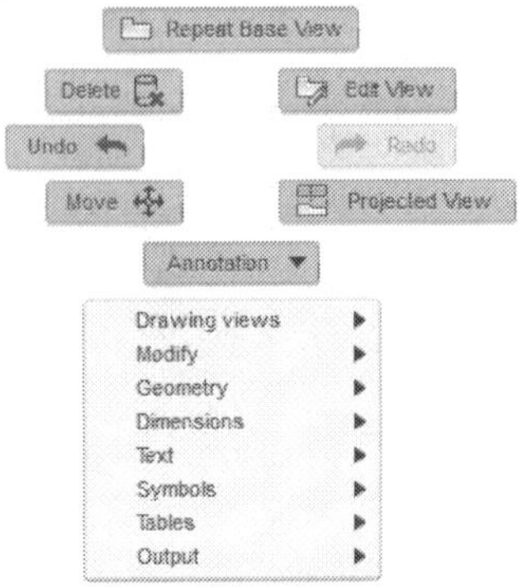

Figure 1-23 Marking menu available in the Drawing workspace

Constraints

Constraints are the logical operations that are performed on the selected design to make it more accurate or to define its position with respect to some other design entity. These constraints are called geometric constraints and are applied on the basic sketch entities to relate them to the standard properties like collinearity, concentricity, perpendicularity, and so on. Autodesk Fusion 360 automatically applies these geometric constraints to the sketch entities at the time of their creation. You do not have to use an extra command to apply these constraints on to the sketch entities. However, you can also manually apply these geometric constraints on to the sketch entities. There are twelve types of geometric constraints.

Perpendicular Constraint

This constraint is used to make the selected line segment normal to another line segment.

Parallel Constraint

This constraint is used to make the selected line segments parallel.

Coincident Constraint

This constraint is used to make two points or a point and a curve coincident.

Concentric Constraint

This constraint forces two selected curves to share the same center point. The curves that can be made concentric are arcs, circles, or ellipses.

Collinear Constraint

This constraint forces two selected line segments or ellipse axes to be placed in the same line.

Horizontal Constraint
This constraint forces the selected line segment to become horizontal.

Vertical Constraint
This constraint forces the selected line segment to become vertical.

Tangent
This constraint is used to make the selected line segment or curve tangent to another curve.

Equal
This constraint forces the selected line segments to become equal in length to another line segment. It can also be used to force two curves to become equal in radius.

Smooth
This constraint adds a smooth constraint between a spline and another entity so that at the point of connection, the line is tangent to the spline.

Fix/ Unfix
This constraint fixes the selected point or curve to a particular location with respect to the coordinate system of the current sketch.

Symmetric
This constraint forces the selected sketched entities to become symmetrical about a sketched line segment which may or may not be a center line.

Assembly Joints
The assembly joints are the logical operations performed on the components in order to join them together to create an assembly. These joints allow motion between the connected components or in the assembly. There are seven types of assembly joints in Autodesk Fusion 360 and they are discussed next.

Rigid
The Rigid joint removes all the degrees of freedom from the component. As a result, the components after applying rigid joints can not move in any direction. The Rigid joint is used to fix two parts rigidly. All the DOFs between the selected parts get eliminated and act as a single component when any motion will be applied to any of the direction.

Revolute
The Revolute joint is used to create a joint between two parts such that one part rotates with respect to the fixed part about a common axis.

Slider
The Slider joint allows the movement of a component along a specified path. The component will be joined to translate in one direction only. You can specify only one translation degree of freedom in slider joint. Slider joint are used to simulate the motion in linear direction.

Cylindrical

The Cylindrical joint allows a component to translate along the axis of a cylindrical component as well as rotate about the axis. You can specify one translation degree of freedom and one rotational degree of freedom in the Cylindrical joint.

Planar

The Planar joint is used to connect the planar faces of two components. The components can slide or rotate on the plane with two translation and one rotational degree of freedom.

Ball

The Ball joint is used to create a joint between two components such that both the components remain in touch with each other and at the same time the movable component can freely rotate in any direction. To create a ball joint between two components, you need to specify one point from each component. The joints thus created will generate three undefined rotational DOFs and restrict the other three DOFs at a common point.

Pin-Slot

Pin-Slot Joints allow the component to rotate about an axis and translate about a different Axis. With the Pin-Slot joint, you allow motion along one translational axis and a rotation about a different axis.

HOTKEYS

As mentioned earlier, there is no command prompt in Autodesk Fusion 360. However, you can use the keys on the keyboard to invoke some tools. The keys that can be used to invoke the tools are called hotkeys. Remember that the working of the hotkeys will be different for different environments. The use of hotkeys in different environments is given next.

Modeling Workspace

The hotkeys that can be used in the **MODEL** Workspace and their functions are given next.

Hotkey	Function
R	2-point Rectangle
A	Appearance
Shift + J	As-built Joint
C	Center Diameter Circle
Ctrl + B	Compute All
E	Extrude
F	Fillet
2	Freeform Selection
H	Hole

X	Normal/ Construction
J	Joint
L	Line
I	Measure
S	Model Toolbox
M	Move
O	Offset
3	Paint Selection
Q	Press Pull
P	Project
Shift + S	Scripts and Add-ins
D	Sketch Dimension
Shift + N	Toggle Component Color Cycling
V	Toggle Visibility
T	Trim
1	Window Selection

Animation Workspace

In addition to the hotkeys of the modeling tool, the following hot keys can also be used in the **Animation** workspace:

Hotkey	**Function**
U	Auto Explode All Levels
E	Manual Explode
P	Publish Video
M	Transform Components
C	View

Drawing Workspace

The hotkeys that can be used in the **Drawing** workspace are given next.

Hotkey	**Function**
B	Balloon
C	Center Mark
Delete	Delete
D	Dimension

M	Move
P	Projected View
T	Text

Edit Form Commands

Alt + Drag	Add geometry
Alt + Control + Drag | Alt + Command + Drag	Add geometry and keep creases

In addition to these keys, you can also use some other keys for the ease of designing. Note that you will have to hold some of these keys down and use them in combination with the pointing device. These hotkeys are given next.

Sculpting Workspace

Hotkey	Function
Shift + Up Arrow	Grow selection
Alt + N | Command + N	Invert selection
Alt + O | Control + O	Loop grow selection
Alt + P | Control + P	Loop selection
Alt + Right Arrow | Control + Command + Right Arrow	Next U
Alt + Up Arrow | Control + Command + Up Arrow	Next V
Alt + Left Arrow | Control + Command + Left Arrow	Previous U
Alt + Down Arrow | Control + Command + Down Arrow	Previous V
Alt + M | Command + M	Range selection
Alt + K | Control + K	Ring grow selection
Alt + L | Control + L	Ring selection
Alt + J | Control + J	Ring shrink selection
Double-click an edge	Select edge ring
Select two faces then double-click a third face	Select face ring
Shift + Down Arrow	Shrink selection
Ctrl + 1 | Control + 1	Toggle box mode
Ctrl + 2 | Control + 2	Toggle control frame mode
Ctrl + 3 | Control + 3	Toggle smooth mode

CAM Workspace

Ctrl + D | Command +D	Duplicate
Ctrl + G | Command + G	Generate Toolpath
Shift+S	Scripts and Add-Ins
Ctrl + L | Command + L	Show Log

Render Workspace

A	Appearance

Simulation Workspace

Ctrl + D | Command + D	DOF View
F	Force
Ctrl + G | Command + G	Groups View
Ctrl + L | Command + L	Model View
N	New Simulation Study
Ctrl + R | Command + R	Results View
E	Settings
C	Structural Constraint
L	Structural Loads

Canvas Selection

Hotkey	**Function**
Ctcrl + C | Command + C	Copy
Ctrl + X | Command + X	Cut
Hold Shift + Hold Middle Mouse Button	Orbit
Hold Shift + Click then Hold Middle Mouse Button	Orbit around point
Hold Middle Mouse Button	Pan
Ctrl + V | Command + V	Paste
Ctrl + Y | Command + Y	Redo
Ctrl + Z | Command + Z	Undo
Roll Middle Mouse Button	Zoom

Self-Evaluation Test

Answer the following questions and then compare them to those given at the end of this chapter:

1. You can invoke the **Line** tool by using the _________ hotkey.

2. Press _________ hot key to invoke **2-Point Rectangle** tool.

3. _________ tool is used to select entities by creating a selection trap.

4. _________ workspace create photo-realistic renderings.

5. _________ workspace is used for the documentation of the parts or assemblies in the form of drawing views.

Review Questions

Answer the following questions:

1. You can invoke the **Model Toolbox** by pressing the _________ key.

2. The _________ tool is used to select an entity by dragging the mouse to draw a free form sketch across it.

3. There are twelve types of geometric constraints in Autodesk Fusion 360. (T/F)

4. You can invoke the **Trim** tool by pressing the X key. (T/F)

5. You can set units at any time by using the _________ dialog box.

Answers to Self-Evaluation Test

1. L, **2.** R, **3. Window Selection, 4. RENDER**, **5. DRAWING**

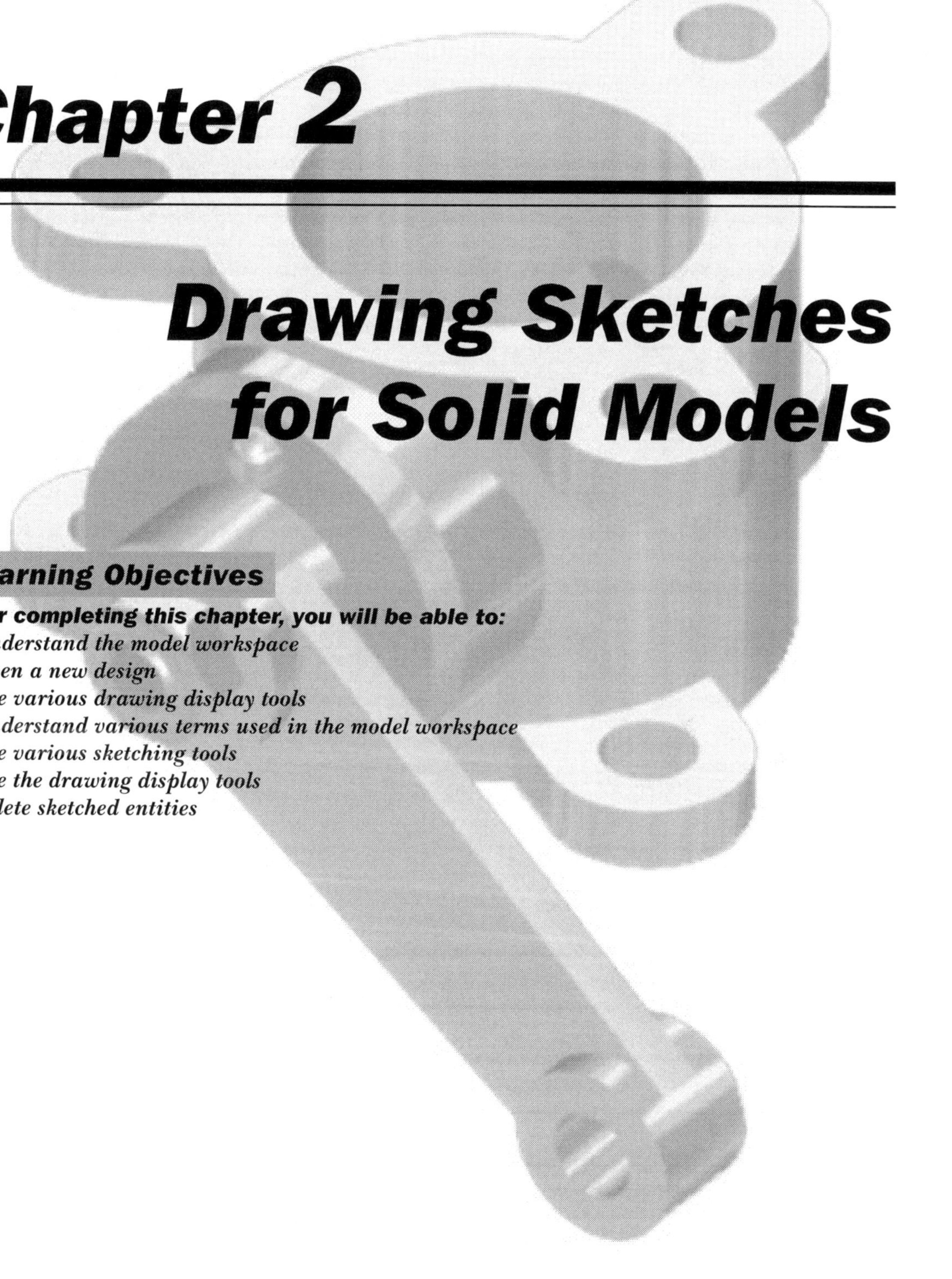

Chapter 2

Drawing Sketches for Solid Models

Learning Objectives

After completing this chapter, you will be able to:

- *Understand the model workspace*
- *Open a new design*
- *Use various drawing display tools*
- *Understand various terms used in the model workspace*
- *Use various sketching tools*
- *Use the drawing display tools*
- *Delete sketched entities*

INTRODUCTION

Sketches are foundation for creating important three-dimensional features. Sketches are typically comprises two-dimensional sketch entities created either on a plane or an existing planar face of a part. Once a sketch is created, it can be used as a profile to create a 3D feature. These profiles can be used to create a unique geometry by pulling them along a path.

Most of the designs created in Autodesk Fusion 360 consist of sketched and placed features. A sketch is a combination of a number of two-dimensional (2D) entities such as lines, arcs, circles, and so on. The features such as extrude, revolve, and sweep that are created by using 2D sketches are known as sketch based features. The features such as fillet, chamfer, thread, and shell that are created without using a 2D sketches are known as placed features. In a design, the base feature or the first feature is always a sketch based feature. For example, the sketch shown in Figure 2-1 is used to create the solid model shown in Figure 2-2. Once you have drawn the basic sketch, refer to Figure 2-1, you can convert it into a solid model using solid modeling tools.

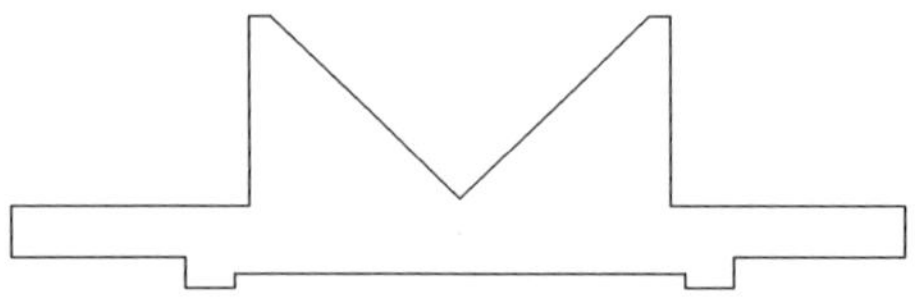

Figure 2-1 *The basic sketch for the solid model*

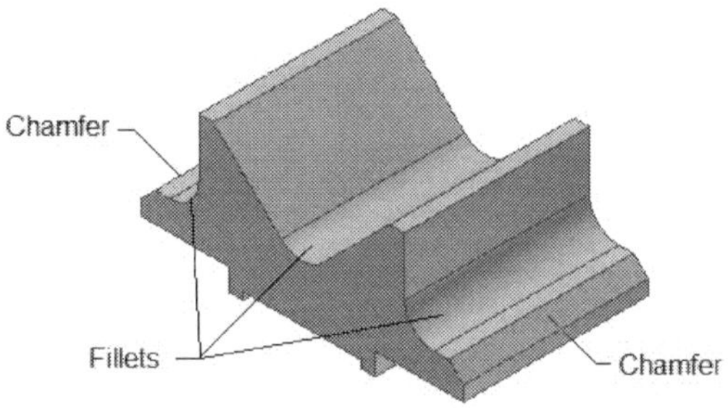

Figure 2-2 *A solid model created using the sketched and placed features*

You can create sketches in the Model workspace by using the tools available in the **SKETCH** panel. This workspace in Autodesk Fusion 360 is invoked by default. Autodesk Fusion 360 enables fast and easy exploration of design ideas with an integrated Product Innovation Platform. The options in the model workspace will be discussed later in this chapter.

TUTORIALS

Tutorial 1

In this tutorial, you will start a new design in Autodesk Fusion 360 and then create a sketch of the solid model shown in Figure 2-3. The sketch of the model is shown in Figure 2-4. Before drawing the sketch, you will modify the snap, grid, and unit settings for the active document. Note that the solid model and the dimensions shown in Figures 2-3 and 2-4 are for your reference only.

(Expected time: 30 min)

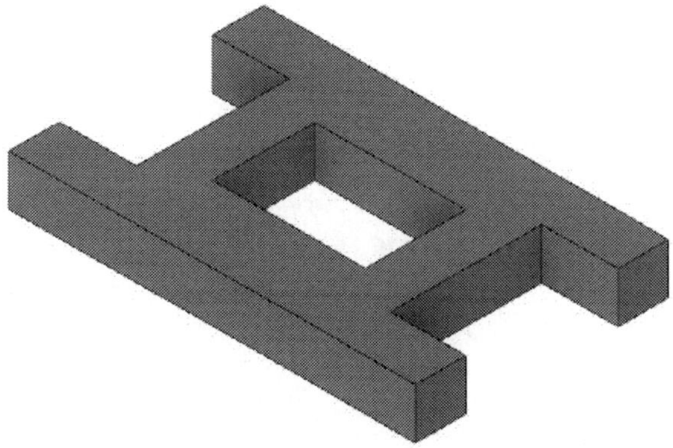

***Figure 2-3** Model for Tutorial 2*

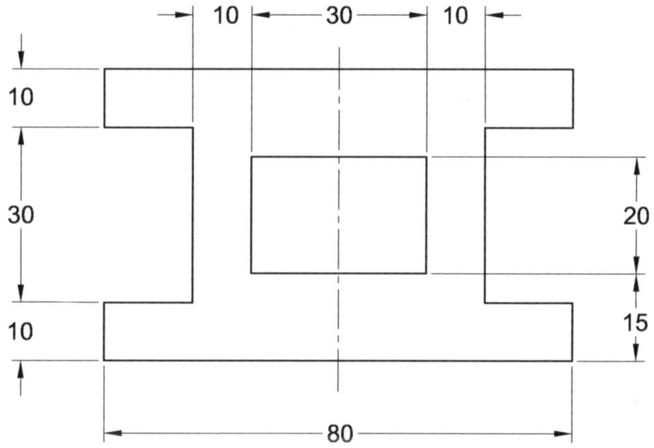

***Figure 2-4** Dimensioned sketch for Tutorial 2*

The following steps are required to complete this tutorial:

a. Start Autodesk Fusion 360 and then start a new design.
b. Modify the snap, grid, units settings.
c. Invoke the **Create Sketch** tool.
d. Create the sketch using the **Line** tool.
e. Save the sketch and then close the file.

Starting Autodesk Fusion 360

When you install the Autodesk Fusion 360, an icon is displayed on the desktop and a folder is added to the **Start** menu.

1. Choose **Autodesk Fusion 360** from the **Start** menu or double-click on the icon of Autodesk Fusion 360 on the desktop of your computer; the Autodesk Fusion 360 window is displayed, as shown in Figure 2-5.

The interface of Autodesk Fusion 360 window consists of Application Bar, Profile and Preferences, Help, Ribbon, ViewCube, Browser bar, Timeline, and Navigation Bar and Display Settings, refer to Figure 2-5.

Application bar: The Application bar is displayed at the top in the initial interface of Autodesk Fusion 360. This bar consists of various options that are used to access the Data Panel, File operations such as save, undo, and redo.

Help: The help option is used to access various help topics.

Ribbon: Ribbon is a dynamic interface that provides access to many of the tools and options available in Autodesk Fusion 360.

Profile and Preference: User can control their profile and account settings. Also, you can set the desired preference.

ViewCube: Use the ViewCube to orbit your design or view the design from standard view positions.

BROWSER bar: The browser bar lists objects in your design. Use the browser to make changes in objects and control the visibility of objects.

Timeline: The timeline lists the actions performed on your design. Right-click a feature in the timeline to make changes. Drag the Rolling back slider to roll back the modification done in the model.

Navigation bar and Display Settings: The navigation bar contains commands to zoom, pan, and orbit your design. The display settings control the appearance of the interface and the design.

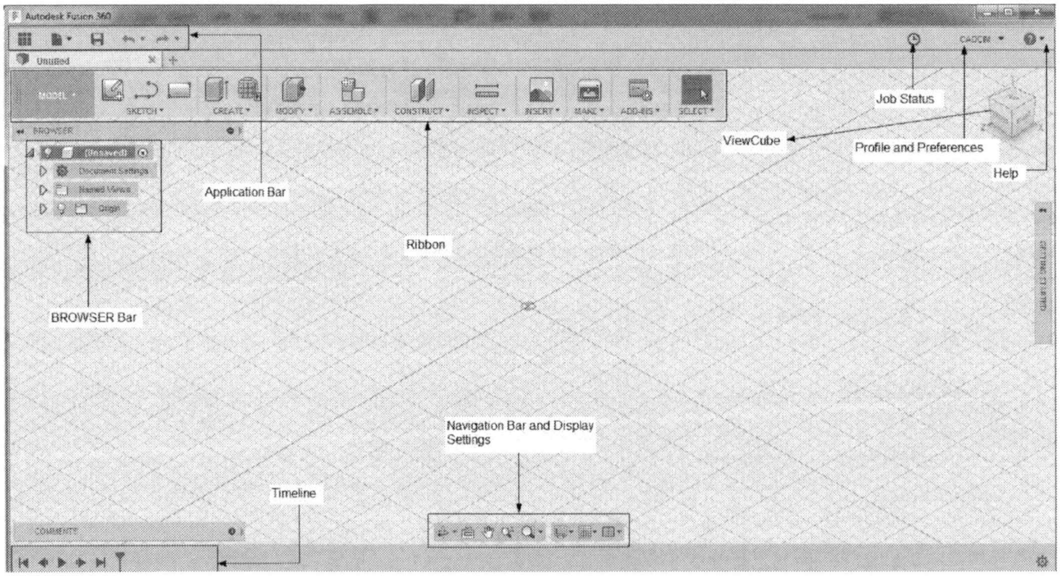

Figure 2-5 *The initial interface of Autodesk Fusion 360*

Whenever you start a new design, by default you switch to the Model workspace. In order to create the sketch of the base feature, you need to invoke the **Create Sketch** tool.

2. Choose the **Create Sketch** tool from the **SKETCH** panel of the Ribbon; the three default planes are displayed in the graphics window, as shown in Figure 2-6.

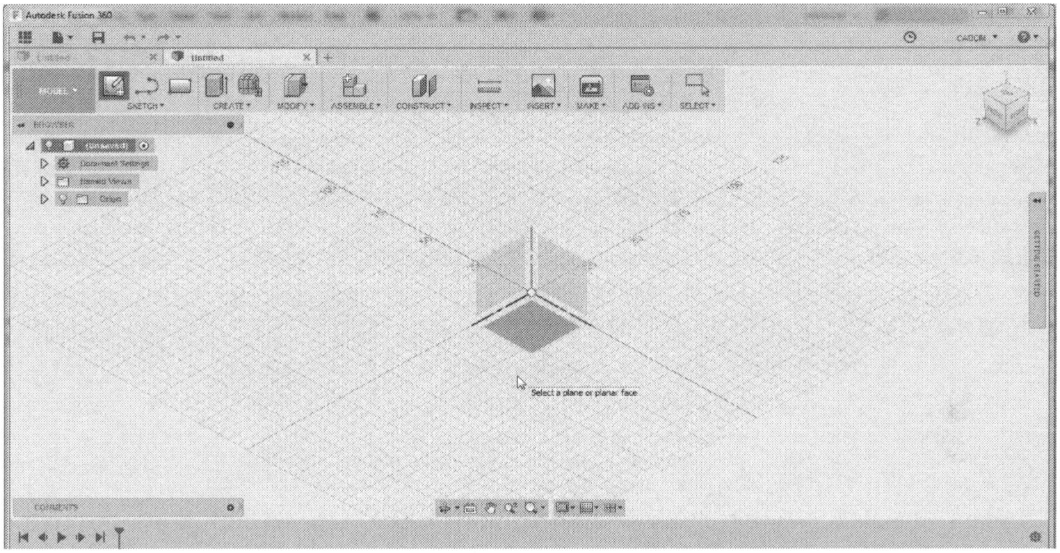

Figure 2-6 *The sketching interface with three default planes*

Depending on the requirement of your design, you can select a plane to draw the sketch of the base feature. The selected plane will be automatically oriented normal to the view so that you can easily create the sketch.

3. Choose the desired plane from the graphics window. Alternatively, you can also select the plane from the **BROWSER** bar. Select the **XZ** plane under the **Origin** node from the **BROWSER** bar; the selected plane gets oriented normal to the view.

   On selecting the plane, you will notice that the **STOP SKETCH** tool gets added at the end of the Ribbon. Also, ViewCube becomes normal to the screen and the **SKETCH PALETTE** window appears in the graphics window, as shown in Figure 2-7.

   By default, the **SKETCH PALETTE** window contains only the **Options** and **Constraints** nodes. The **Options** node controls the sketch properties for the design and the **Constraint** node is used to apply desired constraints to the sketch. Also, an additional node, called the **Feature Options** node, gets added for the tool that is active or the entity that is selected. For example, when the **Slot** tool is active, the **Center to Center slot**, **Center Point slot**, **Overall slot, 3-Point Arc slot and Center Point Arc** options get listed under the **Feature Options** palette.

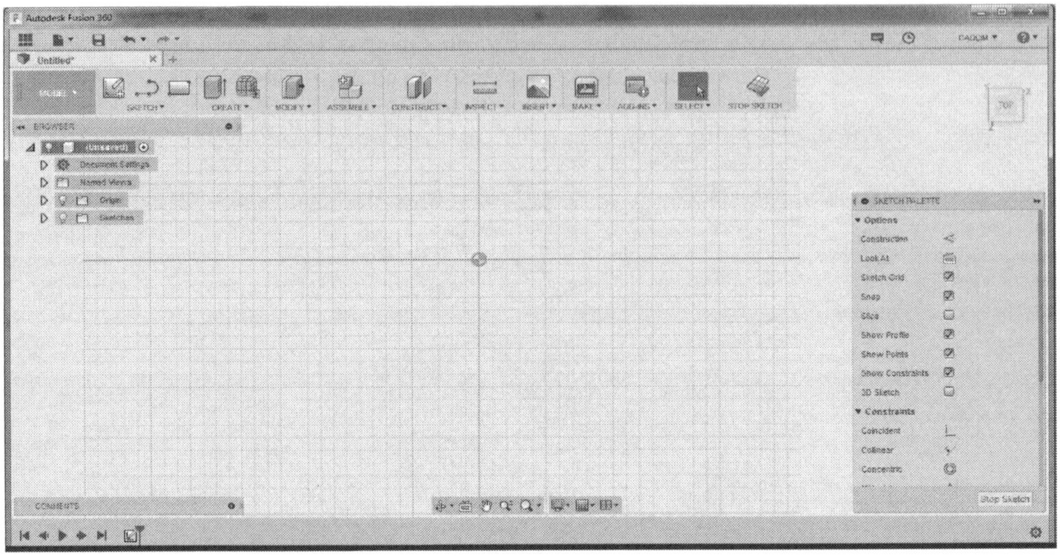

Figure 2-7 *Autodesk Fusion 360 interface along with the* ***SKETCH PALETTE*** *window*

Modifying the Units, Snap, and Grid

In Autodesk Fusion 360, you can modify the unit, snap, and grid settings. For this tutorial, you need to modify the grid and snap settings so that the cursor jumps through a distance of 5 mm.

1. Move the cursor on the **Units: mm**  sub node of the **Document Settings** in the **BROWSER** bar in the graphics window; the **Change Active Units** button is displayed on the right in the graphics window.

2. Choose the **Change Active Units** button; the **CHANGE ACTIVE UNITS** dialog box is displayed, as shown in Figure 2-8.

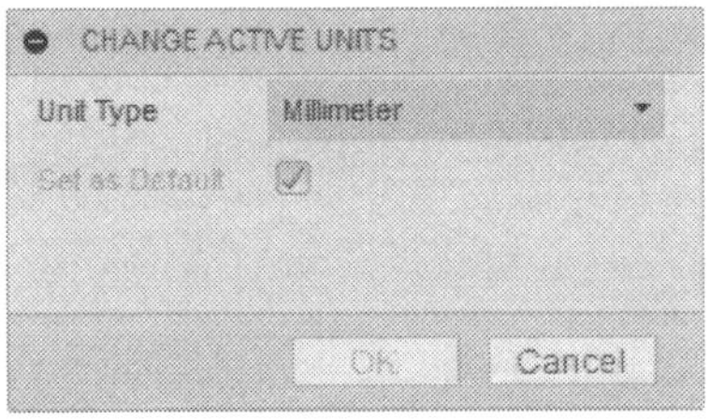

Figure 2-8 *The* ***CHANGE ACTIVE UNITS*** *dialog box*

3. Select **Millimeter** from the **Unit Type** drop-down list if not selected by default, and choose the **OK** button from the dialog box.

Note

In Autodesk Fusion 360, mm is set as the default unit and the ***Set as Default*** *check box remains selected but disabled. However, you can change the default set unit. To do so, select the desired unit from the Unit Type drop-down list; the Set as Default check box will be activated Select the* ***Set as Default*** *check box to make the selected unit as a default unit for the software. Do not select the check box if you only need to change the unit of the current design file.*

As evident from Figure 2-4, the dimensions in the sketch are multiple of 5. Therefore, you need to modify the grid and snap settings so that the cursor jumps through a distance of 5 mm.

4. Choose the **Grid Settings** option from the **Grid and Snaps** flyout at the bottom of the graphics window; the **GRID SETTINGS** dialog box is displayed, as shown in Figure 2-9.

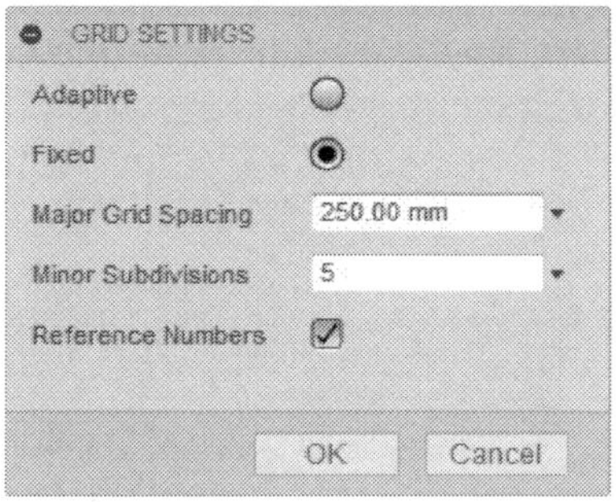

Figure 2-9 *The* ***GRID SETTINGS*** *dialog box*

In this dialog box, two radio buttons, **Adaptive** and **Fixed,** are available. The **Adaptive** radio button is used to adjust the size of the grid as the view zooms in and out. This radio button is selected by default. If you select the **Fixed** radio button, the view remains constant. The **Major Grid Spacing** edit box, the distance between two major grid line is entered. The **Minor-Subdivision** is number of division between two major grid lines. Note that major grid lines are thicker and darker than minor grid lines. The **Reference Numbers** check box is selected to display reference number along X-axis and Y-axis.

5. Select the **Fixed** radio button from the **GRID SETTINGS** dialog box; the dialog box gets expanded. Enter **100** in the **Major Grid Spacing** edit box and **20** in the **Minor Subdivision** edit box and choose the **OK** button.

6. Select the **Snap to Grid** check box from the **Grid and Snaps** flyout in the **Display Settings** bar if it not selected by default.

The **Snap to Grid** check box is available in the **Grid and Snaps** flyout at the bottom of the graphics window. When this check box is selected, any sketch cursor in use will snap to the intersection of the grid squares, indicated by a colored square. If the check box is not selected, the cursor will not snap to any point on the grid, and the endpoint can be placed anywhere.

Note

The distance by which the cursor jumps depends on the ratio between the values in the ***Major Grid Spacing*** *and the* ***Minor Subdivisions*** *edit boxes available in the* ***GRID SETTINGS*** *dialog box. For example, if you want the coordinates to increment by 10 mm, you will have to set the ratio of the major and minor lines to 10. This can be done by setting the value in the* ***Major Grid Spacing*** *edit box to 100 and the* ***Minor Subdivisions*** *edit box to 10. Similarly, to make the cursor jump through a distance of 5 mm, set the value in the* ***Major Grid Spacing*** *edit box to 50 and the* ***Minor Subdivisions*** *edit box to 10.*

Drawing the Sketch

The sketch will be drawn using the **Line** tool. As, Autodesk Fusion 360 is parametric in nature, you can draw the sketch from any point in the drawing window. In this tutorial, Dynamic Input has been used to enter dimensions while drawing the sketch.

As evident from Figure 2-4, the sketch consists of nested loop: inner and outer. When you extrude the outer loop, the inner loop will get subtracted and a cavity will be automatically created in the model. This reduces the time and effort required in creating the inner cavity as another feature.

1. Choose the **Line** tool from the **Sketch** panel; you are prompted to place the first point. Place the first point at the origin in the graphics window; you are prompted to specify the next point.

   The line cursor is actually a combination of line and arc. You can also invoke the **Line** tool by pressing the L key. You can create an arc by pressing and dragging the left mouse button from the end point of line without existing the line tool.

   The line is the basic sketching entity available in Autodesk Fusion 360. In general terms, a line is defined as the shortest distance between two points.

   In Autodesk Fusion 360, the **Line** tool is used to draw a chain of continuous lines which is the default method of drawing lines. In this method, you have to specify the start point and the endpoint of the line using the left mouse button. As soon as you specify the end point of the line, a line will be drawn between the two points. Now, when you move the cursor away from the end point of the line, you will notice that another line is attached with the cursor. It means that you can create a chain of continuous lines one after another. You can end the process of drawing the continuous line by pressing the ESC key or by double-clicking on the screen. You can also right-click to display the marking menu and choose the **OK** option to terminate the process of drawing the line.

Note

When you terminate the process of drawing a line by double-clicking on the screen, the current chain ends but the ***Line*** *tool still remains active. As a result, you can draw other lines. However, if you choose the* ***OK*** *option from the marking menu, the* ***Line*** *tool will be deactivated.*

2. Enter **80** in the length input field and press the TAB key; the angle input field gets activated. Enter **0** in the angle input field and then click in the graphics window; first line is created and the Dynamic Input is displayed again.

3. Move the cursor up and then enter **10** in the length input field. Next, press the TAB key and enter **90** in the angle input field and then click in the graphics window; second line is created and the Dynamic Input is displayed again.

4. Move the cursor toward the left and then enter **15** in the length input field. Next, press the TAB key and enter **90** in the angle input field. Next, click in the graphics window; third line is created and the Dynamic Input is displayed again.

5. Move the cursor up and then enter **30** in the length input field. Next, press the TAB key and enter **90** in the angle input field. Next, click in the graphics window; fourth line is created and the Dynamic Input is displayed again.

6. Move the cursor toward the right and then enter **15** in the length input field. Next, press the TAB key and enter **90** in the angle input field. Next, click in the graphics window; fifth line is created and the Dynamic Input is displayed again.

7. Move the cursor up and then enter **10** in the length input field. Next, press the TAB key and enter **90** in the angle input field. Next, click in the graphics window; sixth line is created and the Dynamic Input is displayed again.

8. Move the cursor toward the left and then enter **80** in the length input field. Next, press the TAB key and enter **90** in the angle input field. Next, click in the graphics window; seventh line is created and the Dynamic Input is displayed again.

9. Move the cursor down and then enter **10** in the length input field. Next, press the TAB key and enter **90** in the angle input field. Next, press click in the graphics window; the eighth line is created and the Dynamic Input is displayed again.

10. Move the cursor toward the right and then enter **15** in the length input field. Next, press the TAB key and enter **90** in the angle input field. Next, click in the graphics window; ninth line is created and the Dynamic Input is displayed again.

11. Move the cursor down and then enter **30** in the length input field. Next, press the TAB key and enter **90** in the angle input field. Next, click in the graphics window; tenth line is created and the Dynamic Input is displayed again.

12. Move the cursor toward the left and then enter **15** in the length input field. Next, press the TAB key and enter **90** in the angle input field. Next, click in the graphics window; eleventh line is created and the Dynamic Input is displayed again.

13. Move the cursor down and then enter **10** in the length input field. Next, press the TAB key and enter **90** in the angle input field. Next, click in the graphics window; twelfth line is created and the Dimension Input is displayed again.

14. Right-click anywhere in the graphics window to display the Marking menu and choose **OK** from it to exit the **Line** tool. The sketch of the outer loop is shown in Figure 2-10.

    You can arrange the positions of the dimensions in the sketch by dragging them to view the sketch clearly.

    Next, you need to draw the inner loop. You can draw the loop by using the **Center Rectangle** tool.

    In Autodesk Fusion 360, the tools that are used to draw a rectangle are grouped under the **Rectangle** option of the **SKETCH** panel in the Ribbon. The **Center Rectangle** tool is used to draw a rectangle by specifying two points to define the center and corner. To create a rectangle using two points for the diagonal corners, use **2-Point Rectangle**. To create a rectangle using three points to define the width, direction, and height, use **3-Point Rectangle**.

15. Choose **Sketch > Rectangle > Center Rectangle** from the ribbon; you are prompted to specify the center of the rectangle.

16. Starting from the origin, move the cursor horizontally by 8 units and then vertically up by 5 units in the graphics window. At this point, click in the graphics window to set the center of the rectangle;you are prompted to specify the size of the rectangle.

17. Enter **30** and **20** in the horizontal and vertical input fields, respectively, in the Dimension Input, and then click in the graphics window.

18. Right-click in the graphics window; a Marking menu is displayed. Choose **OK** from the Marking menu; the sketch is created, as shown in Figure 2-11.

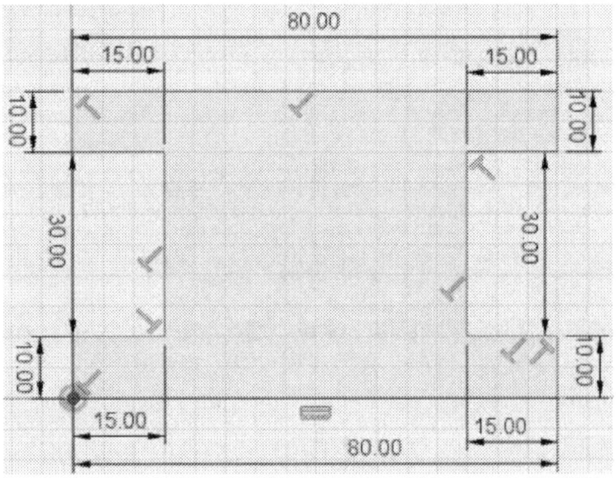

Figure 2-10 *Sketch of the outer loop*

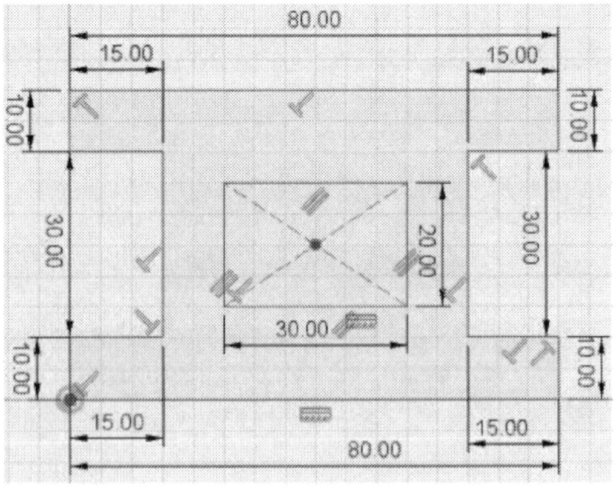

Figure 2-11 *Completed sketch for Tutorial 1*

Saving the Sketch

1. Choose the **Save** option from the File menu or Application menu; the **Save** dialog box is displayed.

2. Enter **c02_Tut_01** in the **Name** edit box.

3. Next click on the down arrow on the extreme right side of the **Location** selection box, the **Save** dialog box gets expanded, refer to Figure 2-12. 

4. Select the existing project from the **PROJECT** column or you can create new project by choosing the **New Project** button from the **Save** dialog box.

5. Select the **Save** button to save the sketch and exit the **Save** dialog box.

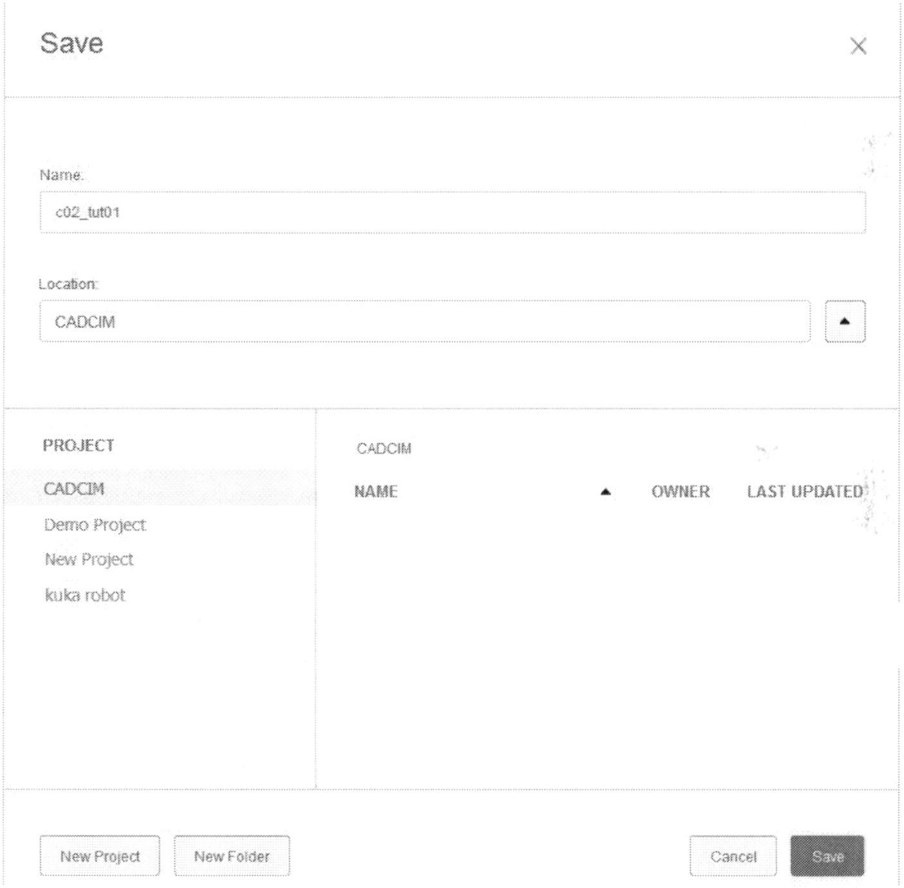

***Figure 2-12** The expanded **Save** dialog box*

Tutorial 2

In this tutorial, you will draw the sketch of the model shown in Figure 2-13. The sketch of the model is shown in Figure 2-14. Do not dimension the sketch as the dimensions are given only for reference.

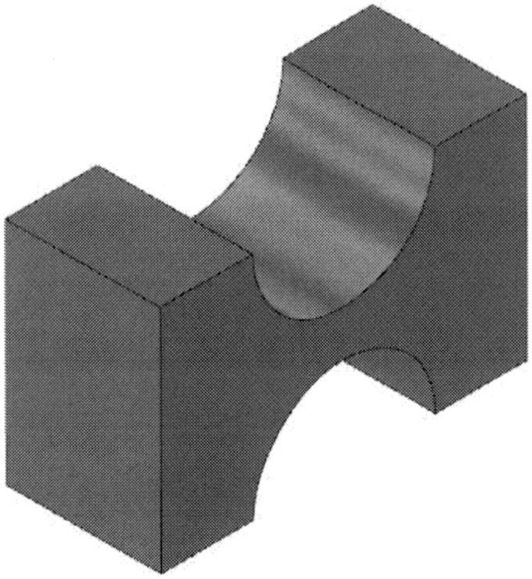

Figure 2-13 Model for Tutorial 3

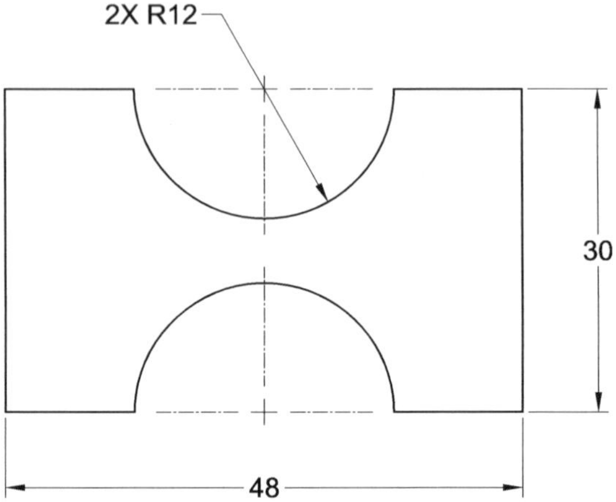

Figure 2-14 Sketch for Tutorial 3

a. Start a new design.
b. Invoke the model workspace.
c. Modify the snap, grid, and units settings .
d. Create the sketch using the **Line** and **Centerline** tools, refer to Figure 2-14.
e. Save the sketch and then close the file.

Starting a New File

1. Choose the **New Design** option from the **File** menu to start new design.

   Whenever you start a new part design, by default you are in the **Model** workspace. You need to start the design by first creating the sketch of the base feature.

2. Choose the **Create Sketch** tool from the **SKETCH** panel of the Ribbon; the three default planes are displayed in the graphics window and you are prompted to select a plane or a planar face.

3. Choose **RIGHT** from the **ViewCube** and select the plane to orient it normal to the view.

   Alternatively, expand the **Origin** node in the **BROWSER** bar and then select the **YZ** Plane to orient it normal to the view.

Modifying the Units, Snap, and Grid

In Autodesk Fusion 360, you can modify the units, snap, and grid. In this tutorial, you need to modify the grid and snap settings such that the cursor jumps through a distance of 3 mm.

1. Move the cursor on the **Units: mm** sub node of the **Document Settings** in the **BROWSER** bar in the graphics window; the **Change Active Units** button is displayed.

2. Choose the **Change Active Units** button; the **CHANGE ACTIVE UNITS** dialog box is displayed.

3. Select the **Millimeter** option from the **Unit Type** drop-down list in the **CHANGE ACTIVE UNITS** dialog box and then choose the **OK** button. If this option is already selected then, choose the **Cancel** button from the dialog box.

   As evident from Figure 2-14, the dimensions in the sketch are multiples of 3. Therefore, you need to modify the grid and snap settings such that the cursor jumps through a distance of 3 mm.

4. Choose the **Grid Settings** option from the **Grid and Snaps** flyout at the bottom of the graphics window; the **GRID SETTINGS** dialog box is displayed.

5. Select the **Fixed** radio button from the **GRID SETTINGS** dialog box; the dialog box expands. Now you need to enter **30** in the **Major Grid Spacing** edit box and **10** in the **Minor Subdivisions** edit box.

6. Choose the **OK** button.

Drawing the Sketch

The upper arc of the sketch can be drawn by specifying its center, start, and end points. Therefore, you need to use the **Center Point Arc** tool to draw it.

1. Choose **SKETCH > Arc > Center Point Arc** from the Ribbon; you are prompted to place the center point of the arc.

2. Move the cursor vertically up from the origin by 5 units and place the center point of the arc; you are prompted to specify the start point of the arc.

3. Move the cursor horizontally toward left from the center of the arc in the graphics window and enter **12** in the length input field and click in the graphics window; the first point of the arc is defined and you are prompted to specify the end point.

4. Move the mouse cursor in a counter-clockwise direction and enter **180** in the angle input field and click in the graphics window; the upper arc is drawn.

   Next, you need to draw lines in the sketch.

5. Choose the **Line** tool from the **SKETCH** panel; you are prompted to specify the first point of the line. 

6. Move the cursor close to the start point of the arc; the square snaps symbol appears indicating that the cursor has snapped to the end point of the arc. Press the left mouse button to select this point as the start point of the line, refer to Figure 2-15.

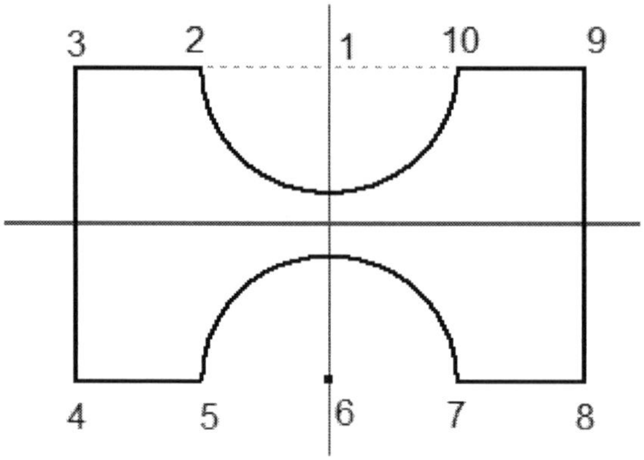

Figure 2-15 *Sketch showing different points for Tutorial 3*

7. Move the cursor toward left in the graphics window. Enter **12** in the length input field and press the TAB key; the angle input field activated. Enter **180** in the angle input field and then click in a graphics window; the line between points 2 and 3 is created. Refer to Figure 2-15 for point numbers.

8. Move the cursor downward in the graphics window. Enter **30** in the length input field and **90** in the angle input field in the Dynamic Input. Next, click in the graphics window; a line between points 3 and 4 is created.

9. Move the cursor toward right in the graphics window. Enter **12** in the length input field and **90** in the angle input field of the Dynamic Input. Next, click in the graphics window; a line is created between points 4 and 5.

   Now, you need to draw the lower arc of the sketch.

10. Choose **SKETCH > Arc > Center Point Arc** from the Ribbon; you are prompted to place the center point of the arc.

11. Move the cursor vertically downward from the origin by 5 units and click to place the center point of the arc.

12. Move the cursor horizontally toward left from the center of the arc in the graphics window and enter **12** in the length input field. Next, click in the graphics window; the first point of the arc is defined and you are prompted to specify the end point.

13. Move the cursor in the clockwise direction and enter **180** in the angle input field and click in the graphics window. The lower arc is drawn.

    Next, you need to draw the remaining lines to complete the sketch.

14. Choose the **Line** tool from the **SKETCH** panel; you are prompted to specify the start point of the line. 

15. Specify the start point of line at the end point of the lower arc; you are prompted to specify the next point of line.

16. Enter **12** in the length input field and **0** in the angle input field. Next click in the graphics window; a line between points 7 and 8 is created.

17. Move the cursor upward in the graphics window. Enter **30** in the length input field and **90** in the angle input field. Next click in the graphics window; a line between points 8 and 9 is created.

18. Move the cursor toward left in the graphics window. Enter **12** in the length input field and **90** in the angle input field in the Dynamic Input. Next click in the graphics window; a line between points 9 and 10 is created.

19. Exit the **Line** tool.

    The final sketch for Tutorial 3 is shown in Figure 2-16.

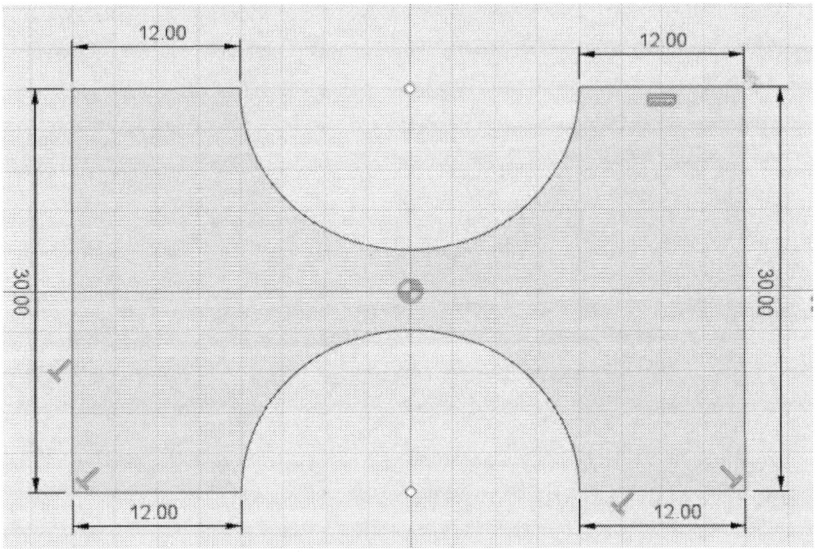

***Figure 2-16** Completed sketch for Tutorial 3*

Saving the Sketch

1. Choose the **Save** option from the **File** menu or Application menu; the **Save** dialog box is displayed.

2. Enter **c02_Tut_02** in the **Name** edit box.

3. Next, click on the down arrow on the extreme right side of the **Location** selection box; the **Save** dialog box expands, refer to Figure 2-17. 

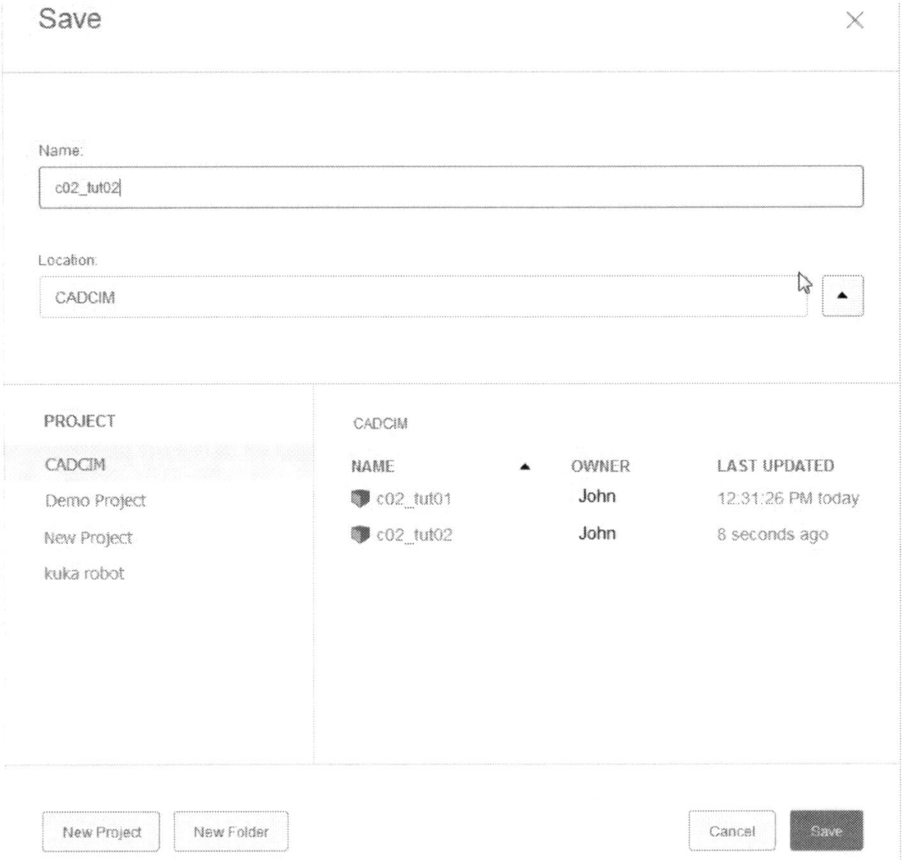

Figure 2-17 *The expanded **Save** dialog box*

4. Select an existing project from the **PROJECT** column or you can create a new project by choosing the **New Project** button from the **Save** dialog box.

5. Select the **Save** button to save the sketch and exit the **Save** dialog box.

Tutorial 3

In this tutorial, you will draw the basic sketch of the revolved solid model shown in Figure 2-18. The sketch for creating this model is shown in Figure 2-19. Do not dimension the sketch as the dimensions are given only for reference. Use Dynamic Input to draw the feature.

(Expected time: 30 min)

The following steps are required to complete this tutorial:

a. Start a new design.
b. Invoke the Model workspace.
c. Modify the snap, grid, and units settings.
d. Draw the sketch with the help of the **Line** tool.
e. Apply fillets.
f. Save the sketch with the name *c03_tut03* and close the file.

Figure 2-18 *Revolved model for Tutorial 4*

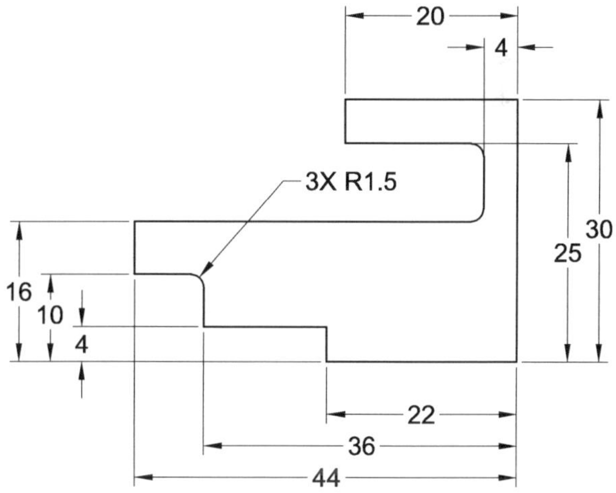

Figure 2-19 *Sketch for the revolved model*

Starting a New Design

1. Choose the **New Design** option from the **File** menu to create a new design.

   Whenever you start a new part design, by default, you are in the Model workspace. You need to start the design by first creating the sketch of the base feature.

2. Choose the **Create Sketch** tool from the **SKETCH** panel of the Ribbon; three default planes are displayed in the graphics window and you are prompted to select the plane.

3. Choose **RIGHT** from the ViewCube and select the plane to orient it normal to the view.

   Alternatively, expand the **Origin** node in the **BROWSER** bar and then select the **YZ** Plane to orient it normal to the view.

Modifying the Units, Snap, and Grid

In Autodesk Fusion 360, you can modify the units, snap, and grid. For this tutorial, you need to modify the grid and snap settings such that the cursor jumps through a distance of 2 mm.

1. Move the cursor on the **Units: mm** sub node of the **Document Settings** in the **BROWSER** bar in the graphics window; the **Change Active Units** button is displayed.

2. Choose the **Change Active Units** button; the **CHANGE ACTIVE UNITS** dialog box is displayed.

3. Choose the **Millimeter** option from the **Unit Type** drop-down list in the **CHANGE ACTIVE UNITS** dialog box, if it is not selected by default and then choose the **OK** button from this dialog box.

4. Choose the **Grid Settings** option from the **Grid and Snaps** flyout in the **Display Setting** at the bottom of the graphics window; the **GRID SETTINGS** dialog box is displayed.

5. Select the **Fixed** radio button from the **GRID SETTINGS** dialog box; the dialog box gets expanded. Enter **20** in the **Major Grid Spacing** edit box and **10** in the **Minor Subdivisions** edit box. Next, choose **OK**.

Drawing the Sketch

1. Choose the **Line** tool from the **SKETCH** panel or right- click in the graphics window; a Marking menu is displayed. Choose **SKETCH>Line**; you are prompted to select the first point for the line to be created. Click on the origin to specify the first point of the sketch.

Now, you need to draw the line1, refer to Figure 2-20

2. Move the cursor toward left, enter **22** in the length input field. Press TAB and enter **180** in the angle input field. Next, click in the graphics window; the line 1 is drawn.

3. Move the cursor upward in the graphics window. Next, enter **4** in the length input field, press TAB, and enter **90** in the angle input field. Next, click in the graphics window; the line2 is drawn, as shown in Figure 2-20.

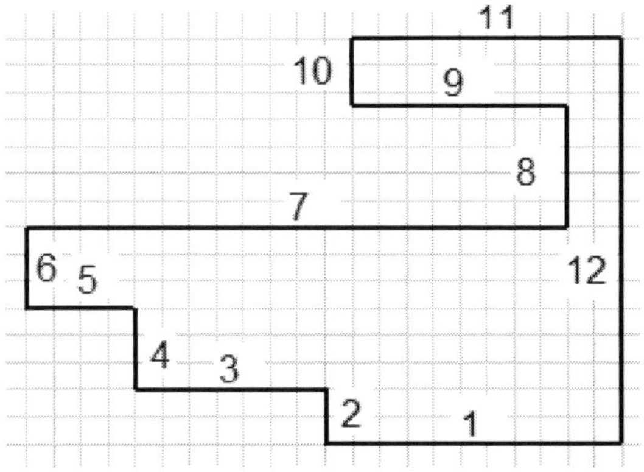

Figure 2-20 *Sketch after drawing the lines*

4. Move the cursor toward left in the graphics window. Next, enter **14** in the length input field, press TAB, and enter **90** in the angle input field. Click in the graphics window; line 3 is drawn.

5. Move the cursor upward in the graphics window. Next, enter **6** in the length input field, press TAB, and enter **90** in the angle input field; line 4 is drawn.

6. Move the cursor toward the left in the graphics window. Next, enter **8** in the length input field, press TAB, and enter **90** in the angle input field. Click in the graphics window; line 5 is drawn.

7. Move the cursor upward in the graphics window. Next, enter **6** in the length input field, press TAB, and enter **90** in the angle input field. Click in the graphics window; line 6 is drawn.

8. Move the cursor toward the right in the graphics window. Next, enter **40** in the length input field, press TAB, and enter **90** in the angle input field. Click in the graphics window; line 7 is drawn.

9. Move the cursor upward in the graphics window. Next, enter **9** in the length input field, press TAB, and enter **90** in the angle input field. Click in the graphics window; line 8 is drawn.

10. Move the cursor toward left in the graphics window. Next, enter **16** in the length input field, press TAB, and enter **90** in the angle input field. Click in the graphics window; line 9 is drawn.

11. Move the cursor upward in the graphics window. Next, enter **5** in the length input field, press TAB, and then enter **90** in the angle input field. Click in the graphics window; line 10 is drawn.

12. Move the cursor toward right in the graphics window. Next, enter **20** in the length input field, press TAB, and enter **90** in the angle input field. Click in the graphics window; line 11 is drawn.

13. Click on the origin, which was the start point of the sketch; the initial sketch is drawn.

14. Exit the **Line** tool by right-clicking and choosing **OK** from the Marking menu.

Applying Fillets

1. Choose the **Fillet** tool from the **SKETCH** panel; you are prompted to select the lines or arc.

2. Select the line 8 and then line 9, refer to Figure 2-20; a fillet is created between these lines and the **Radius** edit box is displayed.

3. Enter **1.5** in the **Radius** edit box; a warning message is displayed informing you that constraint and/ or dimension were removed during this operation. Ignore it.

4. Similarly, select lines 7 and 8 and then lines 4 and 5 to create a fillet between these lines. Next, right-click, and choose **OK** from the Marking menu to exit the **Fillet** tool after all the fillets are created.

   As all the lines are filleted with the same radius value, the radius of the fillet is not displayed on other fillets. This completes the sketch. The final sketch for this tutorial after filleting all the sketches is shown in Figure 2-21.

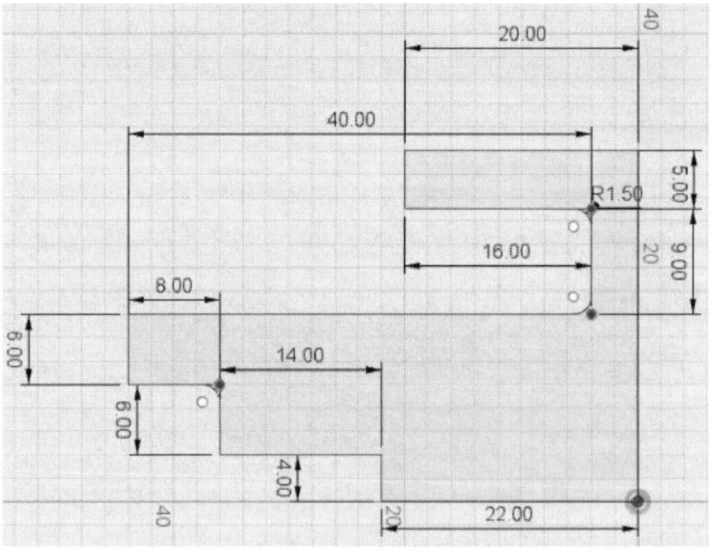

***Figure 2-21** Final sketch after filleting*

Saving the Sketch

Next, you need to save the sketch.

1. Choose the **Save** option from the **File** menu or Application menu; the **Save** dialog box is displayed.

2. Enter **c02_Tut_03** in the **Name** edit box.

3. Click on the down arrow on the extreme right side of the **Location** selection box; the **Save** dialog box gets expanded, refer to Figure 2-22. 

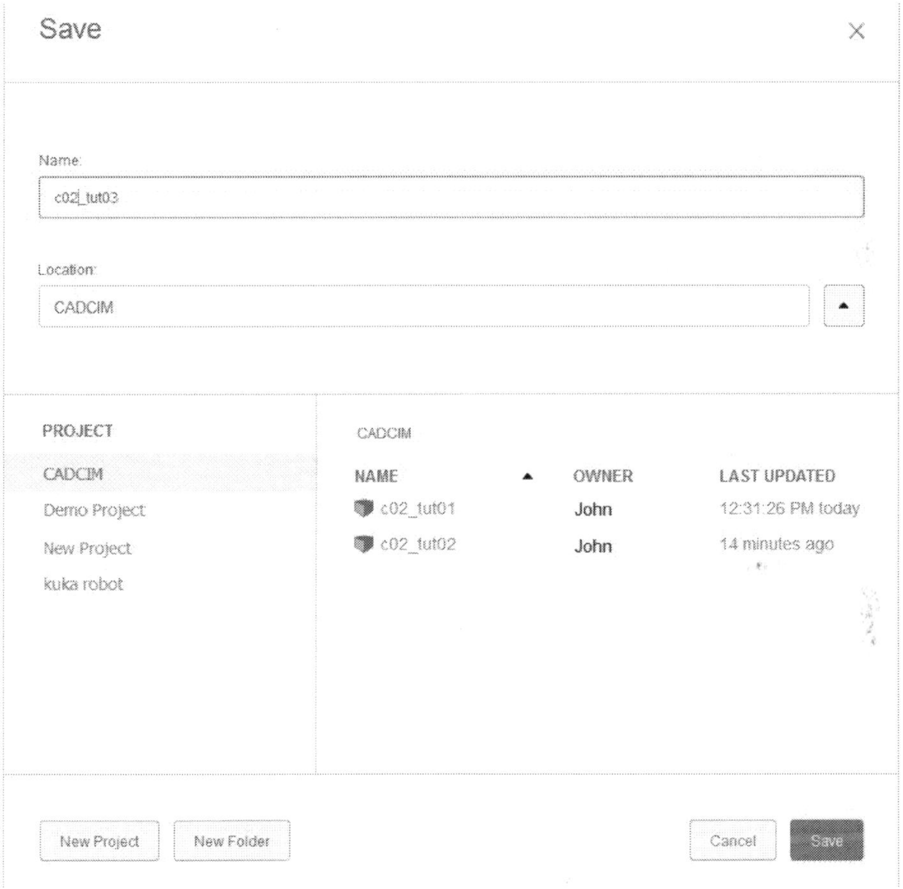

Figure 2-22 *The expanded **Save** dialog box*

4. Select an existing project **CADCIM** from the **PROJECT** column or you can a create new project by using the **New Project** button from the **Save** dialog box.

5. Select the **Save** button to save the sketch and exit the **Save** dialog box.

Tutorial 4

In this tutorial, you will draw the sketch of the model shown in Figure 2-23. The sketch of the model is shown in Figure 2-24. Do not dimension the sketch as the solid model and the dimensions are given for your reference only. **(Expected time: 30 min)**

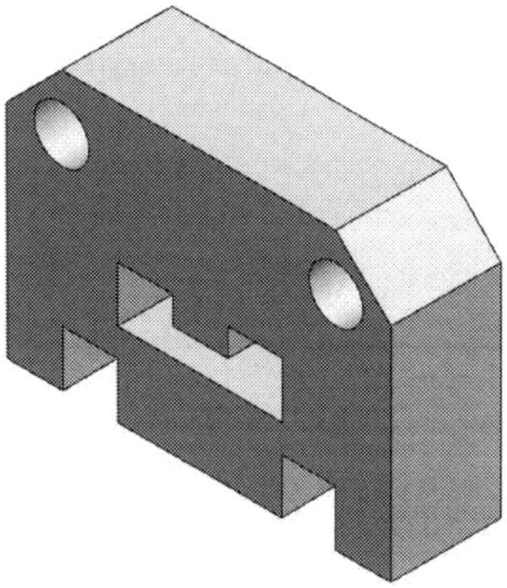

Figure 2-23 Solid model for Tutorial 4

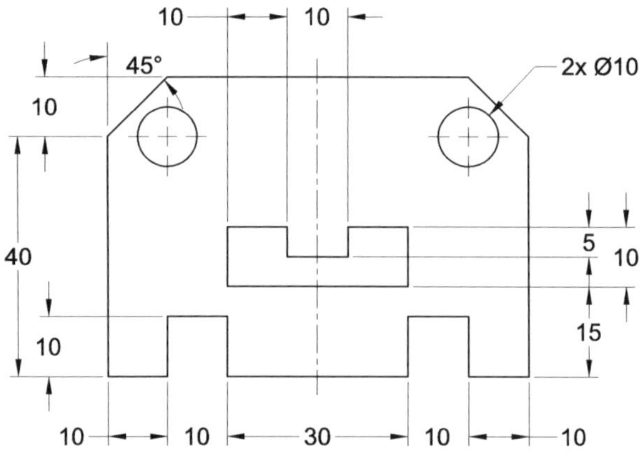

Figure 2-24 Sketch of the model for Tutorial 4

The following steps are required to complete this tutorial:

a. Start Autodesk Fusion 360 and then start a new design.
b. Invoke the model workspace.
c. Draw the outer loop of the sketch, refer to Figure 2-28.
d. Draw circles and sketch of inner cavity, refer to Figure 2-29.
e. Save the sketch and then close the file.

Starting a New File

1. Choose the **New Design** option from the **File** menu to create a new design.

   Whenever you start a new part design, by default, you are in the Model workspace. You need to start the design by first creating the sketch of the base feature.

2. Choose the **Create Sketch** tool from the **SKETCH** panel of the Ribbon; three default planes are displayed in the graphics window and you are prompted to select the plane.

3. Choose **FRONT** from the ViewCube and select the plane to orient it normal to the view. Alternatively, expand the **Origin** node in the **BROWSER** bar and then select the **XY** Plane to orient it normal to the view.

Modifying the Unit and Grid Settings

In Autodesk Fusion 360, you can modify the Units, Snap, and Grid settings. For this tutorial, you need to modify the grid and snap settings such that the cursor jumps through a distance of 5 mm.

1. Move the cursor on the **Units: mm** sub node of the **Document Settings** in the **BROWSER** bar in the graphics window; the **Change Active Units** button is displayed.

2. Choose the **Change Active Units** button; the **CHANGE ACTIVE UNITS** dialog box is displayed.

3. Choose the **Millimeter** option if not chosen by default from the **Unit Type** drop-down list in the **CHANGE ACTIVE UNITS** dialog box, and choose the **OK** button from this dialog box.

   As evident from Figure 2-24, the dimensions in the sketch are multiples of 5. Therefore, you need to modify the grid and snap settings such that the cursor jumps through a distance of 5 mm.

4. Choose the **Grid Settings** option from the **Grid and Snaps** flyout in the **Display Setting** at the bottom in the graphics window; the **GRID SETTINGS** dialog box is displayed.

5. Select the **Fixed** radio button from the **GRID SETTINGS** dialog box; the dialog box gets expanded. Now you need to enter **50** in the **Major Grid Spacing** edit box and **10** in the **Minor Subdivision** edit box. Choose **OK**.

> **Tip**
> *If a grid is displayed on the screen when you initially invoke the model workspace, you can turn off the grid display by clearing the* ***Sketch Grid*** *check box under the* ***Options*** *node in the* ***SKETCH PALATTE*** *window.*

Drawing the Outer Loop of the Sketch

The sketch of the model consists of an outer loop that has two circles and a cavity inside it. You will first draw the outer loop and then the inner entities. The sketch will be drawn by using the **Line** and **Circle** tools.

The outer loop will be drawn using continuous lines. You will start drawing the sketch from the lower left corner of the sketch.

1. Choose the **Line** tool from the **SKETCH** panel or right-click in the graphics window; a Marking menu is displayed. Choose **Sketch > Line**. On doing so, you are prompted to select the first point of the line to be created. Click on the origin to specify the first point of the sketch.

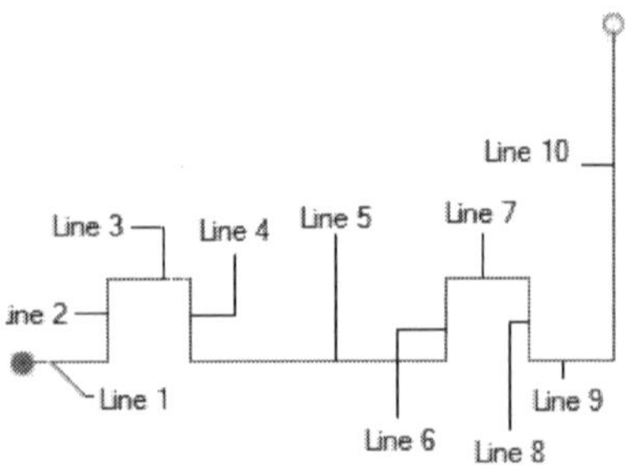

Figure 2-25 *Partial outer loop of the sketch*

2. Now, you need to draw the line 1, refer to Figure 2-25. Move the cursor toward right, enter **10** in the length input field, press TAB and enter **0** in the angle input field of the Dynamic Input. Click in the graphics window; the line 1 is drawn.

3. Move the cursor upward in the graphics window. Next, enter **10** in the length input field, press TAB, and enter **90** in the angle input field. Click in the graphics window; the line 2 is drawn, refer to Figure 2-25.

4. Move the cursor toward right in the graphics window. Next, enter **10** in the length input field, press TAB, and enter **90** in the angle input field in the Dynamic Input. Click in the graphics window; line 3 is drawn.

5. Move the line cursor vertically downward. Next, enter **10** in the length input field, press TAB, and enter **90** in the angle input field in the Dynamic Input. Click in the graphics window; line 4 is drawn.

6. Move the line cursor horizontally toward the right. Next, enter **30** in the length input field, press TAB, and enter **90** in the angle input field in the Dynamic Input. Click in the graphics window; line 5 is drawn.

7. Move the line cursor vertically upward. Next, enter **10** in the length input field, press TAB, and enter **90** in the angle input field in the Dynamic Input. Click in the graphics window; line 6 is drawn, refer to Figure 2-25.

8. Move the line cursor horizontally toward the right Next, enter **10** in the length input field, press TAB, and enter **90** in the angle input field in the Dynamic Input. Click in the graphics window; line 7 is drawn, refer to Line 7 in Figure 2-25.

9. Move the line cursor vertically downward. Next, enter **10** in the length input field, press TAB, and enter **90** in the angle input field in the Dynamic Input. Click in the graphics window; line 8 is drawn, refer to Line 8 in Figure 2-25.

10. Move the line cursor horizontally toward the right. Next, enter **10** in the length input field, press TAB, and enter **90** in the angle input field in the Dynamic Input. Click in the graphics window; line 9 is drawn, refer to Line 9 in Figure 2-25.

11. Move the line cursor vertically upward. Next, enter **40** in the length input field, press TAB, and enter **90** in the angle input field in the Dynamic Input. Click in the graphics window; line 10 is drawn, refer to Line 10 in Figure 2-25.

Now you need to draw a line inclined at an angle of 135 degrees. To draw this line,you need to move the cursor at an angle of 135 degrees with respect to line 10.

12. Move the line cursor diagonally left in an upward direction. Next, enter **14.14** in the length input field, press TAB, and enter **135** in the angle input field in the Dynamic Input. Click in the graphics window; the inclined line is drawn.

13. Move the cursor toward left, enter **50** in the length input field, press TAB and enter **135** in the angle input field of the Dynamic Input. Click in the graphics window; a line of 50 mm length is drawn.

14. Move the line cursor diagonally in downward direction. Next, enter **14.14** in the length input field, press TAB, and enter **135** in the angle input field in the Dynamic Input. Click in the graphics window; the inclined line is drawn.

15. Move the cursor vertically downward to the start point of the first line, which is the origin, and then click to complete the sketch and exit the Line tool.

    This completes the sketch of the outer loop. Since the display of the sketch is small, you need to modify the drawing display area using the **Fit** tool, such that the sketch fits the screen.

16. Choose the **Fit** tool from the **Navigation Bar and Display Settings** to fit the current sketch into the screen. The outer loop of the sketch is completed and is shown in Figure 2-26.

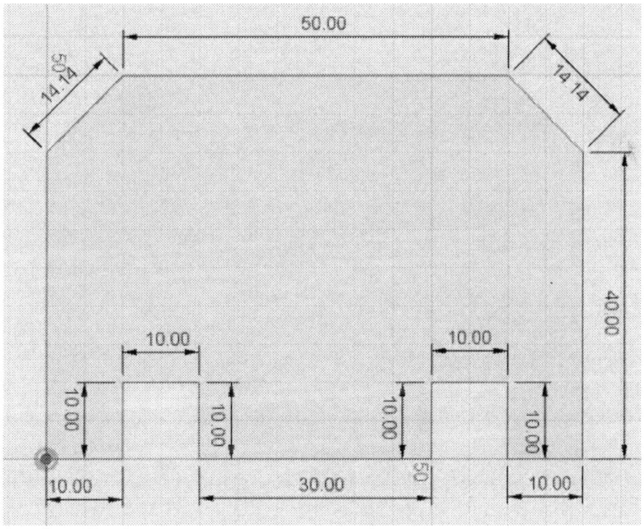

***Figure 2-26** Outer loop of the sketch*

Drawing Circles

In this section, you will invoke the **Circle** tool by using the Marking menu to draw two circles. You will use the inferencing lines originating from the start points and end points of the inclined lines to specify the centerpoint of circles. At a given time, you can either snap to grid or use inferencing lines to draw sketches. In this tutorial, you will use inferencing lines to draw the sketch. So, you need to turn off the snap to grid option.

1. Clear the **Snap to Grid** check box from the **Grid and Snap** flyout.

2. Right-click in the drawing area to invoke the Marking menu and then choose the **Center Diameter Circle** tool from the **Sketch > Circle** ; you are prompted to place the center point of the circle.

   When you invoke the **Circle** tool, the arrow cursor is replaced by the circle cursor.

3. Move the circle cursor close to the lower endpoint of the right inclined line and then move it toward left. Remember that you should not press the left mouse button at this moment. An inferencing line, which is a blue dashed line is displayed originating from the lower end point of the right inclined line. Now, move the cursor to the upper end point of the right inclined line and then move the cursor vertically down; a vertical inferencing line starts following the cursor. Click on the point where the vertical and horizontal inferencing meets, refer to Figure 2-27.

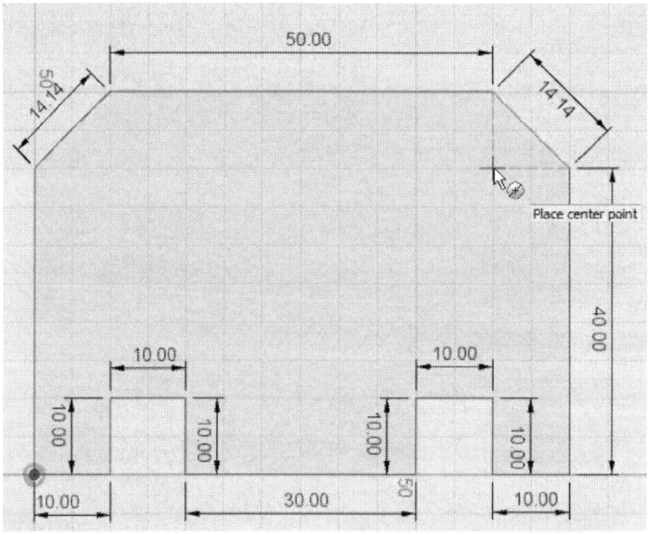

Figure 2-27 *Drawing a circle with the help of inferencing lines*

4. Click at the point where the inferencing lines from both the endpoints of the inclined lines intersect; the center point of the circle is specified and you are prompted to specify the diameter of the circle

5. Move the circle cursor toward left to define a circle and enter **10** in the diameter input field and press ENTER twice to exit the **Circle** tool.

6. Similarly, draw a circle on the left using the inferencing lines generating from the end points of the left inclined line. The sketch after drawing the two circles inside the outer loop is shown in Figure 2-28.

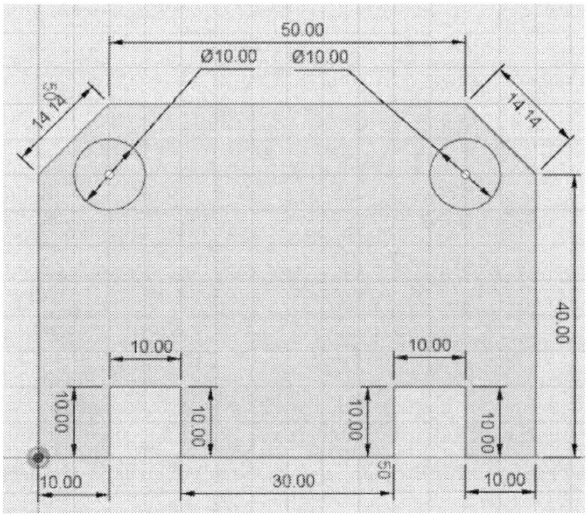

Figure 2-28 Sketch after drawing the two inner circles

Drawing the Sketch of the Inner Cavity

Next, you will draw the sketch of the inner cavity. You will start drawing the sketch with the lower horizontal line.

1. Turn on **Snap to Grid** from the **Grid and Snap** flyout and select the **Sketch Grid** check box from the **SKETCH PALETTE** window.

2. Invoke the **Line** tool by pressing the L key; the arrow cursor is replaced by the line cursor.

3. Starting from the origin, move the line cursor by 4 units to the right and 3 units up and then click to specify the start point of line.

4. Move the cursor horizontally toward the right. Click again when the length of the line above the line cursor is displayed as 30.

5. Move the line cursor vertically upward and click when the length of the line on the line cursor is displayed as 10.

6. Move the line cursor horizontally toward the left and click when the length of the line on the line cursor is displayed as 10.

7. Move the line cursor vertically down and click when the length of the line on the line cursor is displayed as 5.

8. Move the line cursor horizontally toward the left and click when the length of the line on the line cursor is displayed as 10.

9. Move the line cursor vertically up and click when the length of the line on the line cursor is displayed as 5.

10. Move the line horizontally toward the left and click when the length of the line on the line cursor is displayed as 10.

11. Move the line cursor vertically downward to the start point of the first line. The length of the line at this point is displayed as 10.

12. Right-click and then choose the **OK** option from the shortcut menu; the complete sketch is drawn.

13. Choose the **Fit** button from the Navigation Bar and Display Setting to fit the display of the sketch into the screen. The final sketch for Tutorial 4 after hiding the constraint for more clarity, is shown in Figure 2-29.

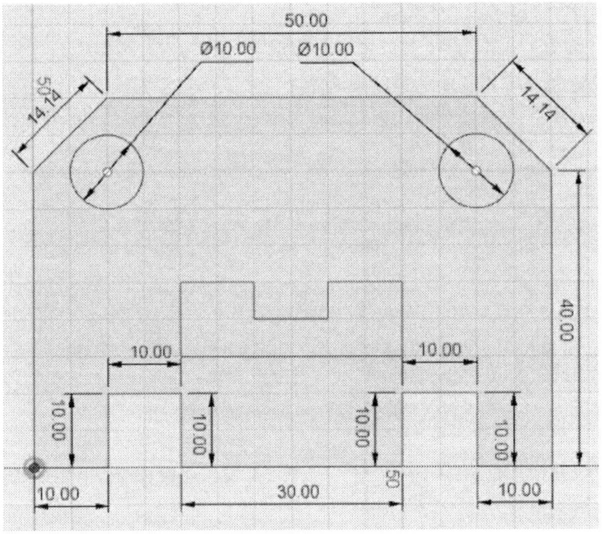

***Figure 2-29** The final sketch for Tutorial 4*

Saving the Sketch

Next, you need to save the sketch.

1. Choose the **Save** option from the **File** menu or **Application** menu; the **Save** dialog box is displayed.

2. Enter **c02_Tut_04** in the **Name** edit box.

3. Next, click on the down arrow on the extreme right side of the **Location** selection box; the **Save** dialog box gets expanded.

4. Select an existing project **CADCIM** from the **PROJECT** column or create new project by choosing the **New Project** button from the **Save** dialog box.

5. Choose the **Save** button to save the sketch and exit the **Save** dialog box.

Self-Evaluation Test

Answer the following questions and then compare them to those given at the end of this chapter:

1. Which of the following is required to begin the creation of a sketch?

   (a) Sketch Plane (b) Sketch References
   (c) Constraints (d) Dimensions

2. You can convert a sketched entity into a construction entity by selecting the _________ option from the Marking Menu.

3. The _________ tool is used to create a rectangle to define width, direction, and height.

4. _________ are temporary lines that are used to track a particular point on the screen.

5. You can also delete the sketched entities by pressing the _________ key.

6. In Autodesk Fusion 360, a rectangle is considered as a combination of individual _________.

7. You can press the ESC key to exit the sketching tool. (T/F)

8. The base feature of any design is a sketched feature and is created by drawing a sketch. (T/F)

9. You can invoke the arc mode using the **Line** tool. (T/F)

10. You cannot turn off the display of grid lines. (T/F)

Review Questions

Answer the following questions:

1. In Autodesk Fusion 360, which of the following entities combine to form a polygon?

   (a) Lines (b) Arcs
   (c) Splines (d) None of these

2. Which of the following key drawing display options is used to interactively zoom in a drawing?

   (a) F5 (b) F6
   (c) F7 (d) F4

3. Which of the following drawing display options will make the selected entity normal to screen?

   (a) **Zoom** (b) **Pan**
   (c) **Look At** (d) None of these

4. Choose the __________ button from the **Navigation Bar and Display Setting** to fit the display of the sketch into the screen.

5. You can specify the position of entities dynamically by using the Dynamic Input. (T/F)

6. In Autodesk Fusion 360, you can start a new file by choosing the **New Design** option from the **File** menu. (T/F)

EXERCISES

Exercise 1

Draw a basic sketch of the model shown in Figure 2-30. The sketch to be drawn is shown in Figure 2-31. Do not dimension it as the dimensions are given only for reference.

(Expected time: 30 min)

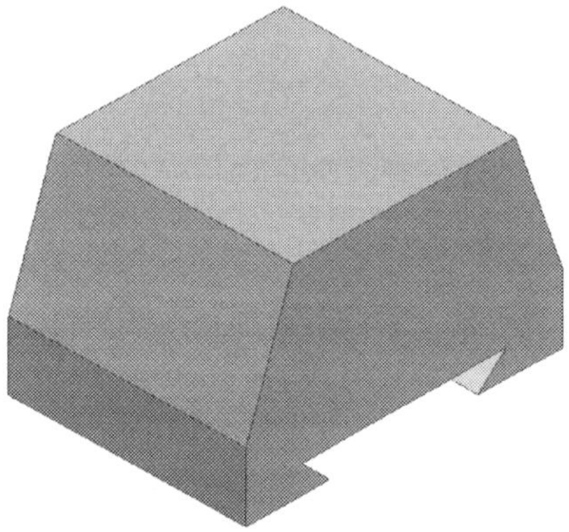

Figure 2-30 Model for Exercise 1

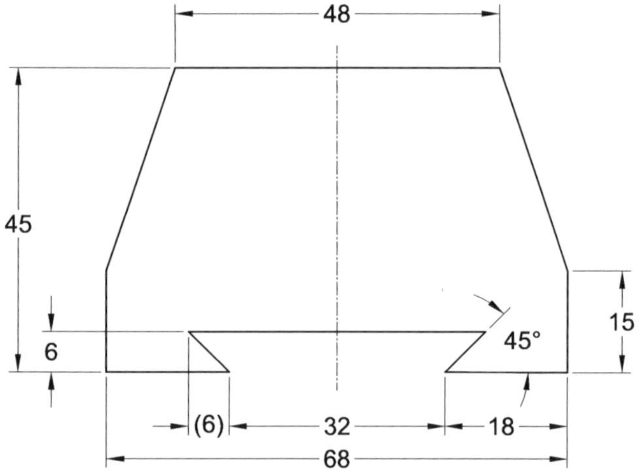

Figure 2-31 Sketch for Exercise 1

Exercise 2

Draw a basic sketch of the model shown in Figure 2-32. The sketch to be drawn is shown in Figure 2-33. Do not dimension it as the dimensions are given only for reference.

(Expected time: 45 min)

Figure 2-32 Model for Exercise 2

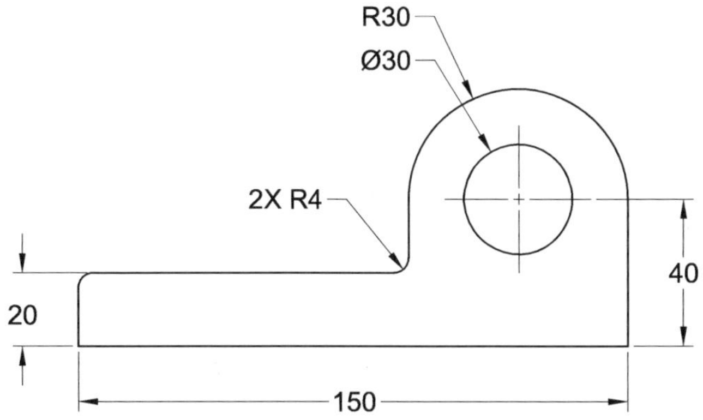

Figure 2-33 Sketch for Exercise 2

Exercise 3

Draw a sketch of the model shown in Figure 2-34. The sketch to be drawn is shown in Figure 2-35. Do not dimension it as the dimensions are given only for reference.

(Expected time: 45 min)

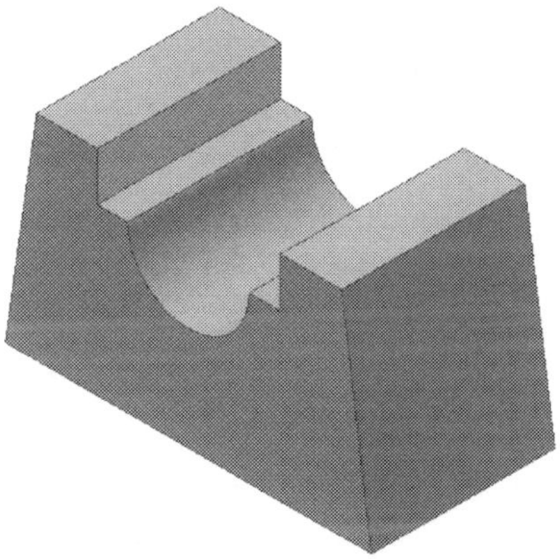

Figure 2-34 Model for Exercise 3

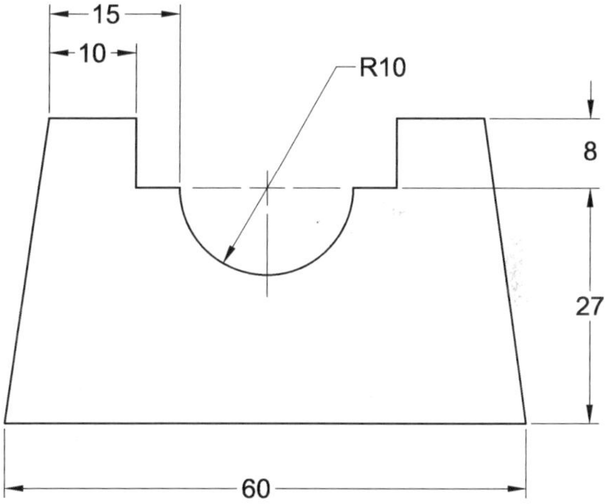

Figure 2-35 Sketch for Exercise 3

Answer to Self-Evaluation Test

1. Sketch Plane, 2. **Normal / Construction**, 3. **3-Point Rectangle**, 4. Inferencing lines, 5. DELETE, 6. lines, 7. T, 8. T, 9. T, 10. F

Chapter 3

Adding Constraints and Dimensions to Sketches

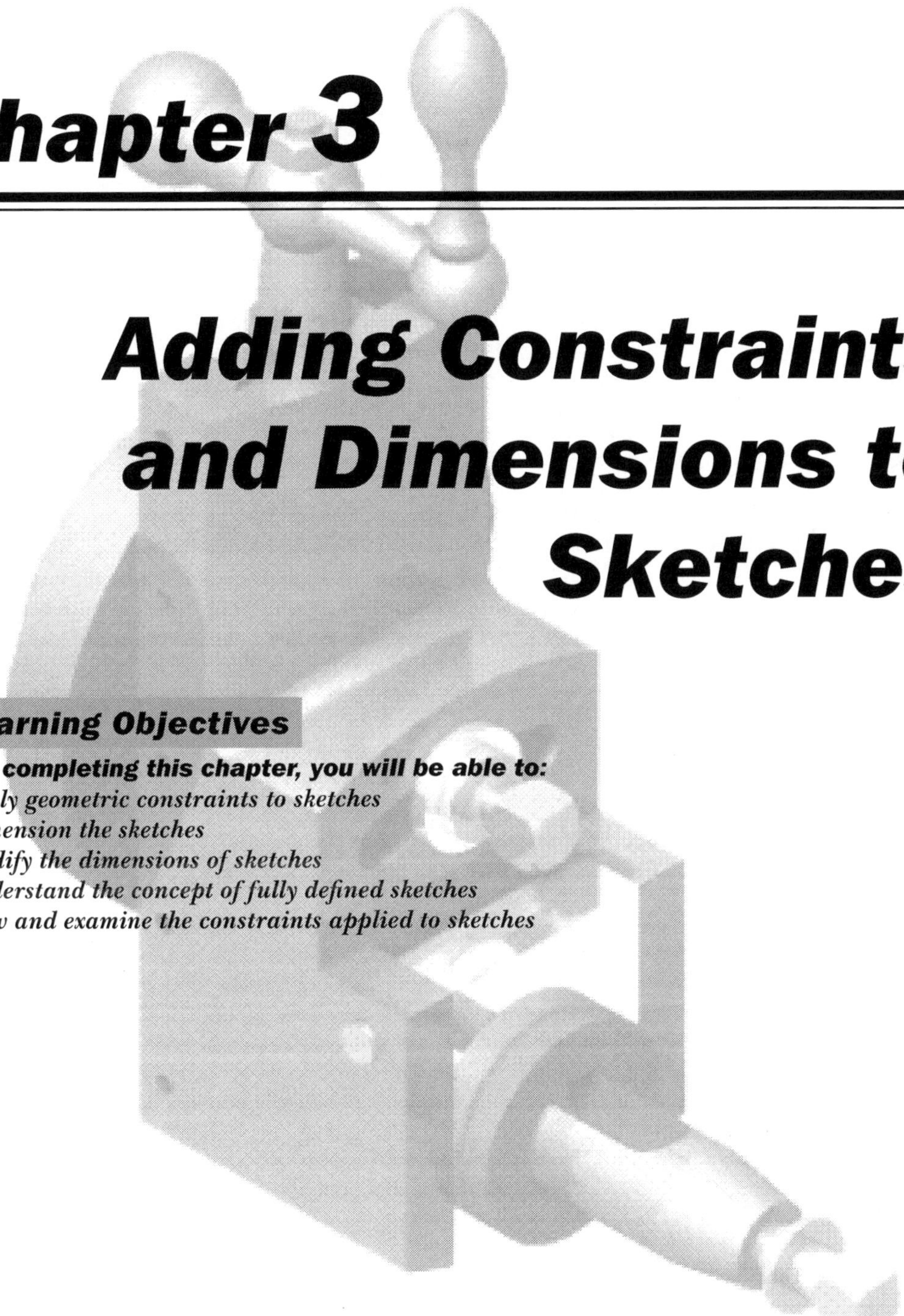

Learning Objectives

After completing this chapter, you will be able to:

- *Apply geometric constraints to sketches*
- *Dimension the sketches*
- *Modify the dimensions of sketches*
- *Understand the concept of fully defined sketches*
- *View and examine the constraints applied to sketches*

INTRODUCTION

Constraints are applied to the sketched entities to define their size and position with respect to other elements. Also, they are useful for capturing the design intent. As mentioned in Chapter 1, there are twelve types of geometric constraints that can be applied to the sketched entities. These constraints restrict their degrees of freedom and make them stable. Most of the constraints get automatically applied to the entities while drawing. However, sometimes you may need to apply some additional constraints to the sketched entities. These constraints are discussed later in this chapter.

After drawing a sketch and adding constraints to it, dimensioning is the next most important step in design creation. As mentioned earlier, Autodesk Fusion 360 is a parametric solid modeling package. The parametric property of the software ensures that irrespective of its original size, the selected entity is driven by the specified dimension value. Therefore, whenever you modify or apply dimension to an entity, it is forced to change its size with respect to the specified dimension value. The type of dimension to be applied varies according to the type of entity selected.

TUTORIALS

Tutorial 1

In this tutorial, you will draw the sketch shown in Figure 3-1. This sketch is the same as the one drawn in Tutorial 1 of Chapter 2. After drawing the sketch, you will add the required constraints and then dimension it. **(Expected time: 30 min)**

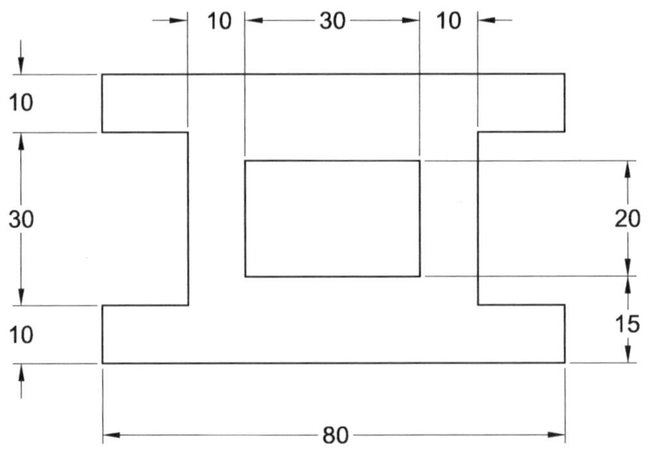

Figure 3-1 *Dimensioned sketch for Tutorial 1*

The following steps are required to complete this tutorial:

a. Start a new design file and invoke the model workspace.
b. Draw the initial sketch by using the **Line** and **2-Point Rectangle** tools.
c. Add required constraints to the sketch.
d. Dimension the sketch by using the origin to fully constrained it.
e. Save the sketch and then close the file.

Starting Autodesk Fusion 360 and Invoking the Model Workspace

1. Start Autodesk Fusion 360 and then invoke the **Model** Workspace by choosing the **New Design** option from the **File** menu.

   Next, you need to invoke the sketching interface.

2. Choose the **Create Sketch** tool from the **SKETCH** panel in the Ribbon; the three default planes are displayed in the graphics window and you are prompted to select a plane or a planar face.

3. Select the **YZ** plane from the **BROWSER** bar by expanding the **Origin** node; the selected plane gets oriented normal to the view.

   You do not need to draw the sketch to the exact length. Use the temporary tracking option for drawing the sketch. For your reference, all lines in the sketch are numbered, see Figure 3-1.

4. Clear the **Sketch Grid** check box from the **SKETCH PALETTE** window to hide the grid, if grid is displayed by default.

5. Expand the **Document Settings** node, move the cursor to the **Units** sub node in the **BROWSER** bar in the graphics window; the **Change Active Units** button is displayed.

6. Click on the **Change Active Units** button; the **CHANGE ACTIVE UNITS** dialog box is displayed.

7. Select the **Millimeter** option if not selected from the **Unit Type** drop-down list in the dialog box and choose the **OK** button from this dialog box.

   Note that if **Millimeter** is already selected then choose the **Cancel** button to exit this dialog box.

Drawing the Initial Sketch

1. Using the **Line** and **Center Rectangle** tools, draw the required sketch, as shown in Figure 3-1.

   You do not need to draw the sketch to the exact dimensions. For your reference, all lines in the sketch are numbered, as shown in Figure 3-2.

Applying Constraints to the Sketch

After drawing the sketch, you need to add geometric constraints.

Geometric constraints are logical operations that are performed to add relationships (such as tangent or perpendicular) between the sketched entities, planes, axes, edges, or vertices. You can apply constraints to a sketch by using the options under the **Constraints** of the **SKETCH PALETTE** window.

In Autodesk Fusion 360, some of the constraints get automatically applied to the sketched entities while drawing. For example, when you specify the start point of a line and move the cursor horizontally toward right or left, you will notice that horizontal line symbol is displayed below the line cursor. This symbol indicates that horizontal constraint is applied to the line while drawing. If you move the cursor vertically downward or upward, a vertical line symbol for the vertical constraint will be displayed below the line cursor. If you move the cursor to the intersection of two or more sketched entities, an intersection symbol will appear on the cursor. Similarly, other constraints are also automatically applied to the sketch when you draw it. These automatically applied relations are called Automatic Constraints.

It is evident from Figure 3-2 that some of the lines need to be of the same length. For example, lines 1 and 7, lines 2 and 6, lines 8 and 12, and so on. You can draw lines of same length in two ways. First by assigning dimensions to the lines. However, this will increase the number of dimensions in the sketch. Second is by applying the equal constraint to all the lines having the same length. Now, if you dimension any one of the constrained lines, all other lines will be forced to acquire the same dimension. The equal constraint is applied in pairs. Note that automatically applied constraints have been hidden for better visibility in this figure.

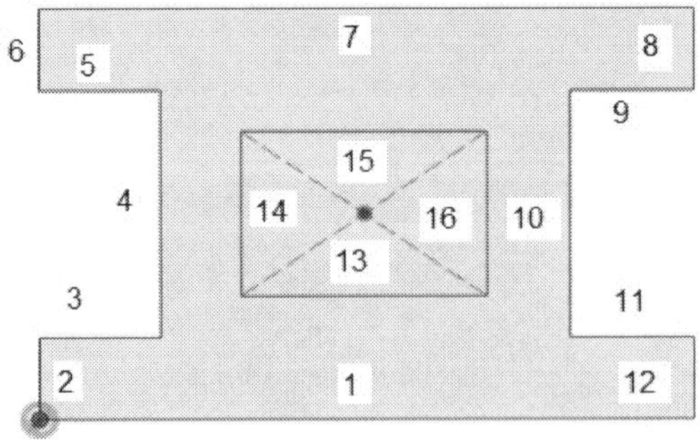

***Figure 3-2** Initial sketch drawn using the sketching tools*

The constraints that can be applied automatically are listed next:

Horizontal	Vertical	Coincident	Midpoint
Tangent	Perpendicular	Concentric	Parallel

1. Choose the **Equal** option under the **Constraints** node in the **SKETCH PALETTE** window, as shown in Figure 3-3 to invoke the **Equal** constraint .

   When you invoke this constraint, you are prompted to select the sketch object or change the constraint type.

2. Select line 2; the selected line gets highlighted and you are prompted to select other geometries. Select line 6; the equal constraint is applied to lines 2 and 6. Again, you are prompted to select the sketch object or change the constraint type.

   Select line 6 as the first line and then line 8 as the second line; the equal constraint gets applied.

3. Similarly, select lines 8 and 12, 1 and 7, 3 and 5, 5 and 9, 9 and 11; the equal constraint gets applied.

   Note that while applying the equal constraint between line 9 and 11, an error message conveying that the sketch geometry is over constrained might be displayed. It means those two geometries are already drawn with equal length with the help of auto trace.

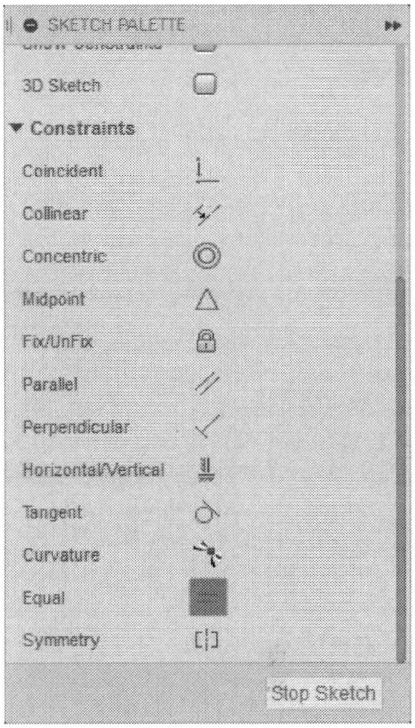

Figure 3-3 *The* ***SKETCH PALETTE*** *window*

4. Right-click in the graphics window and then choose **Cancel** from the Marking menu displayed.

Note

While drawing the sketch of the model, sometimes the constraints get automatically applied to the sketch. In such a case, whenever you apply constraint to the sketch, an error message box is displayed.

5. Apply the horizontal and vertical constraints to the horizontal and vertical lines of the sketch, respectively, if required.

   The constraints available in Autodesk Fusion 360 are discussed next.

Coincident

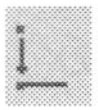

The coincident constraint forces the selected point to become coincident with the selected line, arc, circle, or ellipse. To apply this constraint, select the required entity and then choose the **Coincident** option from the **Constraints** node of the **SKETCH PALETTE** window.

Collinear

The collinear constraint forces the selected lines to lie on the same infinite line. To apply this relation, select the lines to which the colinear constraint is to be applied. Next, choose the **Collinear** option under the **Constraints** node.

Concentric

The concentric constraint forces selected arc or circle to share the same center point with the other arc, circle, point, vertex, or circular edge. To apply this constraint, select the required entity and then choose the **Concentric** option under the **Constraints** node in the **SKETCH PALETTE** window.

Midpoint

The midpoint constraint forces a selected point to move to the midpoint of a selected line. To apply this constraint, select the point and the line. Next, choose the **Midpoint** option under the **Constraint** node in the **SKETCH PALETTE** window.

Fix/UnFix

The fix/unfix constraint forces the selected entity to be fixed or unfixed at the specified position. If you apply this constraint to a line or an arc, its location will be fixed or unfixed but you can change its size by dragging the endpoints. To apply this constraint, select the required entity and choose the **Fix/UnFix** option. Note that only an open sketch profile can be changed not of the closed profile.

Perpendicular

The perpendicular constraint forces the selected lines to become perpendicular to each other. To apply this constraint, select two lines and choose the **Perpendicular** option under the **Constraints** node in the **SKETCH PALETTE** window. Figure 3-4 shows two lines before and after applying the Perpendicular constraint.

Parallel

The **parallel** constraint forces the selected lines to become parallel to each other. To apply this constraint, select two lines and choose the **Parallel** option under the **Constraints** node in the **SKETCH PALETTE** window. Figure 3-5 shows two lines before and after applying the parallel constraint.

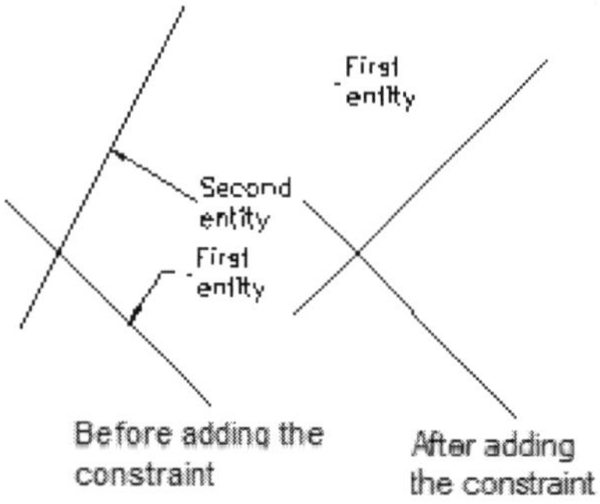

Figure 3-4 Entities before and after applying the perpendicular constraint

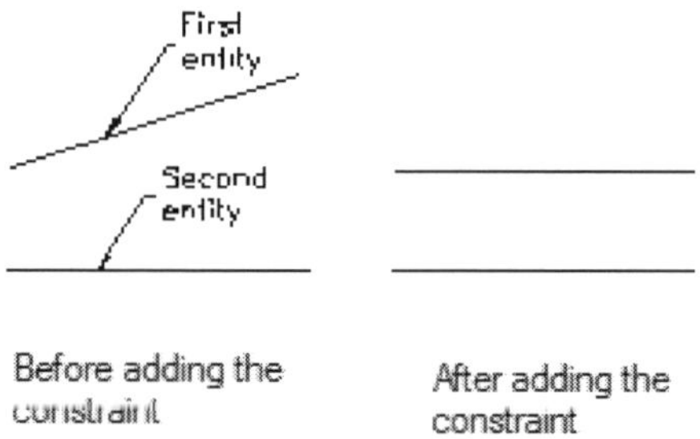

Figure 3-5 Entities before and after applying the parallel constraint

Horizontal/Vertical

The horizontal/vertical constraint forces one or more selected lines or centerlines to become horizontal/vertical. You can also select an external entity such as an edge, plane, axis, or sketch curve on an external sketch that will act as a line to apply this constraint. You can also force two or more points to become horizontal using the horizontal/ vertical constraint. The point can be a sketch point, a center point, an endpoint, a control point of a spline, or an external entity such as origin, vertex, axis, or point in an external sketch.

Tangent

The tangent constraint forces selected arc, circle, or ellipse to become tangent to the another arc, circle, ellipse, line, or edge. To apply this constraint, select the entities and choose the **Tangent** option under the **Constraint** node in the **SKETCH PALETTE** window. Figure 3-6 shows a line and a circle before and after applying the tangent constraint. Figure 3-7 shows two arcs before and after applying the tangent constraint.

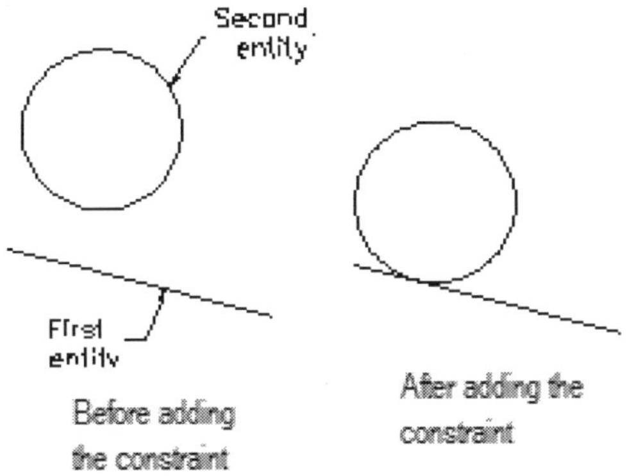

Figure 3-6 Applying the tangent constraint to a line and a circle

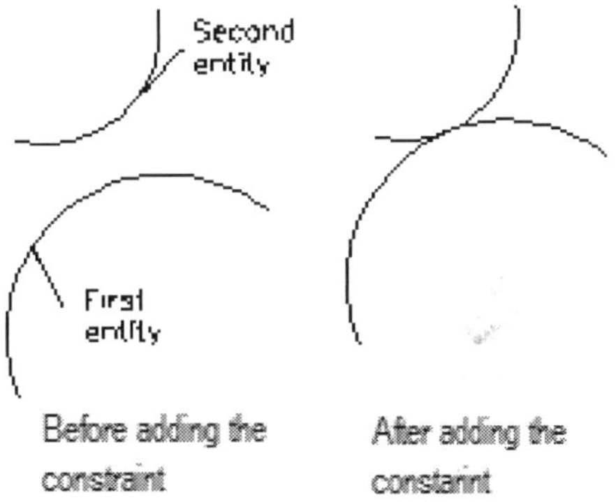

Figure 3-7 Applying the tangent constraint to two arcs

Symmetry

The symmetry constraint forces two selected lines, arcs, points, and ellipses to remain equidistant from the centerline. This constraint also forces the entities to have the same orientation. To apply this constraint, choose the **Symmetry** option from the **Constraints** node in the **SKETCH PALETTE** window. Next, select two lines and an axis or a centerline to apply the symmetry constraint.

Equal

The equal constraint can be used for line segments or curves. If you select two line segments, this constraint will force the length of one of the selected line segments to become equal to the length of the other selected line segment. In case of curves, this constraint should force the radius of one of the selected curves to become equal to that of the other selected curve. Note that if the first selection is a line, the second selection will also be a line. Similarly, if the first selection is a curve, the second selection also needs to be a curve.

Smooth

This constraint is used to apply curvature continuity between a spline and an entity connected to it. The entities that can be selected to apply this constraint include a line, arc, or another spline. Note that these entities should be connected to the spline.

Note that while applying a constraint between two entities, the entity selected first will remain fixed and second entity will adjust with respect to the first one based on the constraint applied.

Dimensioning the Sketch

Once all the required constraints have been applied to the sketch, you can dimension the sketch. As mentioned earlier in this chapter, whenever you modify or apply dimension to an entity, it is forced to change its size with respect to the specified dimension value.

1. Choose the **Sketch Dimension** tool from the **SKETCH** panel. Alternatively, right-click anywhere in the graphics window and then choose **Sketch > Sketch Dimension** from the Marking menu displayed.

2. Select the line 1, refer to Figure 3-2; it gets highlighted and a linear dimension is displayed.

   Note that it is important to modify the value of the dimension after it is placed so that geometries are driven to the required values.

3. Place the dimension below line 1; the **Dimension** edit box is displayed. Enter **80** as the length of line 1 in this edit box and then right-click in the graphics window. Next, choose the **OK** button from the Marking menu.

   You will notice that the length of line 1 as well as line 7 get modified to 80 because of the equal constraint.

   As the **Dimension** tool is still active, you are prompted again to select the objects to dimension.

4. Select line 2 and place the dimension on the left of this line; the **Dimension** edit box is displayed. Change the length of the line to **10** in this edit box and press ENTER.

   You will notice that the length of lines 6, 8, and 12 is also changed to 10 units. This is because the **Equal** constraint has been applied to all these lines.

5. Select line 4 and place it along the previous dimension. Modify the dimension value in the **Dimension** edit box to **30** and press ENTER. Notice that the length of line 10 is also modified.

6. Select line 16 and place the dimension on the right outside the sketch. Modify the dimension value in the **Dimension** edit box to **20** and press ENTER.

7. Select line 15 and place the dimension outside the sketch on the top. Modify the dimension value in the **Dimension** edit box to **30** and press ENTER.

8. Select line 4 and line 14. Now, apply the dimension on the top outside the sketch and then change the dimension value to **10** in the **Dimension** edit box and press ENTER.

9. Similarly, select lines 16 and 10 to dimension the distance between these two lines and place the dimension on the top outside the sketch. Change the dimension value to **10** in the **Dimension** edit box and press ENTER. You will notice that the lengths of lines 5, 9, 3, and 11 also get automatically adjusted because of equal constraint.

   Now, you need to dimension the inner rectangle vertically from the outer loop,

10. Select lines 1 and 13, and then place the dimension on the right of the sketch. Next, modify the dimension value in the **Dimension** edit box to **15** and press ENTER.

    With this step, you have applied all the required constraints and dimensions to the sketch. Now the sketch is ready to be converted into a feature. If you try to add more constraints or dimensions to this sketch, the **Fusion 360** error message box will appear, informing that adding this dimension will over-constrain the sketch, see Figure 3-8. If you still want this dimension to be displayed, choose the **OK** button from this message box; the dimension will be added as a driven dimension. A driven dimension is placed inside parentheses and is not used during the manufacturing process. This dimension is used only for reference. Note that you cannot edit the value of a driven dimension. The sketch after applying all the dimensions and constraints should look similar to the one shown in Figure 3-9.

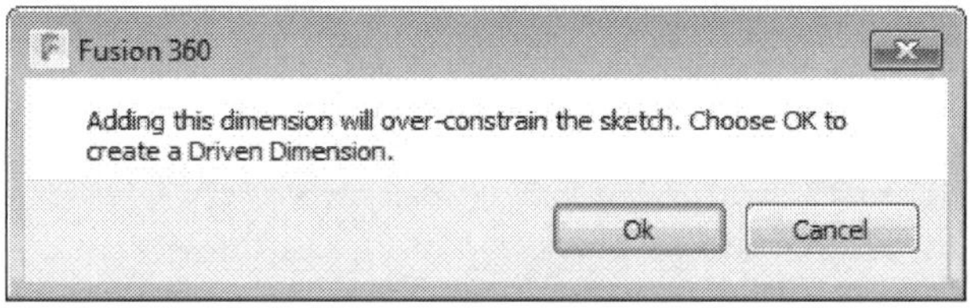

Figure 3-8 *Sketch after adding dimensions*

Even after adding all the dimensions if the color of entities in the sketch is blue, it indicates that the sketch is not fully constrained. In order to fully constrain the sketch, you need to constrain it with respect to the origin which is fixed by default.

11. Choose the **Coincident** option under the **Constraints** node of the **SKETCH PALETTE** window; you are prompted to select the object or change the constraint type. 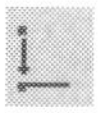

12. Select the lower left vertex of the sketch which is the intersection point of lines1 and 2; you are prompted to select the other geometries.

13. Select the origin; the entire sketch shifts itself such that the lower left vertex of the sketch is now at the origin.

    Figure 3-10 shows the fully constrained sketch for Tutorial 1.

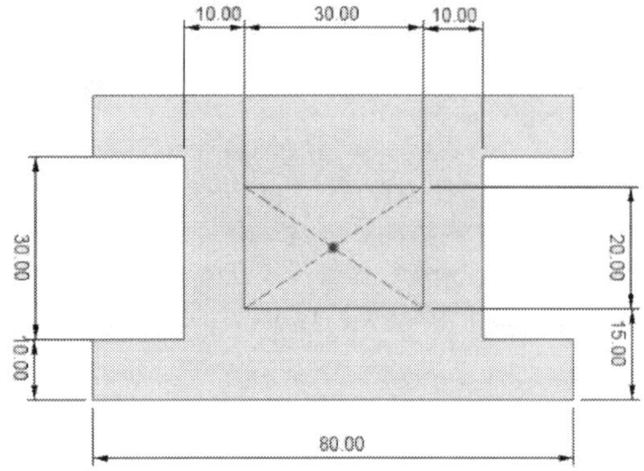

Figure 3-9 Sketch after adding dimensions

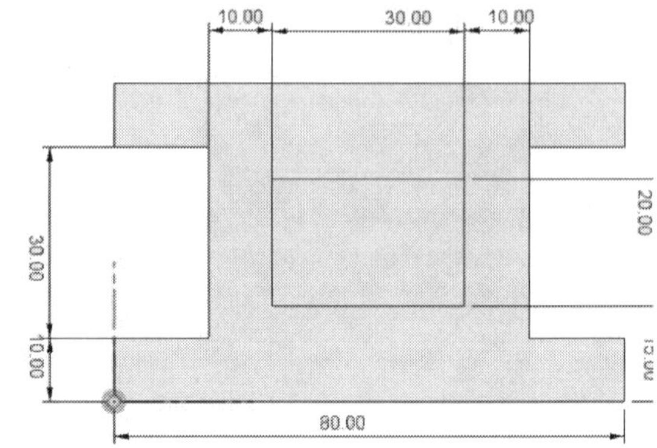

Figure 3-10 Fully constrained sketch for Tutorial 1

Saving the Model

Next, you need to save the sketch.

1. Choose the **Save** option from the **File** menu or the **Application** menu; the **Save** dialog box is displayed.

2. Enter **c03_Tut_01** in the **Name** edit box.

3. Click on the down arrow on the extreme right side of the **Location** selection box; the **Save** dialog box gets expanded. 

4. Select **CADCIM** from the **PROJECT** column or you can create a new project by choosing the **New Project** button from the **Save** dialog box.

5. Choose the **Save** button to save the sketch and exit the **Save** dialog box.

Tutorial 2

In this tutorial, you will draw a sketch of the revolved model shown in Figure 3-11. The dimensioned sketch to be drawn is shown in Figure 3-12. The solid model is given for your reference only. **(Expected time: 30 min)**

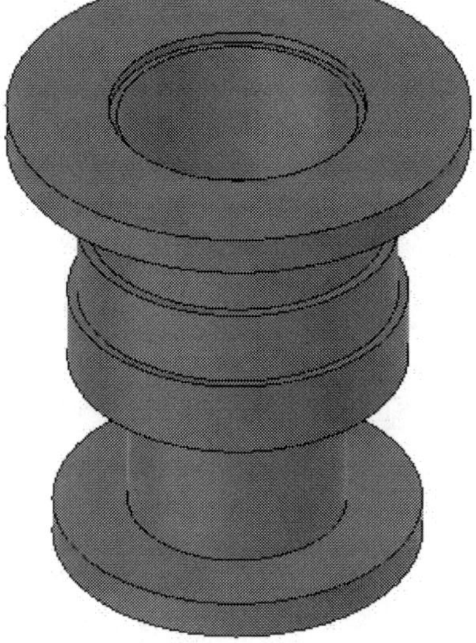

Figure 3-11 Solid model for Tutorial 2

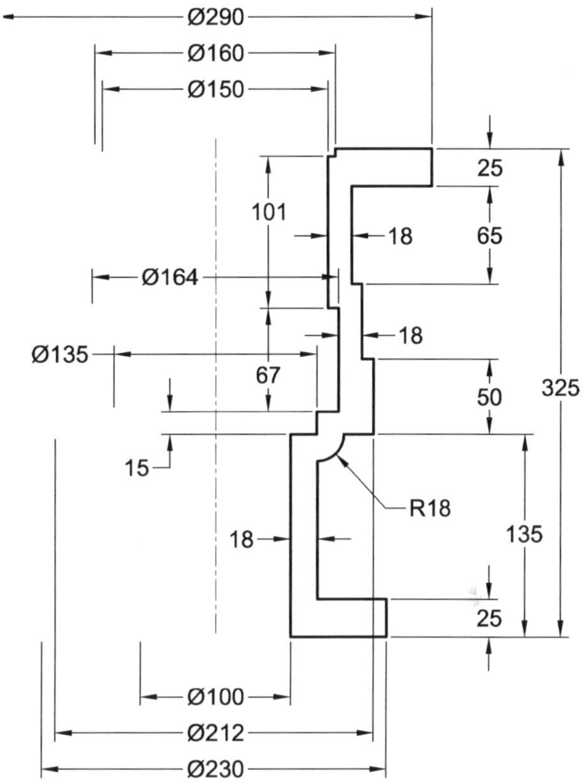

Figure 3-12 Sketch of the model

The following steps are required to complete this tutorial:

a. Start a new design file and invoke the model workspace.
b. Draw a centerline to add linear diameter dimensions to the sketch of the model.
c. Create the sketch by using various sketching tools.
d. Add dimensions to the sketch to fully define it.

Starting Autodesk Fusion 360 and Invoking the Model Workspace

1. Start Autodesk Fusion 360 and then invoke the **Model** workspace by choosing the **New Design** option from the **File** menu.

   Next, you need to invoke the sketching interface.

2. Choose the **Create Sketch** tool from the **SKETCH** panel in the Ribbon; the three default planes are displayed in the graphics window and you are prompted to select a plane or planar face.

3. Select the **YZ** plane from the **BROWSER** bar by expanding the **Origin** node; the selected plane gets oriented normal to the view.

4. Move the cursor to the **Units** sub node of (Unsaved) component in the **BROWSER** bar in the graphics window; the **Change Active Units** button is displayed.

5. Choose the **Change Active Units** button; the **CHANGE ACTIVE UNITS** dialog box is displayed.

6. Select the **Millimeter** option, if not selected, from the **Unit Type** drop-down list in the dialog box and choose the **OK** button from this dialog box. Note that if **Millimeter** is already selected then choose the **Cancel** button to exit this dialog box.

7. Clear the **Sketch Grid** check box from the **SKETCH PALETTE** window to hide the grids, if grid is displayed by default.

Drawing the Sketch

To draw the sketch of the revolved model, you need to draw a centerline around which the sketch of the base feature will be revolved.

1. Invoke the **Line** tool from the **SKETCH** panel.

2. Draw a vertical centerline starting from the origin. Press the ESC key. Then select this vertical centerline and choose the **Construction** option under the **Options** node in the **SKETCH PALETTE** window to make it as central axis.

3. Draw the sketch similar to the one shown in Figure 3-12 by using the **Line** and **3-Point Arc** tools.

   Use the temporary tracking option for drawing the sketch. The sketch after drawing all the entities is shown in Figure 3-13.

   Note that while dimensioning, the shape of sketch might change. To avoid this, try to draw the sketch with approximate dimension.

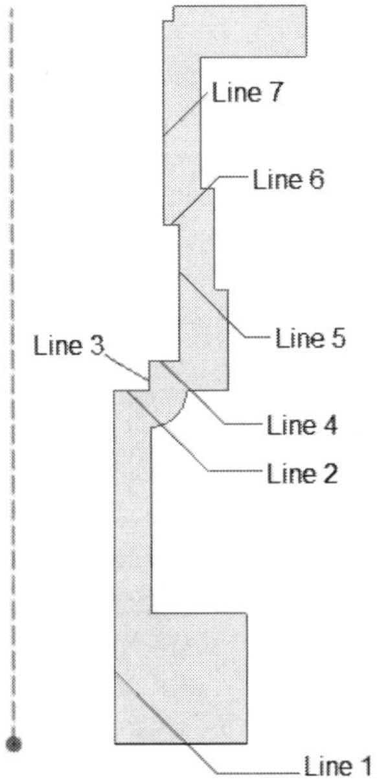

Figure 3-13 *Sketch of the model*

Applying Constraints to the Sketch

Next, you need to add required relations to the sketched entities.

1. Press and hold the CTRL key and select the start point of the Line 1 and then select the origin for applying the horizontal relation, refer to Figure 3-13. Release the CTRL key after the selection. Then, choose the **Horizontal/Vertical** option under the **Constraints** node in the **SKETCH PALETTE** window; the horizontal relation is applied to the selected entities. Click anywhere in the drawing area to clear the selection set.

2. Press and hold the CTRL key and then select the horizontal lines one by one, refer to Figure 3-14. Next, release the CTRL key after selecting the lines. Then, choose the **Collinear** constraint under the **Constraints** node in the **SKETCH PALETTE** window; the collinear constraint is applied to the selected entities.

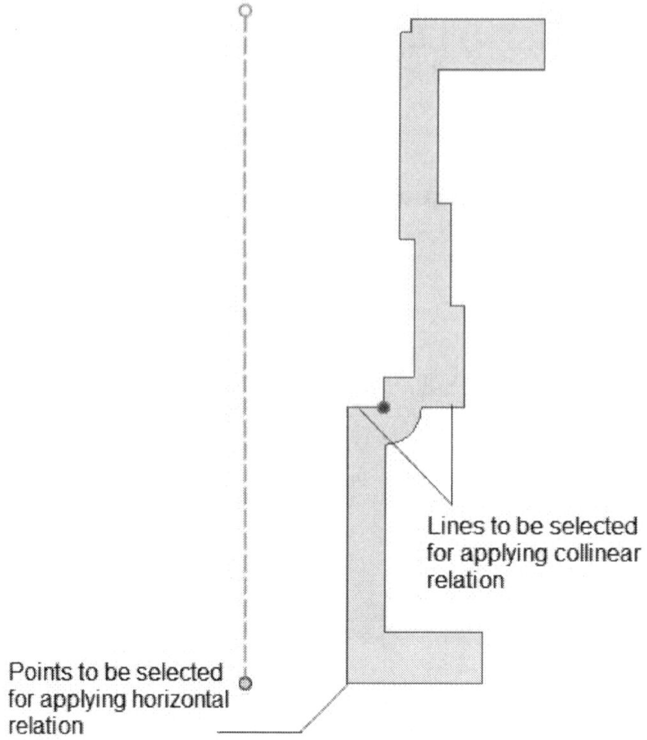

Figure 3-14 *Sketch of the model*

3. Press and hold the CTRL key and select the intersection point of the lines 2 and 3, refer to Figure 3-13. Next, select the center point of the arc and then release the CTRL key.

4. Choose the **Coincident** option under the **Constraints** node in the **SKETCH PALETTE** window; the selected points are made coincident together.

   As discussed earlier, some of the relations such as horizontal, vertical, perpendicular, and so on are automatic relations and are therefore applied automatically to a sketch while drawing. This means, you need not to apply horizontal constraint to the horizontal lines and vertical constraint to the vertical lines of the sketch. However, if it is not applied automatically, you need to apply them manually by expanding the **Constraints** node in the **SKETCH PALETTE** window.

Applying Dimensions to the Sketch

After drawing and applying constraints to the sketch, you need to add dimensions to fully define the sketch. As it is evident from Figures 3-11 and 3-12, the model is made by revolving a sketch, therefore, you need to apply the linear diameter dimensioning.

1. Invoke the **Sketch Dimension** tool from the **SKETCH** panel; you are prompted to select the object to dimension.

2. Select the vertical centerline and line1, refer to Figure 3-13 and then move the cursor to the centerline; a preview of the dimension gets attached to the cursor.

3. Right-click in the graphics window; a Marking menu is displayed. Choose the **Diameter Dimension** option from this Marking menu; the preview of linear diameter dimension gets attached to the cursor and you are prompted to select the location to place the dimension.

4. Click to place the dimension below the sketch, refer to Figure 3-15, for the placement of dimension. As soon as you click the left mouse button to place the dimension, an edit box is displayed.

5. Enter **100** in the edit box and press ENTER; the linear diameter dimension is applied between the selected entities. Also, note that the **Sketch Dimension** tool is still active and you can directly select the other entities of sketch to apply linear diameter dimension with respect to the centerline.

   The linear diameter dimensioning is used to dimension the sketch of a revolved component. If you dimension the sketch of the base feature of the given model using the linear dimensioning method, the same dimensions will be generated in the drawing views. This may be confusing because in the shop floor drawing, you need the diameter dimension of a revolved model. To overcome this problem, it is recommended that you apply the linear diameter dimension to the revolved components.

6. Select the centerline, refer to line 3 in Figure 3-13. As soon as you select the vertical line, a preview of the linear dimension is attached to the cursor between the selected vertical line and the centerline.

7. Right-click in the drawing window; a Marking menu is displayed. Choose the **Diameter Dimension** option from this Marking menu; a preview of linear diameter dimension gets attached to the cursor and you are prompted to select the location to place the dimension.

8. Click to place the dimension below the sketch, refer to Figure 3-12 for the placement of dimension. As soon as you click the left mouse button to place the dimension, an edit box is displayed.

9. Enter the value **135** in the edit box and press ENTER.

10. Similarly, apply linear diameter dimension to the remaining entities of the sketch. Final sketch after applying all the dimensions is shown in Figure 3-15.

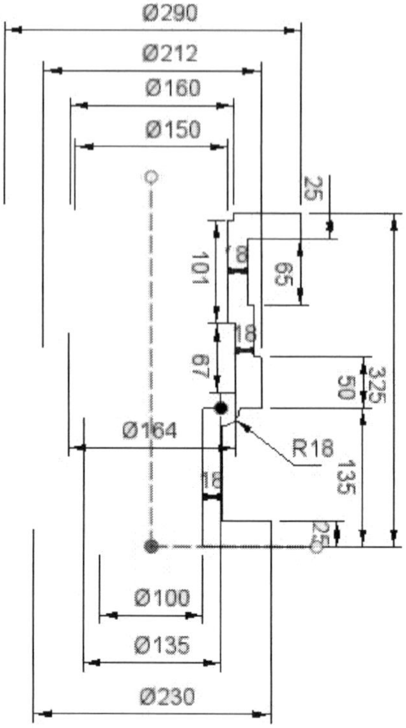

Figure 3-15 *Final sketch for Tutorial 2*

Note

You can also dimension the sketch entities while creating the sketch of the model, as discussed in the tutorials of the previous chapter.

Saving the Sketch

Next, you need to save the sketch.

1. Choose the **Save** option from the **File** menu or the **Application** menu; the **Save** dialog box is displayed.

2. Enter **c03_Tut_02** in the **Name** edit box.

3. Click on the down arrow on the extreme right side of the **Location** selection box; the **Save** dialog box gets expanded. 

4. Select **CADCIM** from the **PROJECT** column or you can create a new project by selecting the **New Project** button from the **Save** dialog box.

5. Choose the **Save** button to save the sketch and exit the **Save** dialog box.

Tutorial 3

In this tutorial, you will draw the sketch for the model shown in Figure 3-16. After drawing the sketch, you will add required constraints and then dimension it. The dimensioned sketch required for this model is shown in Figure 3-17. The solid model shown in Figure 3-16 is only for reference.

The sketch shown in Figure 3-17 is a combination of multiple closed loops: outer loop and inner circles. As the numbers of loops increase, so does the complexity of the sketch. This is because the numbers of constraints and dimensions in a sketch increase in case of multiple loops. Now, to draw sketches without using the Dynamic Input, it is recommended that you first draw the outer loop of the sketch and then add constraints and dimensions to it. This is because once the outer loop has been constrained and dimensioned, the inner circles can be constrained and dimensioned easily with reference to the outer loop.

(Expected time: 30 min)

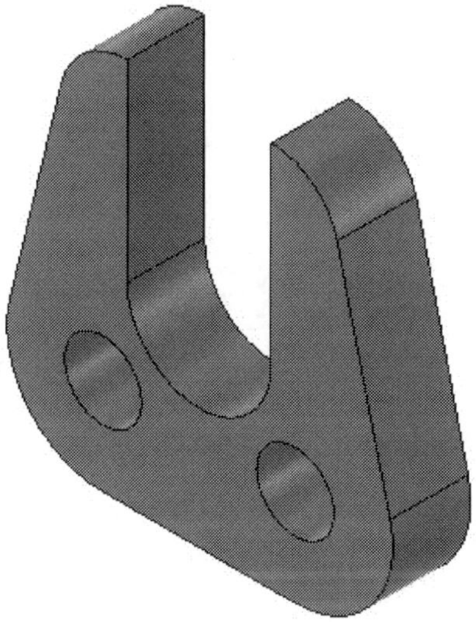

Figure 3-16 Model for Tutorial 3

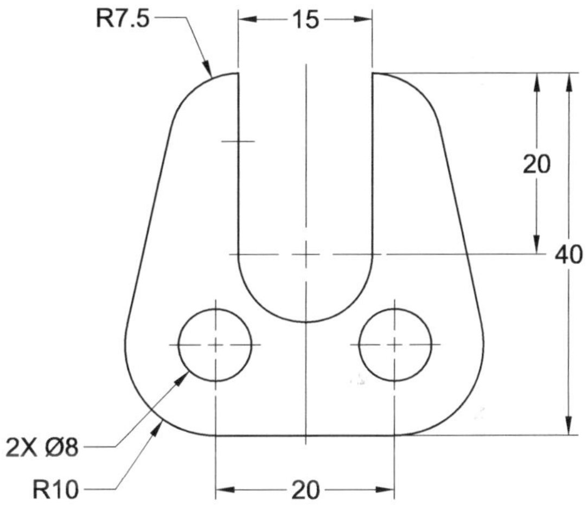

Figure 3-17 Dimensioned sketch of the model

The following steps are required to complete this tutorial:

a. Start a new design file and invoke the model workspace.
b. Draw the outer loop of the sketch.
c. Add required dimensions and constraints to the outer loop.
d. Draw inner circles and add constraints and dimensions to them.
e. Save the sketch with the name **c03_Tut_03** and close the file.

Starting a New Design

1. Choose the **New Design** option from the **File** menu; a new design tab is added to the graphics window.

2. Choose the **Create Sketch** tool from the **SKETCH** panel of the toolbar; the three default planes are displayed in the graphics window and you are prompted to select a plane or planar face.

3. Select the **XY** Plane from the **BROWSER** bar by expanding the **Origin** node; the selected plane gets oriented normal to the view. Alternatively, select the **FRONT** view from the ViewCube and select the plane that gets oriented normal to the view.

4. Clear the **Sketch Grid** check box from the **SKETCH PALETTE** window to hide the grids, if grid is displayed by default.

5. Expand the **Document Settings** node and move the cursor to the **Units** sub node in the **BROWSER** bar in graphics window; the **Change Active Units** button is displayed.

6. Click on the **Change Active Units** button; the **CHANGE ACTIVE UNITS** dialog box is displayed.

7. Select the **Millimeter** option, if not selected, from the **Unit Type** drop-down list in the dialog box and choose the **OK** button from this dialog box. Note that if **Millimeter** is already selected then choose the **Cancel** button to exit this dialog box.

Drawing the Outer Loop

1. Invoke the **Line** tool from the **SKETCH** panel in the **MODEL** workspace and create line1. Next, invoke the **3-Point Arc** tool and create arc 2.

2. Similarly, create rest of the sketch, as shown in Figure 3-18.

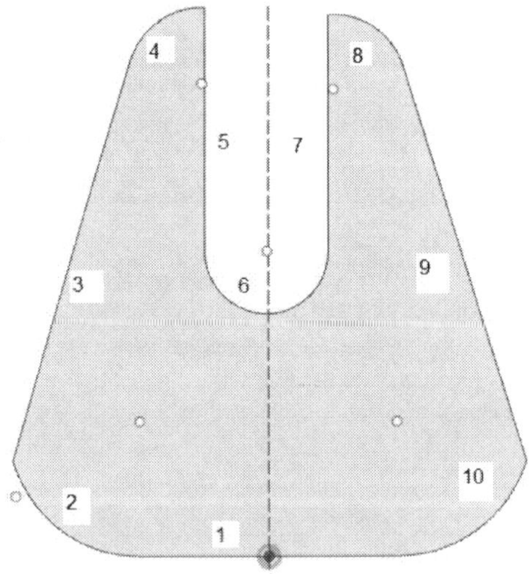

***Figure 3-18** Profile with geometries numbered*

You can also draw tangent arcs while using the **Line** tool. This can be done by invoking the **Line** tool and by pressing the left mouse button and then dragging it in the required direction. Alternatively, move the cursor close to the endpoint of the last line. Next, press and hold the left mouse button and drag the mouse through a small distance in the upward direction.

For your reference, all the geometries in the sketch are numbered. You will draw inner holes in the sketch after dimensioning the outer loop.

Note
The outer loop that you created in the previous step might be different from the one shown in Figure 3-18. You can find the missing constraints by following the next step and applying constraints accordingly.

Applying Constraints to the Sketched Entities

As evident from Figure 3-18, some of the constraints such as tangent and equal are missing in the sketch. The sketch shown in Figure 3-18 may not be symmetrical and all the lines in the sketch may not be tangent to the arcs. Therefore, you need to add these missing constraints manually to the sketch to complete it. You can choose the **Show Constraints** option from the **Options** node in the **SKETCH PALETTE** window.

In Figure 3-18, the tangent constraint is missing between line 9 and arc 10. You need to add the constraint between them.

1. Choose the **Tangent** constraint under the **Constraints** node in the **SKETCH PALETTE** window; you are prompted to select the sketch object or change the constraint type. Select arc 10 as the sketch curve; you are prompted to select the other geometries. Select line 9 as the second curve. Similarly, add this constraint to the arcs and lines wherever required.

   The geometries 5,7, and 3,9, respectively are the lines that must be of equal length. Also, the geometries 2,10, and 4,8, respectively are the arcs that must be of equal radii. Therefore, you need to add the equal constraint between these pairs of geometries.

2. Choose the **Equal** option from the **Constraints** palette of the **SKETCH PALETTE** window; you are prompted to select the sketch object or change the constraint type.

3. Select line 5 as the first line and then line 7 as the second line to apply the **Equal** constraint to them; you are again prompted to select a sketch object.

4. Select line 3 and then line 9 to apply the **Equal** constraint to them; you are prompted to select the sketch object or change the constraint type.

5. Select arc 2 and then arc 10 to apply the **Equal** constraint to them. Applying this constraint to arcs or circles forces their radii or diameters to be equal.

6. Similarly, apply the **Equal** constraint to arc4 and arc8.

7. Apply the **Coincident** constraint between the center point of arc 4 and line 5, and the center point of arc 8 and line 7 if not applied automatically.

8. Choose the **Symmetry** constraint under the **Constraints** node of the **SKETCH PALETTE** window to apply the **Symmetry** constraint between line 3 and line 9 about the centerline.

9. Choose the C**oincident** option under the **Constraints** node in the **SKETCH PALETTE** window; you are prompted to select the sketch object or change the constraint type.

10. Select the origin; you are prompted to select other geometries.

11. Select the center point of arc 6; the entire sketch moves to make the origin coincident with the center point of the arc. The sketch after applying all the constraints is shown in Figure 3-19.

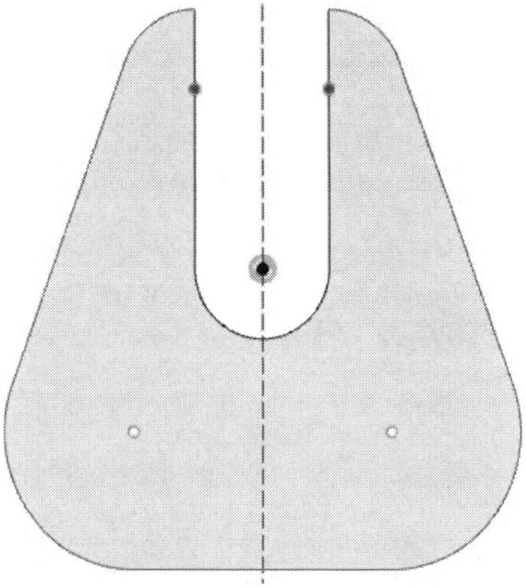

Figure 3-19 *The sketch after applying all constraints*

Note
The shape of the sketch that you have drawn may be a little different from the final sketch at this stage because of the difference in specifying points while drawing the sketch. However, once all the dimensions are applied, the shape of the sketch will be the same as of the final sketch. Also you may need to apply some constraint to make it fully constrain sketch.

Dimensioning the Sketch

1. Choose the **Sketch Dimension** tool from the **SKETCH** panel; you are prompted to select the sketch objects to dimension. Select line 1 and place the dimension below the sketch. Modify the value of this dimension in the **Dimension** edit box to **20**.

2. Select arc4 and place the dimension on the left of the sketch; the radius dimension of the sketch is placed. Modify the dimension value in the **Dimension** edit box to **7.5**. The size of arc8 is also modified because the Equal constraint is applied between these two entities.

Note
As discussed in Chapter 1, you may need to use a combination of hot keys to zoom or pan the model

3. Select arc 2 and place the radius dimension on the left in the sketch. Modify the dimension value in the **Dimension** edit box to **10** and press ENTER; the size of arc 10 is also modified because of the **Equal** constraint applied between these two entities.

4. Select the upper endpoint of line 7 and select line 1, and then place the dimension on the right of the previous dimension. Modify the value of this dimension to **40** and press ENTER.

5. Select line 5 and then line 7, and then place the dimension above the sketch. Modify the value of this dimension in the **Dimension** edit box to **15** and press ENTER.

6. Select line 7 and place the dimension on the right in the sketch. Modify the value of this dimension in the **Dimension** edit box to **20** and press ENTER.

   Figure 3-20 shows the sketch with all the dimensions applied except between the center points of arcs 4 and 6 or arcs 8 and 6 where horizontal dimension is not applied. The need of these dimensions depends on the constraints and dimensions assumed while drawing the sketch. If you still apply the horizontal dimensions then the sketch gets over-constrained and the **Autodesk Fusion 360** message box is displayed. Choose **Cancel** from the message box.

Drawing Circles

Once all the required dimensions and constraints have been applied to the sketch, you need to draw the circles. Figure 3-17 shows that circles are concentric with arcs 2 and 10.

1. Choose **Circle> Center Diameter Circle** from the **SKETCH** panel to draw concentric circles; you are prompted to place the center of the circle. Move the cursor close to the center of arc 2. Specify the center point when the cursor snaps to the center point of arc 2. Next, move the cursor away from the center and specify a point to size the circle.

2. Similarly, create the other circle taking reference of the center of arc 10.

Applying Constraints to Circles

As both the circles have the same diameter, you can apply the equal constraint to them. On applying the dimension to one of the circles, the other circle will automatically be created as per the specified diameter value.

1. Choose the **Equal** option from the **Constraints** node. Select the first circle and then the second circle to apply the equal constraint.

Dimensioning Circles

1. Choose the **Sketch Dimension** tool from the **SKETCH** panel and select the left circle. Place the dimension on the left in the sketch. In the **Dimension** edit box, change the value of the diameter of the circle to **8** and press ENTER.

You will notice that because of the **Equal** constraint, the size of the right circle is automatically modified to match the dimension of the left circle. The final sketch for Tutorial 3 after drawing and dimensioning circles is shown in Figure 3-21.

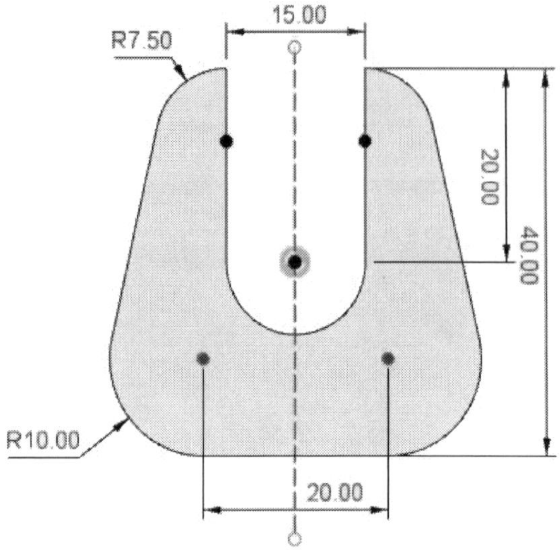

Figure 3-20 *Dimensioned sketch for Tutorial 3*

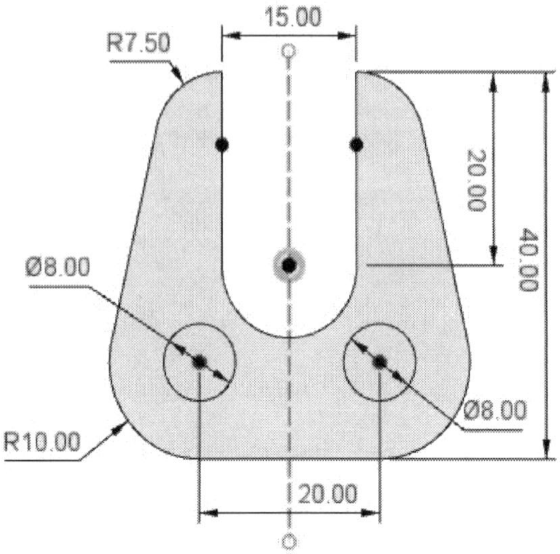

Figure 3-21 *The final dimensioned sketch for Tutorial 3*

Saving the Sketch

Next, you need to save the sketch.

1. Choose the **Save** option from the **File** menu or the **Application** menu; the **Save** dialog box is displayed.

2. Enter **c03_Tut_03** in the **Name** edit box.

3. Click on the down arrow on the extreme right of the **Location** selection box; the **Save** dialog box gets expanded. 

4. Select **CADCIM** from the **PROJECT** column or you can create a new project by choosing the **New Project** button from the **Save** dialog box.

5. Choose the **Save** button to save the sketch and exit the **Save** dialog box.

Tutorial 4

In this tutorial, you will draw the sketch of the model shown in Figure 3-22. The dimensions of the sketch are shown in Figure 3-23. After drawing the sketch, add constraints and dimensions. The solid model is given for reference only. **(Expected time: 30 min)**

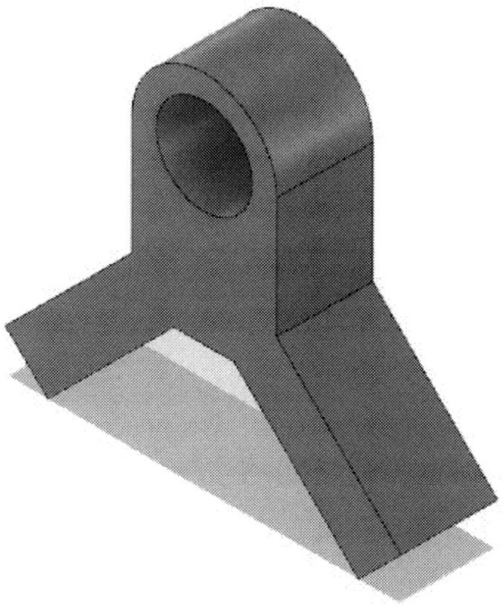

Figure 3-22 Model for the sketch of Tutorial 4

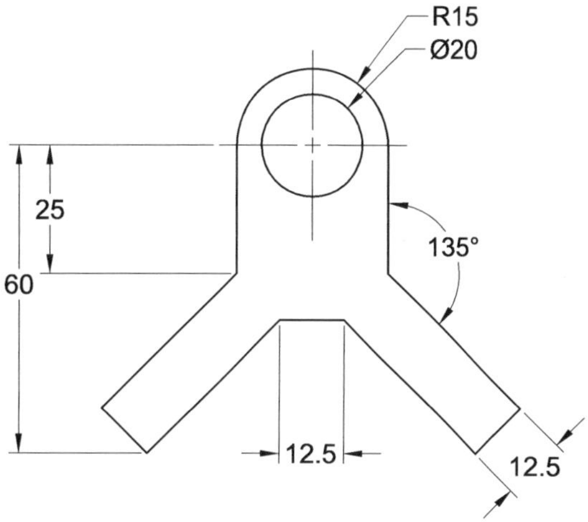

Figure 3-23 The final dimensioned sketch for Tutorial 4

The following steps are required to complete this tutorial:

a. Start a new design file and invoke the model workspace.
b. Draw the outer loop of the sketch.
c. Add the required dimensions and constraints to the sketch.
d. Add the inner circle to the sketch and dimension it
e. Save the sketch with the name **c03_Tut_04** and close the file.

Starting a New Design and invoke the Model Workspace

1. Choose **New Design** option from option the **File** menu; a new design tab is added to the graphics window.

2. Choose the **Create Sketch** tool from the **SKETCH** panel in the Ribbon; the three default planes are displayed in the graphics window and you are prompted to select a plane or planar face.

3. Select the **XY** Plane from the **BROWSER** bar by expanding the **Origin** node; the selected plane gets oriented normal to the view.

   Alternatively, select the **FRONT** view from the View Cube and select the plane that is oriented normal to view.

4. Clear the **Sketch Grid** check box from the **SKETCH PALETTE** window to hide the grids, if grid is displayed by default.

5. Expand the **Document Settings** node and move the cursor to the **Units** sub node in the **BROWSER** bar in graphics window; the **Change Active Units** button is displayed.

6. Click on the **Change Active Units** button; the **CHANGE ACTIVE UNITS** dialog box is displayed.

7. Select the **Millimeter** option, if it is not selected, from the **Unit Type** drop-down list in the dialog box and choose the **OK** button from this dialog box. Note that, if **Millimeter** is already selected then choose the **Cancel** button to exit this dialog box.

Drawing the Outer Loop

1. Invoke the **Line** tool from the **SKETCH** panel to draw the outer loop and the centerline, refer to Figure 3-23. As mentioned earlier, you should draw the inner loop after drawing and dimensioning the outer loop. This is because once the outer loop is dimensioned, you can draw the inner loop by taking the reference of the outer loop.

   You can draw the arc while the **Line** tool is active as described in the previous tutorial. You can also use the temporary tracking option to draw this sketch. For your reference, the geometries in the sketch are numbered, see Figure 3-24.

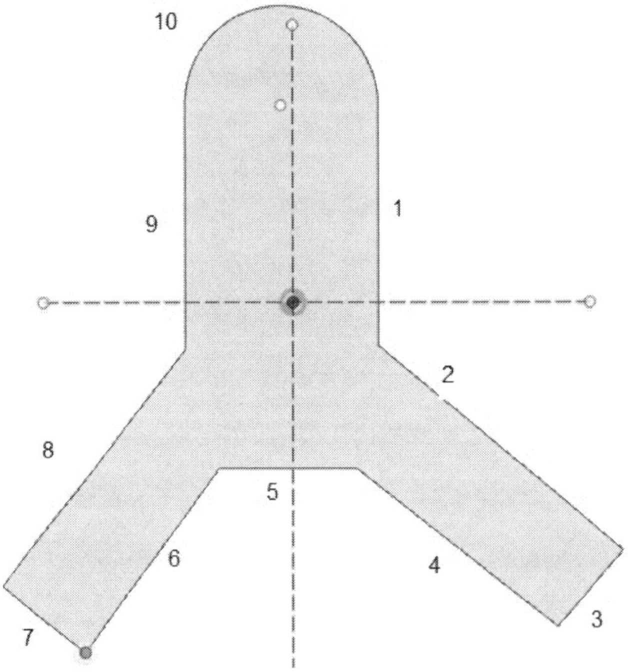

Figure 3-24 *Initial sketch with the geometries numbered*

Adding Constraints to the Outer Loop

1. Choose the **Equal** option from the **Constraints** node of the **SKETCH PALETTE** window; you will be prompted to select sketch objects or change the constraint type.

2. Select line 1; you are prompted to select other geometries. Select line 9; equal length are applied on the line 1 and line 9.

3. Similarly, apply the **Equal** constraint to lines 2 and 8, 3 and 5, 5 and 7, and 4 and 6.

4. Apply the **Perpendicular** constraint to lines 2 and 3 and 7 and 8 if not applied automatically.

5. Choose the **Horizontal/ Vertical** option from the **Constraints** node and select line 5 to apply this constraint if not applied automatically.

6. Add the **Tangent** constraint to lines 1 and 9 which are connected with arc 10, if missing.

7. Next, choose the **Coincident** option under the **Constraints** node in **SKETCH PALETTE** window; you are prompted to select the sketch object or change the constraint type.

8. Select the center of the arc; you are prompted to select the other geometries.

9. Select the origin; the entire sketch shifts itself such that the center of arc of the sketch is at the origin.

   After the sketch is shifted to a new place, it may not be visible completely in the drawing window.

10. Choose the **Fit** tool from the **Navigation Bar** to fit the sketch into the graphics window. You will notice that all the entities in the sketch turn black, indicating that the sketch is fully constrained. Press the ESC key to exit the **Coincident** constraint.

Dimensioning the Outer Loop

1. Choose the **Sketch Dimension** tool from the **SKETCH** panel; you are prompted to select sketch objects to dimension . Select line 9 and place the dimension on the left in the sketch. Modify the dimension value in the **Dimension** edit box to **25** and press ENTER.

2. Select the center of the arc and then the lower endpoint of line 6. Place the dimension on the left of the previous dimension. Modify the dimension value in the **Dimension** edit box to **60** and press ENTER.

3. Select line 3 and then right-click to display the Marking menu. Choose the **Aligned** option from the Marking menu and then place the dimension below the sketch. Modify the dimension value in the **Dimension** edit box to **12.5** and press ENTER.

   You will notice that the length of lines 5 and 7 is also modified because of the equal constraint.

4. Select lines 1 and 2 and then place the angular dimension on the right of the sketch. Modify the value of the angular dimension in the **Dimension** edit box to **135** and press ENTER.

5. Select arc 10 and then place the radius dimension above the sketch. Modify the value of the radius of the arc in the **Dimension** edit box to **15** and press ENTER.

Drawing the Circle

1. Choose **Circle > Center Diameter Circle** from the **SKETCH** panel; you are prompted to place the center of the circle.

2. Move the cursor close to the center of the arc; the cursor snaps to the center point and turns cyan. Select this point as the center of the circle and then move the cursor away from the center to size the circle. Specify a point to give it an approximate size.

Dimensioning the Circle

1. Choose the **Sketch Dimension** tool from the **SKETCH** panel; you are prompted to select sketch objects to dimension. Select the circle and place the diameter dimension below the arc dimension. Enter **20** in the **Dimension** edit box and then press ENTER. This completes the sketch for Tutorial 4. The final dimensioned sketch is shown in Figure 3-25.

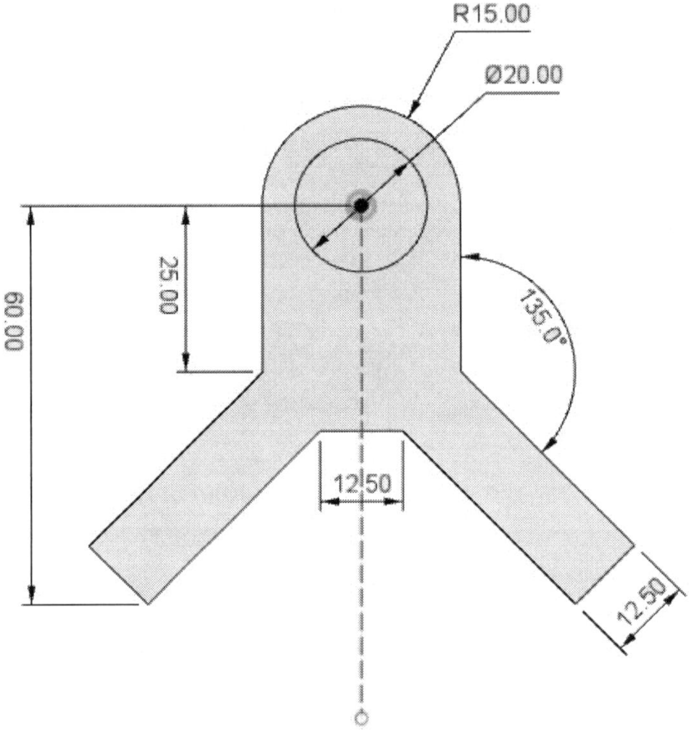

***Figure 3-25** The final dimensioned sketch for Tutorial 4*

Saving the Sketch

Next, you need to save the sketch.

1. Choose the **Save** option from the **File** menu or **Application** menu; the **Save** dialog box is displayed.

2. Enter **c03_Tut_04** in the **Name** edit box.

3. Click on the down arrow on the extreme right side of the **Location** selection box, the **Save** dialog box gets expanded.

4. Select the existing **CADCIM** project from the **PROJECT** column or you can create a new project by choosing the **New Project** button from the **Save** dialog box.

5. Choose the **Save** button to save the sketch and exit the **Save** dialog box.

Self-Evaluation Test

Answer the following questions and then compare them to those given at the end of this chapter:

1. The __________ nature of Autodesk Fusion 360 ensures that a selected entity is driven to a specified dimension value irrespective of its original size.

2. When you select a circle for dimensioning, the __________ dimension is applied to it by default.

3. The __________ dimension has one arrowhead and is placed outside a circle or an arc.

4. The __________ dimension displays the distance between two selected line segments in terms of diameter and the distance shown is twice the original length.

5. The __________ tool is used to measure the dimension of an objects.

6. A __________ constrained sketch is the one whose entities are completely constrained to their surroundings.

7. The perpendicular constraint forces a selected entity to become perpendicular to another specified entity. (T/F)

8. The coincident constraint can be applied to two line segments. (T/F)

9. The collinear constraint can only be applied to line segments. (T/F)

10. If an unnecessary constraint is applied to a sketch, Autodesk Fusion 360 displays a message box informing that adding this constraint will over-constrain the sketch. (T/F)

Review Questions

Answer the following questions:

1. Which of the following constraints forces a selected arc to share the same center point with another arc or point?

   (a) Concentric (b) Collinear
   (c) Coincident (d) Equal

2. In addition to lines, which of the following entities can be selected to apply the colinear constraint?

   (a) Arc (b) Circle
   (c) Ellipse (d) Ellipse axis

3. Which of the following combinations of entities cannot be used to apply the tangent constraint?

   (a) Line, line (b) Line, arc
   (c) Circle, circle (d) Arc, circle

4. The __________ constraint is used to apply curvature continuity between a spline and an entity connected to it.

5. You cannot apply the Concentric constraint between a point and a circle. (T/F)

6. You can use the Horizontal constraint or the Vertical constraint to line up arcs, circles, or ellipses in the respective horizontal or vertical direction. (T/F)

7. You can view all or some of the constraints applied to a sketch. (T/F)

8. There are twelve types of geometrical constraints that can be applied to the sketched entities. (T/F)

9. The linear dimensions are the dimensions that define the shortest distance between two points. (T/F)

10. A sketch in which the number of dimensions or constraints exceeds the required number is called an Over-constrained sketch. (T/F)

EXERCISES

Exercise 1

Create the sketch of the model shown in Figure 3-26. Apply required constraints and dimensions to the sketch and fully define it. The sketch of the model is shown in Figure 3-27. The solid model is given for your reference only. **(Expected time: 30 min)**

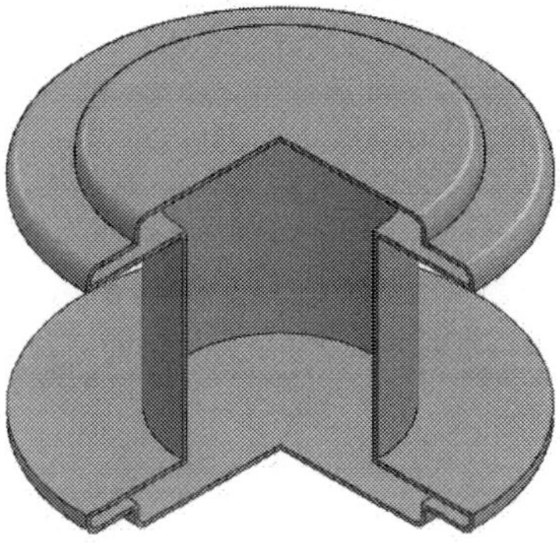

Figure 3-26 Solid model for Exercise 1

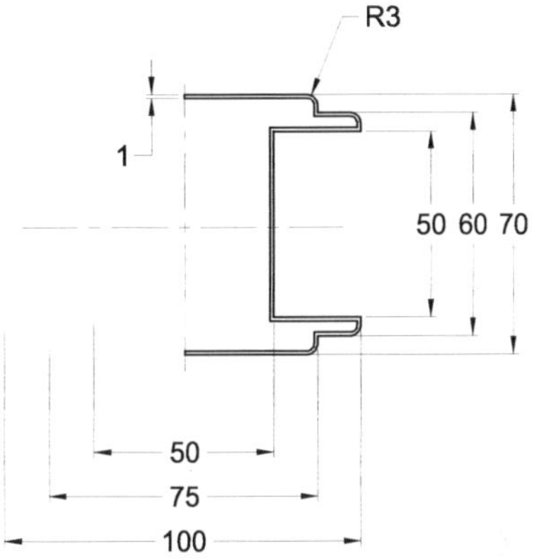

Figure 3-27 Sketch for Exercise 1

Exercise 2

Create the sketch of the model shown in Figure 3-28. Apply required constraints and dimensions to the sketch and fully define it. The sketch of the model is shown in Figure 3-29. The solid model is given for your reference only. **(Expected time: 30 min)**

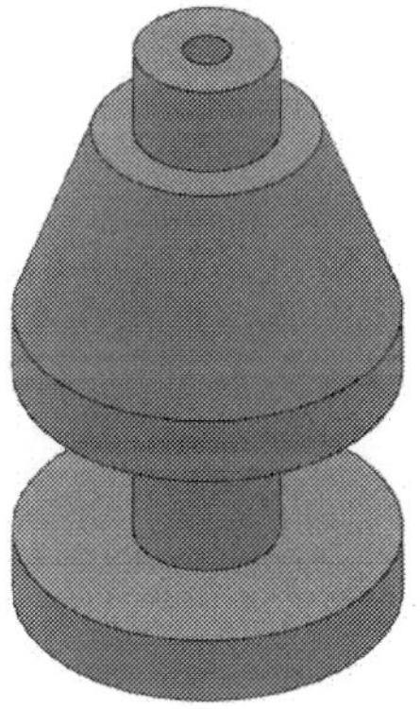

Figure 3-28 Solid model for Exercise 2

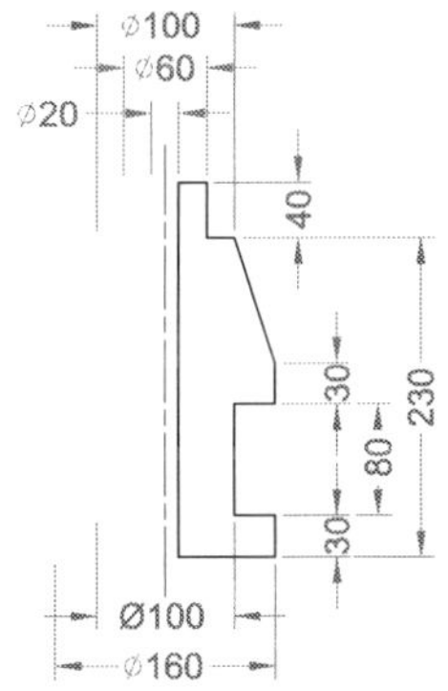

Figure 3-29 Sketch for Exercise 2

Exercise 3

Create the sketch of the model shown in Figure 3-30. Apply required constraints and dimensions to the sketch and fully define it. The sketch of the model is shown in Figure 3-31. The solid model is given for your reference only. **(Expected time: 30 min)**

Figure 3-30 Solid model for Exercise 3

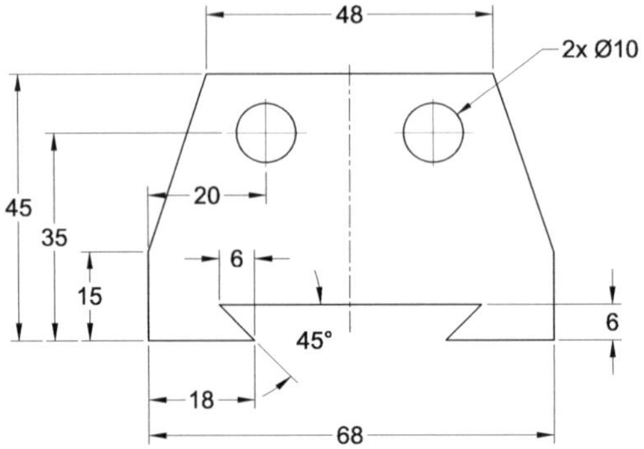

Figure 3-31 Sketch for Exercise 3

Answers to Self-Evaluation Test

1. parametric, 2. diameter, 3. radius, 4. linear diameter, 5. **Measure**, 6. fully, 7. T, 8. F, 9. T, 10. T

Chapter 4

Advanced Modeling-I

Learning Objectives

After completing this chapter, you will be able to:

- *Create solid base extruded features*
- *Create solid base revolved features*
- *Mirror the features*
- *Chamfer the edges of a model*
- *Create circular patterns of features*

INTRODUCTION

In Autodesk Fusion 360, you can apply all possible relations and dimensions to a sketch. In this chapter, you will learn about the advanced sketching techniques that are used to dimension the sketch. Also, you will learn about the tools that are used to convert a sketch into the base feature of a model in the **MODEL** workspace.

TUTORIALS

Tutorial 1

In this tutorial, you will create the model shown in Figure 4-1. Its dimensions are shown in Figure 4-2. The extrusion thickness for the model is 10 mm. After extrusion, you will apply Steel material to the model. **(Expected time:45 min)**

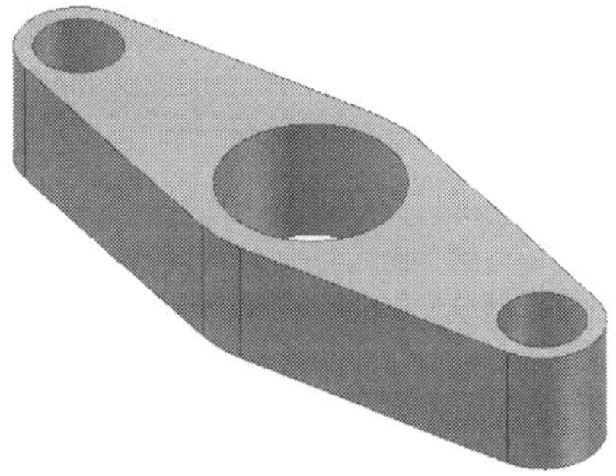

Figure 4-1 Model for Tutorial 1

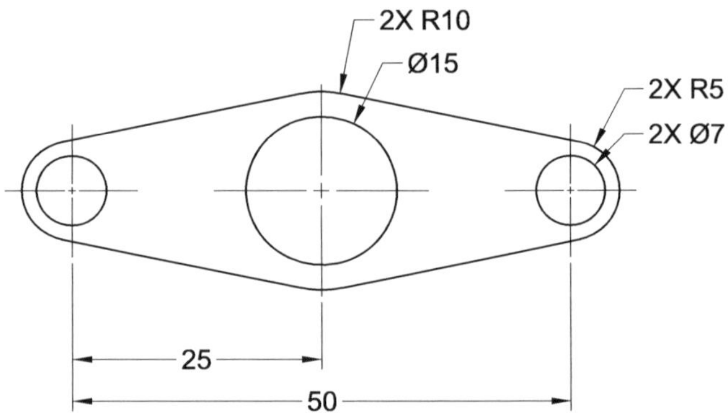

Figure 4-2 Dimensions of the model

The following steps are required to complete this tutorial:

a. Start a new design file. Draw the sketch of the outer loop and add constraints to it.
b. Draw the inner circles and add required constraints. Dimension the complete sketch.
c. Extrude the sketch upto a distance of 10 mm using the **Extrude** tool.
d. Assign materials to the model.
e. Determine the mass properties of the model.

Note

*You can change the units used in Autodesk Fusion 360 file by using the **CHANGE ACTIVE UNITS** dialog box. In this dialog box, choose the desired units from the **Unit Type** drop-down list.*

Starting a New File and Creating the Sketch of the Model

The sketch given in Figure 4-2 is a combination of three circles and an outer loop. First, you will create the outer loop. This outer loop will be created by drawing three circles, two at the ends and one at the center, and then connecting the middle circle with the other two circles through tangent lines. Finally, you will trim the unwanted portions of the circles. The process to create the sketch is discussed next.

1. Choose a **New Design** option from the **File** menu to create new design.

   Whenever you start a new part design, by default, you are in the **Model** workspace but you need to start the design by first creating the sketch of the base feature.

2. Choose the **Create Sketch** tool from the **SKETCH** panel of the **MODEL** workspace; the three default planes are displayed on the graphics window.

3. Choose the Top Plane from the graphics window. Alternatively, you can also select the plane from the **BROWSER** bar. Select the **XZ** plane under the **Origin** node from the **BROWSER** bar; the selected plane gets oriented normal to the view.

4. Draw the sketch consisting of three circles and four tangent lines.

5. Apply the **Equal** constraint to all four lines as well as on the circles on the left and right sides. Also, add the Tangent constraint to the lines wherever missing.

   Note that if the Coincident constraint automatically gets applied while drawing the sketch then one of the Tangent constraint will not be applied.

6. Finally, apply the Horizontal constraint one by one to the center of the both sides circles with respect to the center of the middle circle. The sketch, after drawing and adding constraints, should look similar to the sketch, as shown in Figure 4-3. Note that the applied constraints have been hidden for clarity.

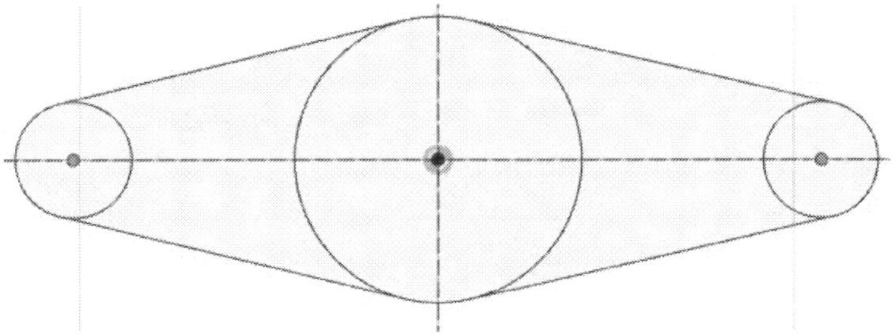

Figure 4-3 *Sketch after drawing and adding constraints*

Next, you need to remove unwanted portions of the circles using the **Trim** tool.

7. Choose the **Trim** tool from the **SKETCH** panel; you are prompted to select the section of the curves to be trimmed. 

   The **Trim** tool is used to trim the unwanted entities in a sketch. You can use this tool to trim a line, arc, ellipse, parabola, circle, spline, centerline intersecting another line, arc, ellipse, parabola, circle, spline, or centerline.

8. Move the cursor close to the right-half of the left circle.

   As you move the cursor close to the circle, the color of the selected curve of the circle turns red.

9. Select the right half of the left circle; the right half of this circle is trimmed. Similarly, select portions of the other circles to trim, as shown in Figure 4-4. On doing so, a warning is displayed stating that constraints and /or dimension were removed during the operation.

10. Draw three circles concentric to the three trimmed arcs and add the Equal constraint between the left and right circles. The sketch after drawing circles and applying the Equal constraint is shown in Figure 4-5.

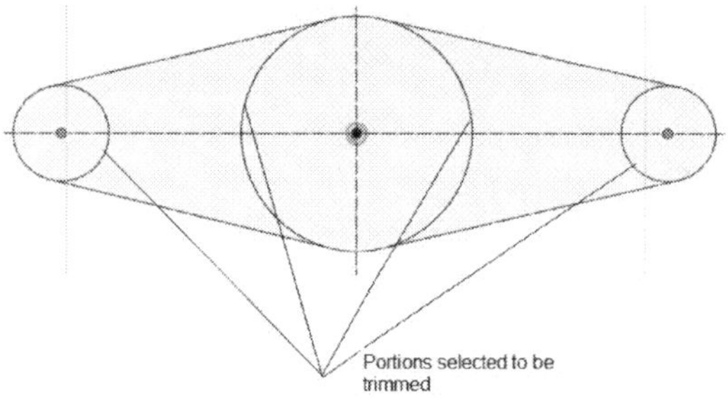

Figure 4-4 *Selecting the portions to be trimmed*

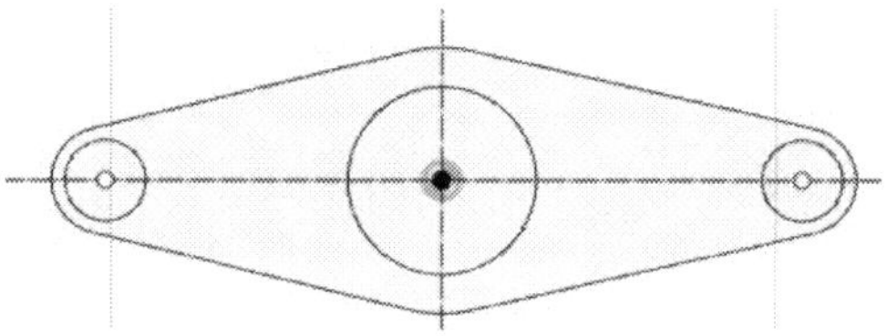

Figure 4-5 *Sketch after creating the outer loop and the three circles*

Dimensioning the Sketch

1. Using the **Sketch Dimension** tool, dimension the sketch. The sketch after dimensioning is shown in Figure 4-6. Note that if the sketch is not getting fully defined, then you need to apply more constraints which were removed during the trim operation.

> **Tip**
> *To avoid distortion in the sketch, it is recommended to draw the sketch of the model with the dimension which are closer to the actual dimension.*

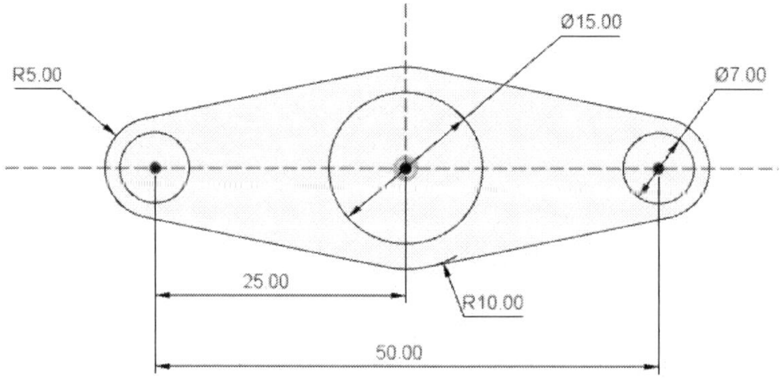

Figure 4-6 *The sketch after adding dimensions*

Extruding the Sketch

After creating the sketch, convert the sketch into a base feature. To do so, invoke the **Extrude** tool and extrude the sketch using parameters given in the tutorial description. The sketch consists of four loops: the outer loop and the three circles. When you extrude the sketch, the three circles will be automatically subtracted from the outer loop. As a result, you will get the required model.

1. Choose the **Extrude** tool from the **CREATE** panel of the Ribbon; the **EXTRUDE** dialog box is displayed, as shown in Figure 4-7, with the **Profile Plane** option selected in the **Start** drop-down list. And, you are prompted to select the profile or planar face to extrude the sketch.

Figure 4-7 *The* ***EXTRUDE*** *dialog box*

The **Extrude** tool is used to add depth to the area defined in a sketch.

In the **EXTRUDE** dialog box, the **Profile Plane** option is selected by default in the **Start** drop-down list. Therefore, the resulting extrude feature will start from the sketching plane on which the sketch is drawn. This option is the most commonly used option while creating an extrude feature.

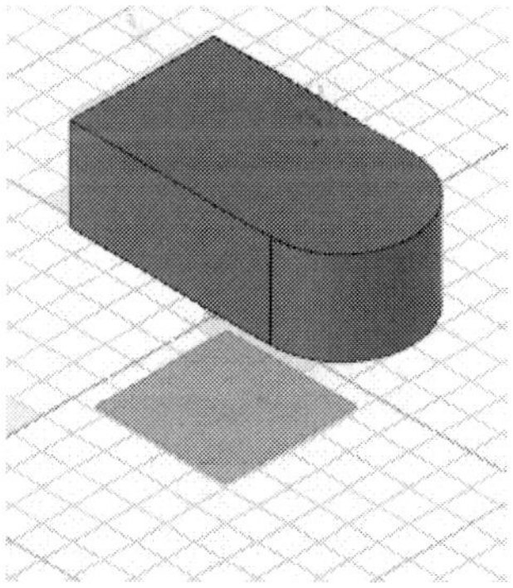

Figure 4-8 *Resulting extruded feature at an offset distance*

In Autodesk Fusion 360, you can also select the **Offset Plane** or **From Object** options from the **Start** drop-down list. The **Offset Plane** option is used to start the extrude feature at an offset distance from the plane on which the sketch is drawn, as shown in Figure 4-8.

The **From Object** option is used to select a face, plane, or vertex to define the start profile position. Figure 4-9 shows the sketch to be extruded and the face of the object selected as the face from which the extrusion will be started. Figure 4-10 shows the resulting extrude feature created on the selected face of the object up to a specified depth.

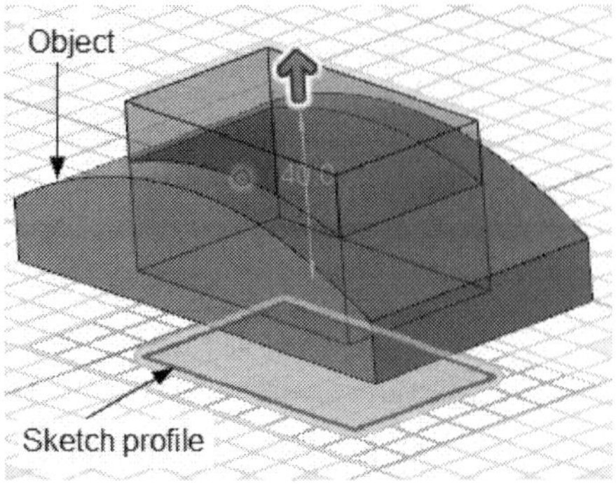

Figure 4-9 Sketch to be extruded and the top face of an object

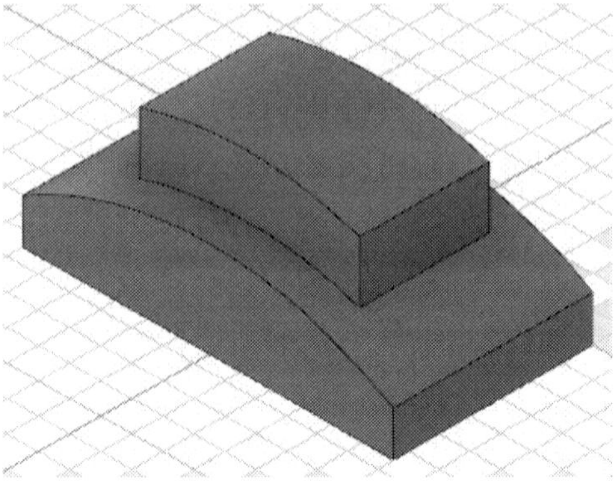

Figure 4-10 Resulting extruded feature

2. Move the cursor anywhere inside the outer loop but outside all three circles and click on the profile; the profile is selected and highlighted in blue color.

   Notice that the area inside any of these circles is not shaded. This shows that the area inside these circles will not be extruded. This is also one of the methods to cross check whether the profile selected is the one you need to extrude or not.

3. Make sure that the **Distance** option is selected in the **Extent** drop-down list of the **EXTRUDE** dialog box.

   The **Extent** drop-down list is used to specify the termination options for the feature to be extruded. The options in this drop-down list are discussed next.

Distance

This option is selected by default in this drop-down list and is used to define the termination of the extrude features by specifying the depth of extrusion. The depth of extrusion is specified in the **Distance** edit box of the dialog box. You can also specify the taper angle in the **Taper Angle** edit box.

To Object

This option is used to terminate the extruded feature on a selected face or plane. On selecting this option, the **Object Select** button will be displayed. This button is used to define the end of the extruded feature by selecting an object.

All

The **All** option is used to extrude the sketch from the sketching plane to all the existing geometric entities. On selecting this option, the **Flip** button becomes available through which you can flip the direction of extrusion. This option gets available in the **Extent** drop-down list only after you create a base feature.

4. Enter **10** in the **Distance** edit box; the sketch gets converted into a solid feature and a preview is displayed in the graphics window, as shown in Figure 4-11.

   Also, you will observe that by default the direction of extrusion is in an upward direction. This direction can be changed by selecting the options in the **Direction** drop-down list such as **One Side**, **Two Sides**, or **Symmetric**.

5. Choose the **OK** button from the **EXTRUDE** dialog box; the final model gets displayed, as shown in Figure 4-12.

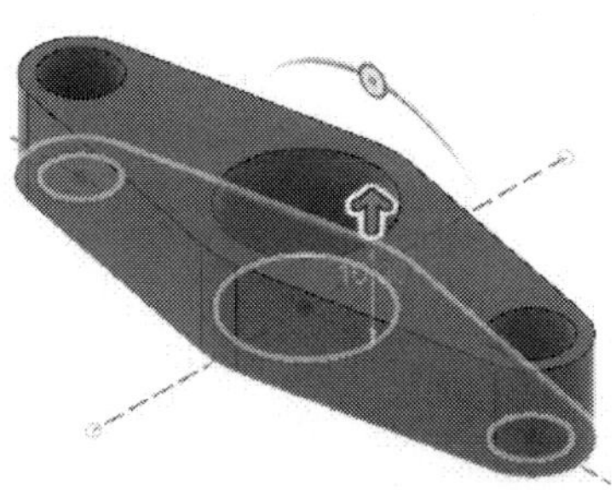

Figure 4-11 *Preview of the solid feature*

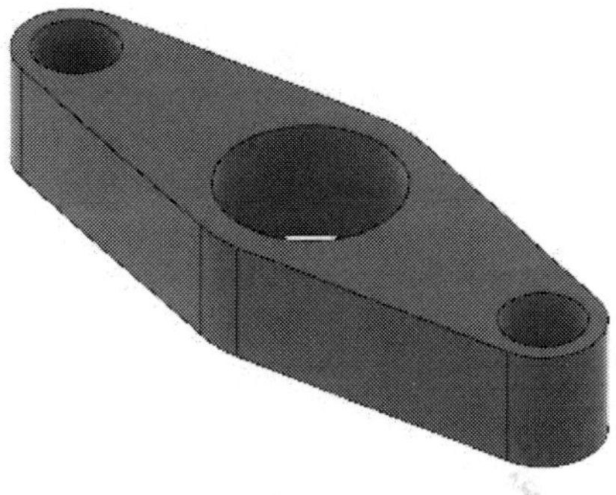

Figure 4-12 *The final model for Tutorial 1*

Assigning Materials to the Model

As mentioned in the tutorial description, you need to apply Steel material to the model. When you apply material to a model, the physical properties such as Density, Mass, Volume and so on of the selected material are assigned to the model. As a result, when you calculate the mass properties of the model, they will be based on the physical properties of the material applied.

1. Expand the **Bodies** node in the **BROWSER** bar and right-click on the **Body1** sub-node; a shortcut menu is displayed, as shown in Figure 4-13.

2. Choose the **Physical Material** option from the shortcut menu; the **PHYSICAL MATERIAL** dialog box is displayed, as shown in Figure 4-14.

3. Expand the **Metal** folder; a list of metals is displayed.

4. Select **Steel** from the displayed list and drag and drop it on the model; the metal is applied to the model.

5. Choose the **Close** button from this dialog box to exit from the dialog box.

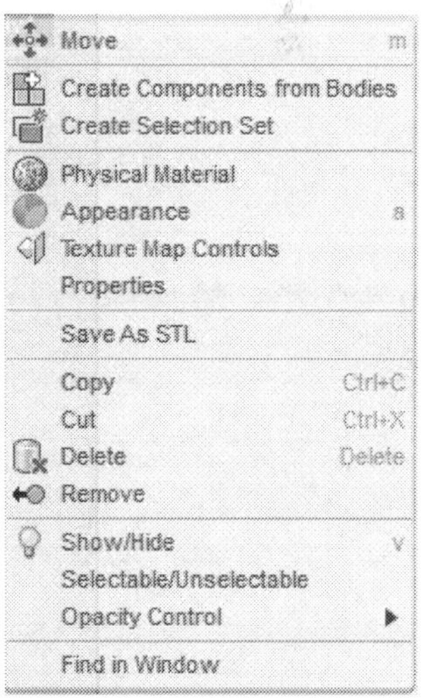

Figure 4-13 *The shortcut menu*

Determining the Mass Properties of the Model

After assigning the material to the model, you need to calculate the mass properties of the model.

1. Expand the **Bodies** node again and right-click on the **Body1** sub-node; a shortcut menu is displayed.

2. Choose the **Properties** option from the shortcut menu; the **PROPERTIES** dialog box is displayed, as shown in Figure 4-15. This dialog box displays the physical properties of the model.

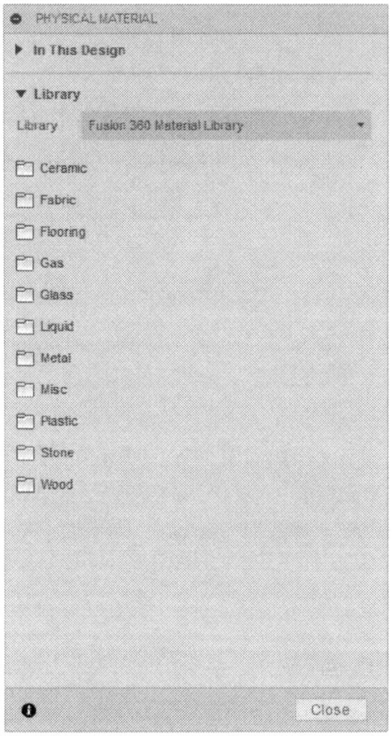

Figure 4-14 *The* ***PHYSICAL MATERIAL*** *dialog box*

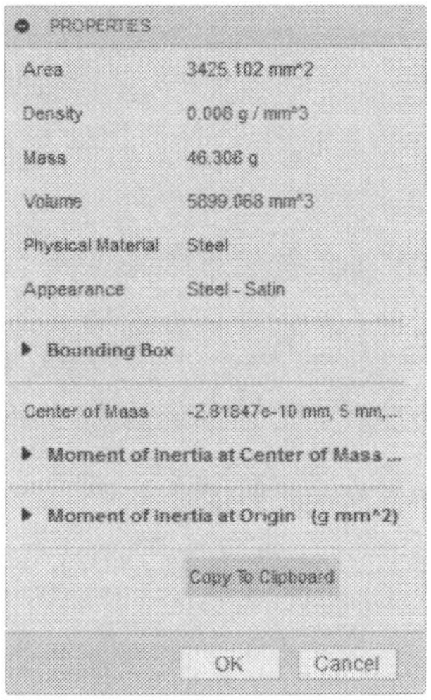

Figure 4-15 *The* ***PROPERTIES*** *dialog box*

Saving the Model

Next, you need to save the sketch.

1. Choose the **Save** option from the **File** menu or the **Application** menu; the **Save** dialog box is displayed.

2. Enter **c04_Tut_01** in the **Name** edit box.

3. Click on the down arrow on the extreme right of the **Location** selection box; the **Save** dialog box gets expanded. 

4. Select an existing project from the **PROJECT** column or you can create a new project by choosing the **New Project** button from the **Save** dialog box.

5. Choose the **Save** button to save the sketch and exit the **Save** dialog box.

Tutorial 2

In this tutorial, you will create the model shown in Figure 4-16. Then you will apply required relations and dimensions to the sketch to make it fully defined. You will also determine the section properties of the sketch and convert it into a revolve feature by revolving the sketch at an angle of 180 degrees, as shown in Figure 4-16. **(Expected time: 45 min)**

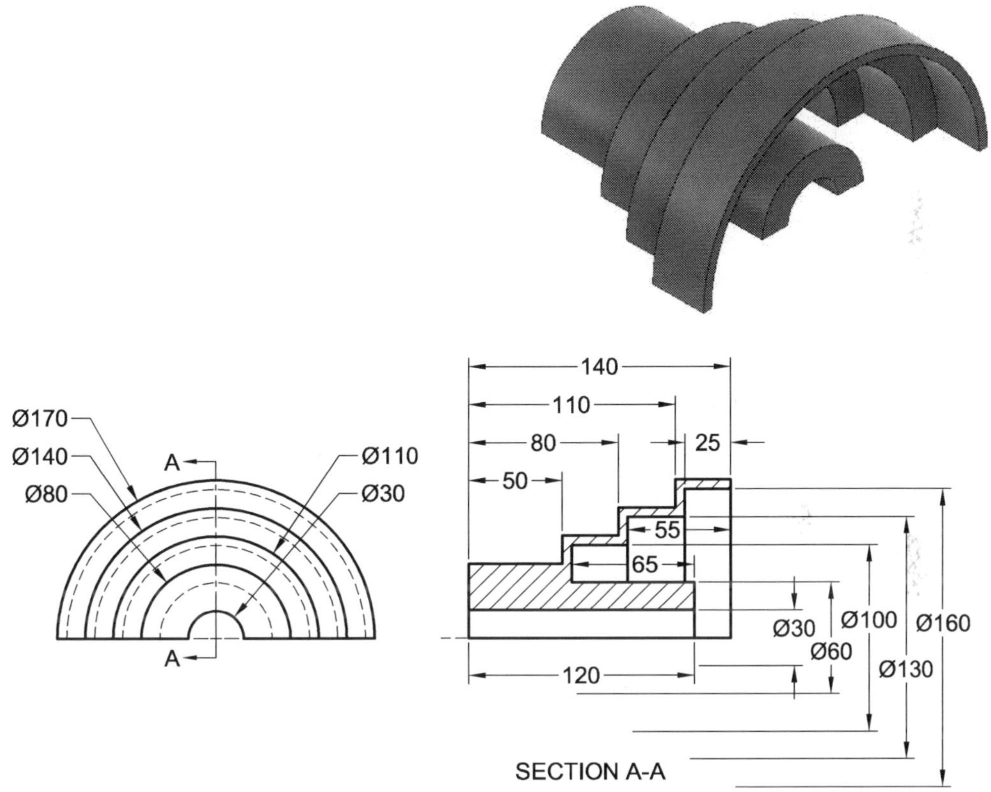

Figure 4-16 *Model and dimensions for Tutorial 2*

The following steps are required to complete this tutorial:

a. Start a new design file and create the sketch.
b. Apply required relations and dimensions to the sketch.
c. Invoke the **Revolve** tool and revolve the sketch at an angle of 180 degrees.
d. Change the projection type and view the model.
e. Save the model.

Starting a New File and Drawing the Sketch of the Base Feature

1. Choose the **New Design** option from the **File** menu to create a new design.

   In this tutorial, you will create the base feature of the model by revolving a sketch around the centerline. To do so, you need to draw the sketch of the base feature and the centerline in the XZ planes and then apply the Geometrical and Dimensional constraints to the sketch to make the sketch fully defined.

2. Choose the **Create Sketch** tool from the **SKETCH** panel of the Ribbon; the three default planes are displayed in the graphics window and you are prompted to select a plane or planar face.

3. Choose the Top view from the ViewCube and select the plane normal to the view. Alternatively, you can also select the plane from the **Browser** bar. Select the **XZ** Plane from the **BROWSER** bar by expanding the **Origin** node; the selected plane gets oriented normal to the view.

4. Create the sketch of the model, as shown in Figure 4-17.

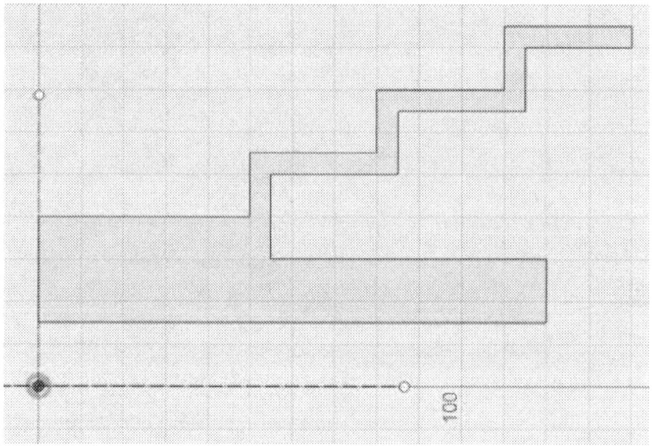

Figure 4-17 *Sketch displayed in the graphics window*

Applying Relations and Dimensions

As mentioned earlier, some of the constraints are automatically applied to a sketch while drawing. If not you need to apply Horizontal and Vertical constraints to horizontal and vertical lines, respectively in the sketch using the **Constraints** node in the **SKETCH PALETTE** window.

1. Apply Equal constraint to equal length entities of the sketch, refer to Figure 4-18. Next, choose the vertical line which is closest to the vertical axis and then apply the coincident constraint between vertical axis and vertical line.

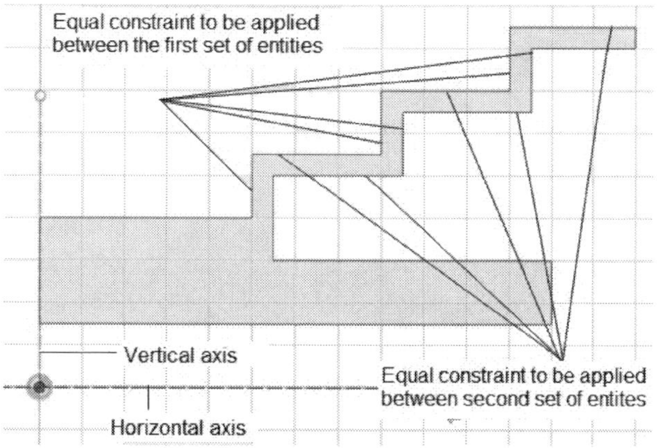

Figure 4-18 Equal constraint to be applied

2. Invoke the **Sketch Dimension** tool from the **SKETCH** panel in the Ribbon and apply the linear diameter dimensions and horizontal dimension to the sketch to make it fully defined, as shown in Figure 4-19.

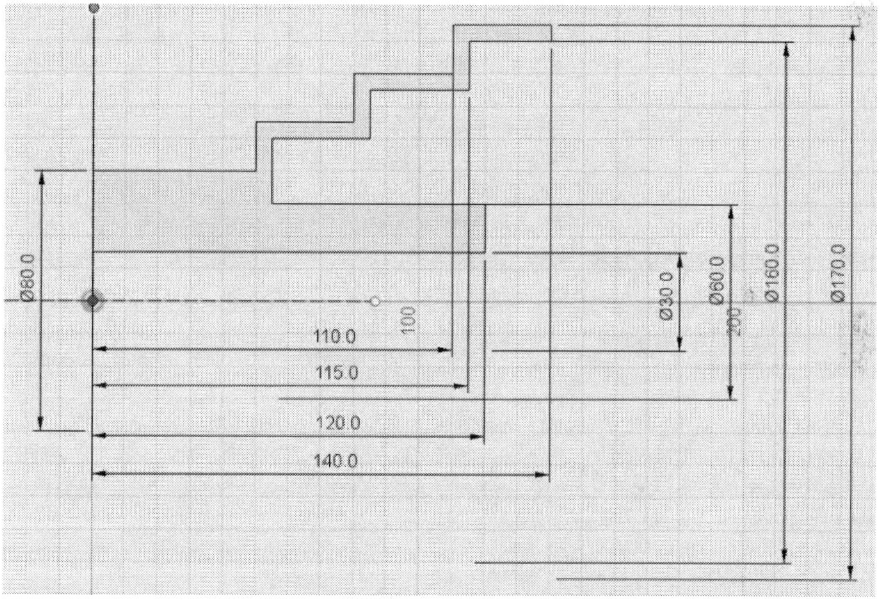

Figure 4-19 Sketch after applying the linear diameter and horizontal dimensions

Revolving the Sketch

After creating the sketch, you need to convert it into a revolve feature by revolving it at an angle of 180 degrees around the horizontal centerline. To do so, you need to invoke the **REVOLVE** dialog box.

1. Choose the **Revolve** tool from the **CREATE** panel in the Ribbon; the **REVOLVE** dialog box is displayed, as shown in Figure 4-20, and you are prompted to select the sketch profiles or planar faces to revolve. Alternatively, you can invoke the **REVOLVE** dialog box from the marking menu. Right-click in the graphics window; a marking menu is displayed. Hover the cursor on the **Create** option and choose the **REVOLVE** tool from the cascading menu.

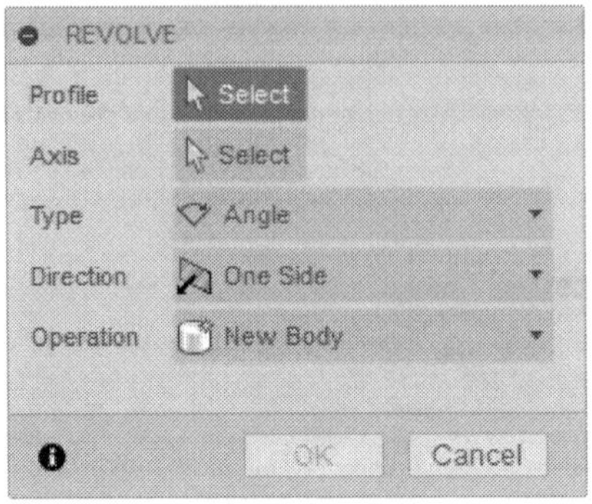

Figure 4-20 *The **REVOLVE** dialog box*

The **Revolve** tool is used to create circular features such as shafts, couplings, pulleys, and so on. You can also use this tool for creating cylindrical cut features. A revolved feature is created by revolving the sketch about an axis. You can use a normal line segment, a center line, or a construction line.

2. Select the closed sketch from the graphics window.

3. Select the **Axis** selection box from the **REVOLVE** dialog box; you are prompted to select the sketch curve, edge, or axis.

4. Select the horizontal centerline as an axis from the graphics window; a preview of the revolved model and the **Angle** edit box is displayed in the graphics window.

   You will also notice that the **Angle** option is selected in the **Type** drop-down list. Note that by default **Angle** is selected in the **Type** drop-down list. As a result, the **Angle** edit box is displayed in the **REVOLVE** dialog box. In this edit box, you can specify the value of the angle. There are also two other options available in the **Type** drop-down list namely, **To** and **Full**.

   The **To** option is used to define the termination of the revolved feature using an extended face, a work plane, or a planar face.

   The **Full** option is selected to create a feature by revolving the sketch through 360-degree.

5. Enter **180** in the **Angle** edit box of the dialog box.

6. Choose the **OK** button from the dialog box to get the revolved model, as shown in Figure 4-21.

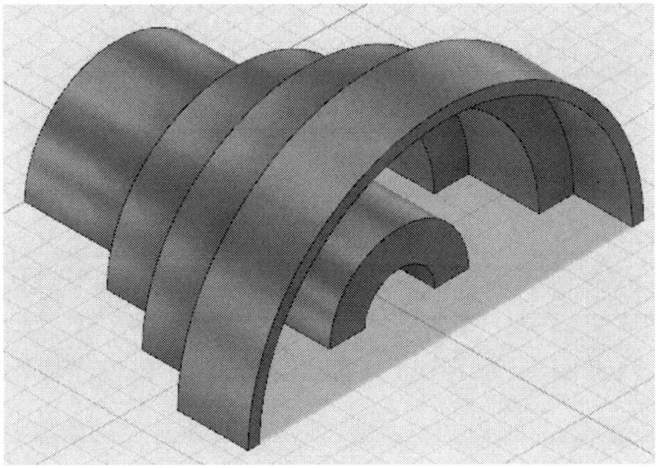

Figure 4-21 *The revolved model at 180-degrees*

Changing the View of the Model

The current view in which the model is displayed, does not show the model properly, refer to Figure 4-21. Therefore, you need to change the view of the model.

1. Click on the top-most corner of the ViewCube, as shown in Figure 4-22.

2. Press and hold the left mouse button on any face of the ViewCube and drag the cursor; the model also reorients to give you a better view of the model.

Changing the Camera Type

By default, the model is displayed in the orthographic camera type. You need to change the camera type from default orthographic to the perspective camera. This is done by choosing the **Perspective** tool from the **Display Settings**. On doing so, the model will be displayed in the perspective camera.

1. Choose **Display Settings > Camera > Perspective** from the Display Settings and Navigation bar; the perspective view of the model is displayed, refer to Figure 4-23.

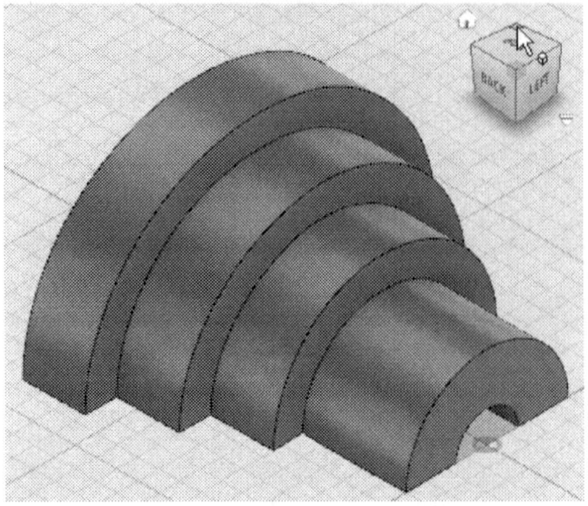

Figure 4-22 *Selecting the corner of the ViewCube*

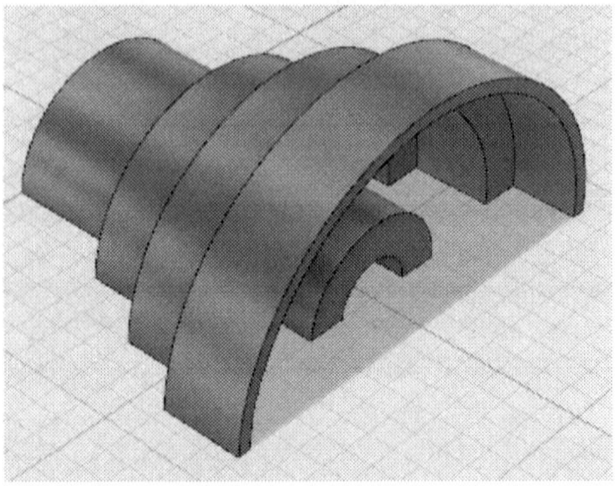

Figure 4-23 *Model displayed using the perspective projection*

Saving the Model

Next, you need to save the sketch.

1. Choose the **Save** option from the **File** menu or the **Application** menu; the **Save** dialog box is displayed.
2. Enter **c04_Tut_02** in the **Name** edit box.
3. Click on the down arrow on the extreme right of the **Location** selection box; the **Save** dialog box gets expanded. 
4. Select an existing project from the **PROJECT** column or you can create a new project by choosing the **New Project** button from the **Save** dialog box.
5. Choose the **Save** button to save the sketch and exit the **Save** dialog box.

Tutorial 3

In this tutorial, you will create the model shown in Figure 4-24. Its views and dimensions are given in the same figure. **(Expected time: 45 min)**

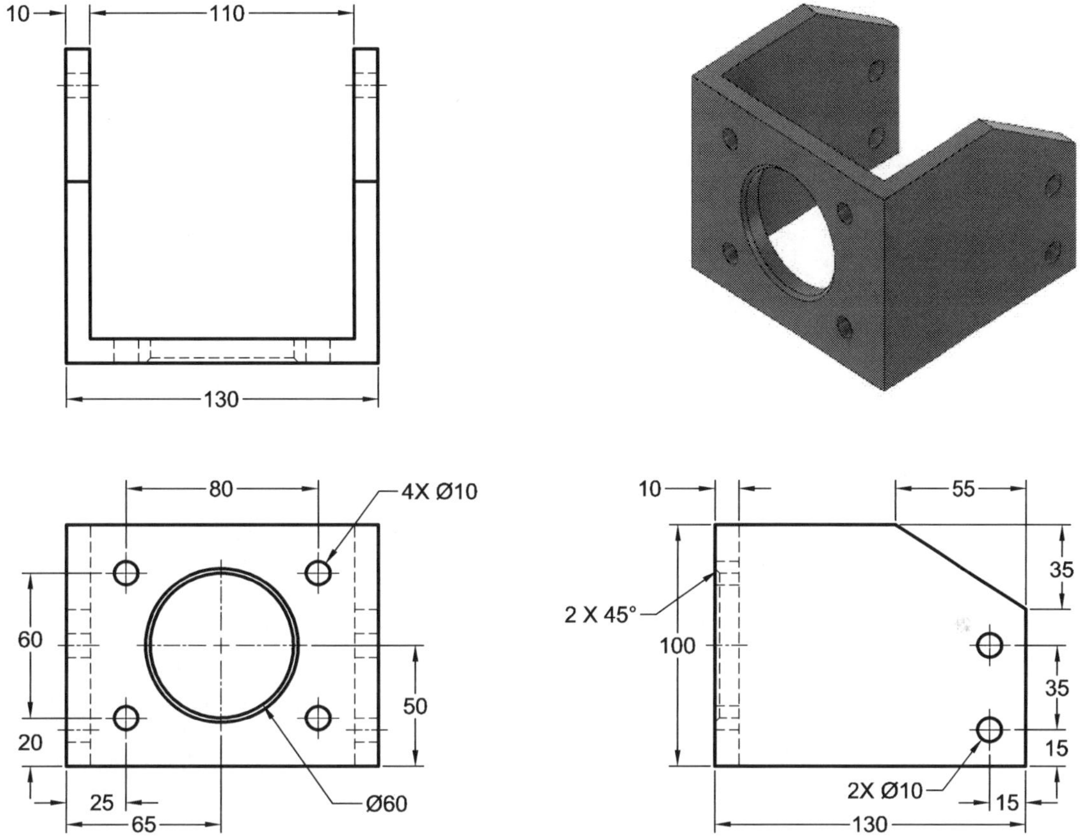

Figure 4-24 *Views and dimensions of the model*

The following steps are required to complete this tutorial:

a. Start a new design and create the base feature of the model on the **XY** plane, refer to Figure 4-24.
b. Create the cut feature on the top planar face of the base feature.
c. Create the third feature, which is an extrude feature, on the right face of the base feature.
d. Create the mirror of the previously created third feature.
e. Apply Fillet and Chamfer to the model.
f. Save the model.

Creating the Base Feature

1. Start a new design using the **New Design** option from the **File** menu or the **Application** menu.

2. Choose the **Create Sketch** tool from the **SKETCH** panel of the Ribbon and then select the **XY** plane from the **BROWSER** bar as the sketching plane; the selected plane gets oriented normal to the screen.

3. Create the sketch of the base feature of the model and apply required constraint to the sketch from the **SKETCH PALETTE** window. Next, dimension the sketch, as shown in Figure 4-25.

4. Choose the **Extrude** tool from the **CREATE** panel; the **EXTRUDE** dialog box is displayed.

5. Click on the sketch profile anywhere outside the circles and select the created sketch from the graphics window such that the inner circles are not selected.

6. Make sure that the **One Side** option is selected in the **Direction** drop-down list of the **EXTRUDE** dialog box.

7. Enter the value **10** in the **Distance** edit box. Next, choose the **OK** button from the **EXTRUDE** dialog box; the base feature of the model is created. The isometric view of the model is shown in Figure 4-26.

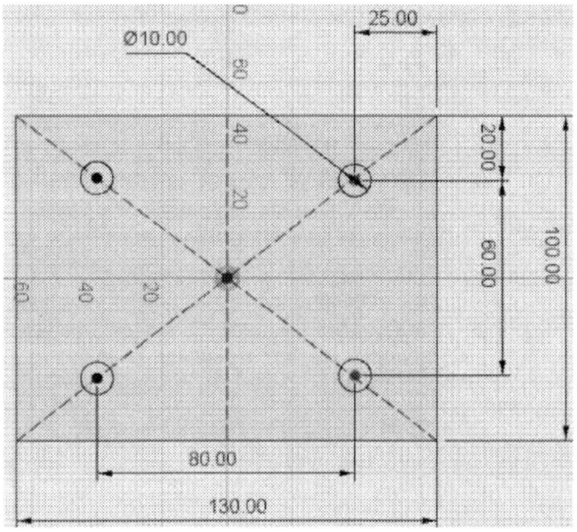

Figure 4-25 Sketch of the base feature

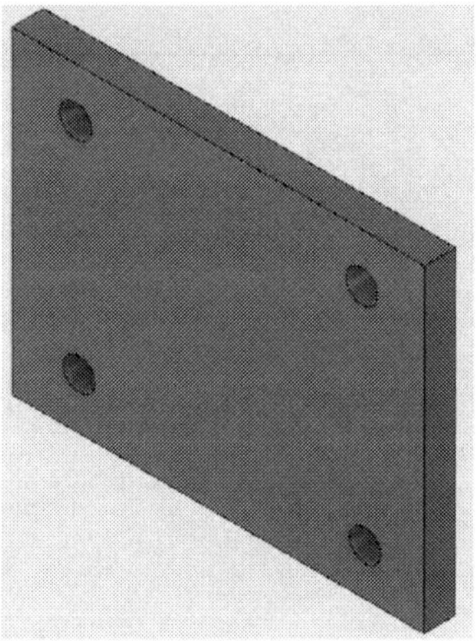

Figure 4-26 Model after creating the base feature

Creating the Second Feature

The second feature of the model is also an extruded feature. The sketch of this extruded feature will be drawn on the right face of the base feature and extruded as a new body.

1. Choose the **Create Sketch** from the **SKETCH** panel and then select the right face as the sketching plane; the selected plane gets oriented normal to the screen.

2. Create the sketch of the second feature of the model and dimension it, as shown in Figure 4-27.

3. Choose the **Extrude** tool from the **CREATE** panel of the Ribbon; the **EXTRUDE** dialog box is displayed.

4. Select the created sketch from the graphics window.

5. Select the **Two Sides** option from the **Direction** drop-down list in the **EXTRUDE** dialog box; the two arrows opposite to each other are displayed in the graphics window. You might have to change the orientation of the model to view them. Also, **Side 1** node and **Side 2** node are displayed in the dialog box along with three drop-down lists, as shown in Figure 4-28.

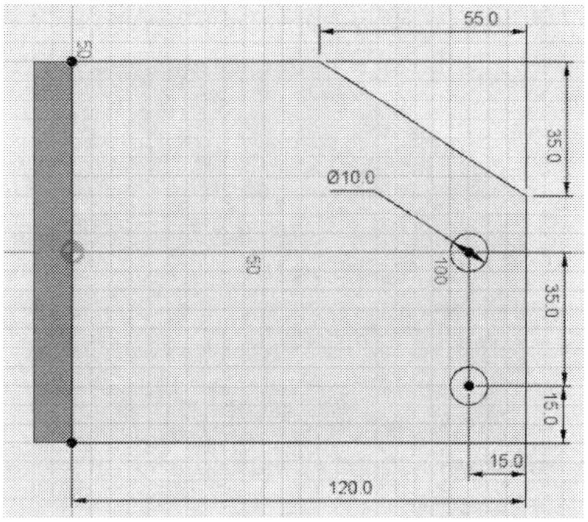

Figure 4-27 *Sketch of the second feature*

In these drop-down lists, individually you can enter the thickness of the feature in two directions. You can also define the taper angle individually in the **Taper Angle** edit boxes.

6. Enter **10** mm extrusion thickness under the **Side 2** node in the **Distance** edit box; the sketch is extruded on the back of the selected right face.

When the mirror feature is applied to the model, make sure that **Join** is selected in the **Operation** drop-down list. This is to ensure that the second feature is not considered as a separate feature.

7. Choose the **OK** button from this dialog box. The isometric view of the model after creating the second feature is shown in Figure 4-29.

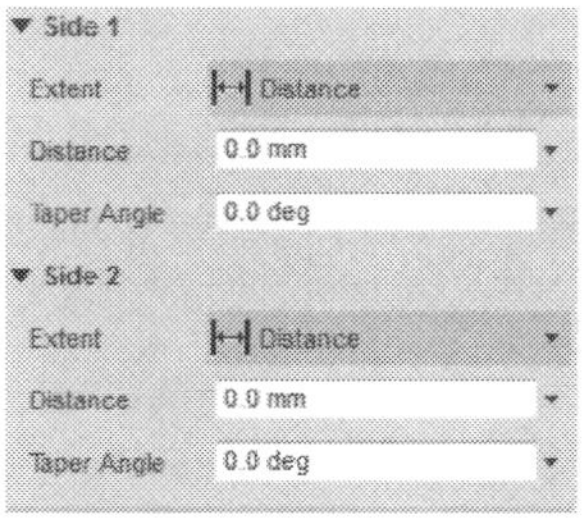

Figure 4-28 *Partial view of the* ***EXTRUDE*** *dialog box*

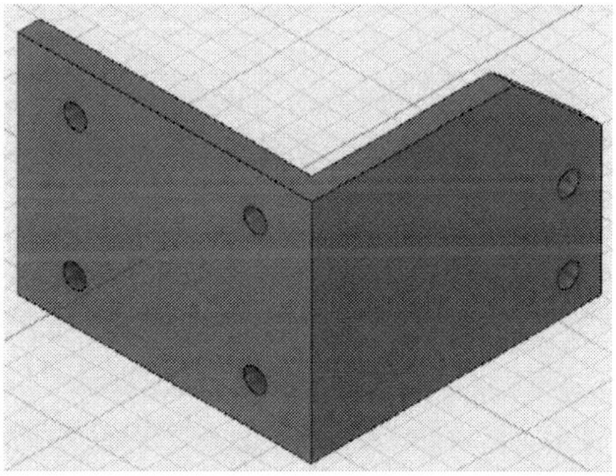

Figure 4-29 *Model after creating the second feature*

Creating the Mirror Feature

Now, you need to create a mirror copy of the last created feature by using the **Mirror** tool.

1. Choose the **Mirror** tool from the **CREATE** panel in the Ribbon; the **MIRROR** dialog box is displayed, as shown in Figure 4-30. Also, you are prompted to select the faces or features to mirror. The **Mirror** tool is used to mirror a selected feature, face, or body about a specified mirror plane which can be a reference plane or a planar face.

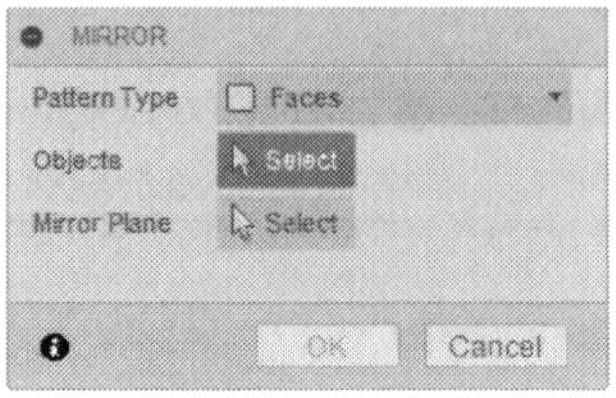

*Figure 4-30 The **Mirror** dialog box*

   The **Object** selection box is used to select the faces, solid feature, bodies, or component to mirror. The **Mirror Plane** selection box is used to select a plane or planar face about which the feature will be mirrored. The **Pattern Type** drop-down list provides different patterning options. The **Faces** option is used to select the face of the model to be mirrored and the **Bodies** option is used to select the complete body to be mirrored. You can also mirror any particular feature of a model about the selected plane or face by selecting the **Feature** option from the **Pattern Type** drop-down list. The **Component** option is selected to mirror the component of an assembly about a selected plane or face.

2. Select the **Feature** option from the **Pattern Type** drop-down list and then select the second feature created from the timeline to mirror it.

3. Click on the **Mirror Plane** selection box to enable the selection box; you are prompted to select the planar face or plane about which the features will be mirrored.

4. Select the **YZ** plane from the **BROWSER** bar under the **Origin** node; preview of the mirrored feature is displayed in the graphics window.

5. Choose the **OK** button from the **MIRROR** dialog box. The isometric view of the model after creating the mirrored feature is shown in Figure 4-31.

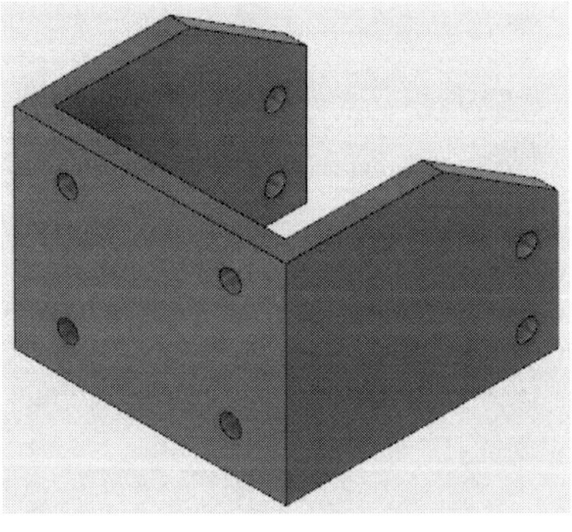

Figure 4-31 Model after creating the mirrored feature

Creating the Cut Feature

In this section, you will create the cut feature using the **All** option. The sketch of the cut feature will be drawn on the front planar face of the base feature.

1. Choose the **Create Sketch** from the **SKETCH** panel and then select the front face as the sketching plane; the selected plane gets oriented normal to the screen.

2. Create sketch of the cut feature of the model, refer to Figure 4-24.

3. Choose the **Extrude** tool from the **CREATE** panel of the Ribbon; the **EXTRUDE** dialog box is displayed. Also, you are prompted to select the sketch profile to extrude.

4. Select the created sketch from the graphics window and select the **All** option from the **Extent** drop-down list. Also, choose the **Flip** button from the **EXTRUDE** dialog box to reverse the direction of extrusion.

5. Choose the **OK** button from the **EXTRUDE** dialog box. The isometric view of the model after creating the cut features is shown in Figure 4-32.

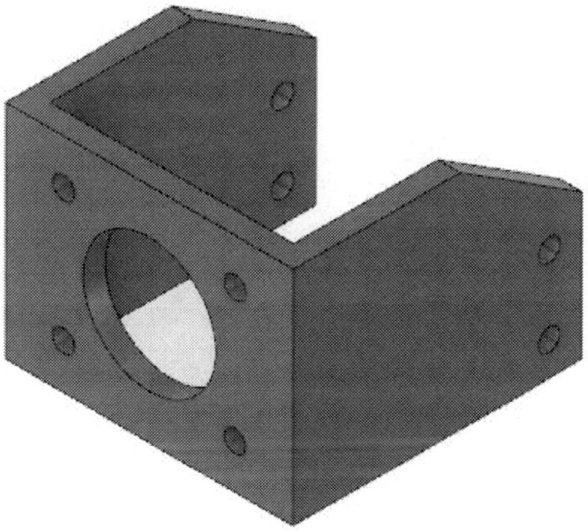

Figure 4-32 *The model after creating the cut feature*

Creating the Chamfer Features

The next feature that you need to add to this model is a chamfer feature.

1. Choose the **Chamfer** tool from the **MODIFY** panel of the Ribbon; the **CHAMFER** dialog box is displayed, as shown in Figure 4-33. Also, you are prompted to select the edges to chamfer.

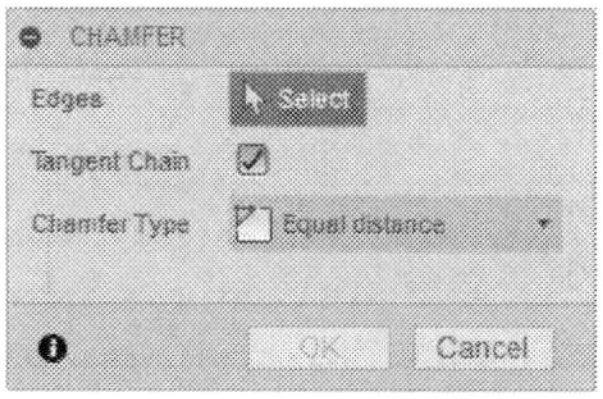

Figure 4-33 *The **CHAMFER** dialog box*

   Chamfering is a process in which the sharp edges of the model are beveled in order to reduce the areas of stress concentration. This process also eliminates undesirable sharp edges and corners of the model.

2. Select the edge of the cut feature from the graphics window and then select the **Distance and angle** option from the **Chamfer Type** drop-down list; the **Distance** and **Angle** edit boxes are displayed.

   By default, the **Equal distance** option is selected in the **Chamfer Type** drop-down list. Therefore, after selecting the edge to be chamfered, the **Distance** edit box is displayed. Now, you can specify the same distance for both sides of the chamfer.

   The **Distance and angle** option in the **Chamfer Type** drop-down list allows you to specify the value of the distance and angle in the **Distance** and **Angle** edit boxes of the **CHAMFER** dialog box or enter values directly in the **Distance** and **Angle** edit boxes in the graphics window. The **Tangent Chain** check box is selected by default. Therefore, the edges that are tangent to the selected edge are selected automatically.

3. Enter **2.5** in the **Distance** edit box and **45** in the **Angle** edit box.

4. Choose the **OK** button from the **CHAMFER** dialog box. The final model after applying the **Chamfer** tool is shown in Figure 4-34.

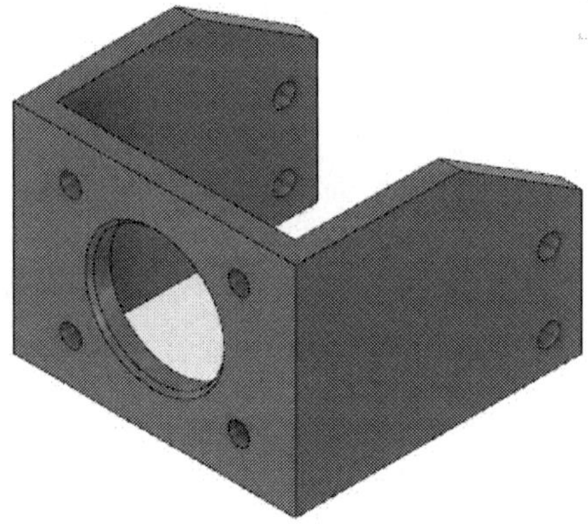

Figure 4-34 *The final model for Tutorial 3*

Saving the Model

Next, you need to save the model.

1. Choose the **Save** option from the **File** menu or **Application** menu; the **Save** dialog box is displayed.

2. Enter **c04_Tut_03** in the **Name** edit box.

3. Click on the down arrow on the extreme right of the **Location** selection box; the **Save** dialog box gets expanded. 
4. Select the existing project from the **PROJECT** column or you can create a new project by choosing the **New Project** button from the **Save** dialog box.

5. Choose the **Save** button to save the sketch and exit the **Save** dialog box.

Tutorial 4

In this tutorial, you will create model of a train wheel by revolving its sketch, refer to Figure 4-35. The dimensions of the model are shown in Figure 4-36. **(Expected time: 45 min)**

Figure 4-35 The model of the train wheel

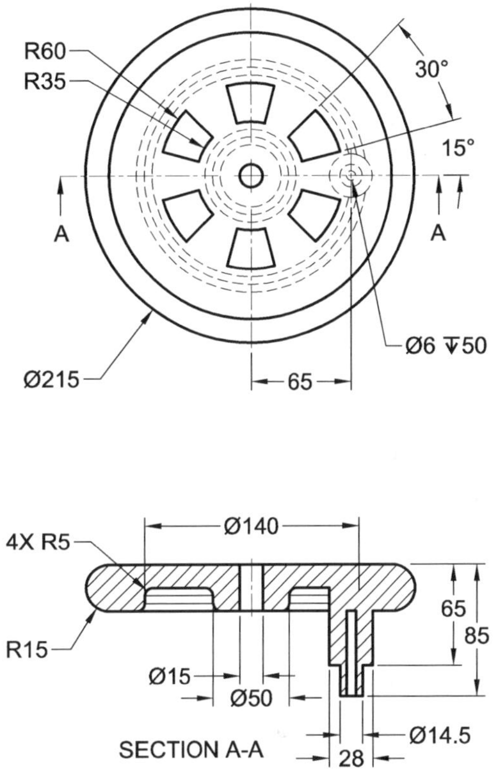

Figure 4-36 The dimension of the model

The following steps are required to complete this tutorial:

a. Create a revolve feature as the base feature of the model.
b. Create an extruded cut feature.
c. Create a circular pattern of the extruded cut feature.
d. Create an extrude feature.
e. Save the model.

Creating the Base Feature

First, you will create base feature of the model, which is a revolved feature.

1. Start a new design using the **New Design** option from the **File** menu or the **Application** menu.

2. Select the **Create Sketch** from the **SKETCH** panel. Next, select the **XY** plane from the **BROWSER** bar as the sketching plane; the selected plane gets oriented normal to the screen.

3. Create the sketch of the revolved feature of the model and apply required constraint and dimensions to the sketch from the **SKETCH PALETTE** window, as shown in Figure 4-37.

4. Invoke the **Revolve** tool and revolve the sketch by selecting its vertical centerline as the axis of revolution. The rotated view of the model after creating the revolved feature is shown in Figure 4-38.

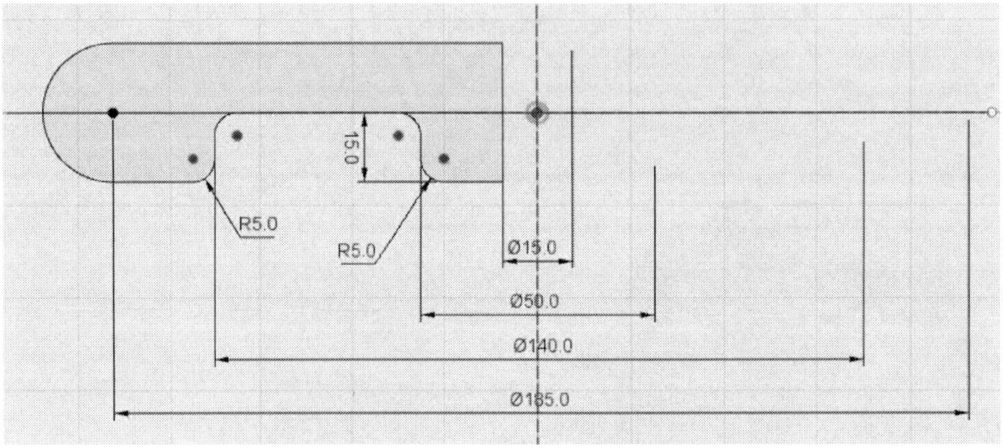

***Figure 4-37** Sketch of the Base feature*

***Figure 4-38** Rotated view of the revolved feature*

Creating the Cut Feature

The second feature to be created is a cut feature. The sketch of this cut feature will be created on the bottom of the revolved feature, refer to Figure 4-39.

1. Invoke the **Create Sketch** tool from the **SKETCH** panel in the Ribbon and select the bottom face of the revolved feature; the selected face gets normal to the screen.

2. Create the sketch of the cut feature, as shown in Figure 4-39.

3. Choose the **Extrude** tool from the **CREATE** panel of the Ribbon; the **EXTRUDE** dialog box is displayed and you are prompted to select the sketch profile to extrude.

4. Select the sketch from the graphics window that corresponds to the cut feature and select the **All** option from the **Extent** drop-down list. Also, choose the **Flip** button from the **EXTRUDE** dialog box to reverse the direction of extrusion.

5. Choose the **OK** button from the **EXTRUDE** dialog box. The rotated view of the model after creating the cut feature is shown in Figure 4-40.

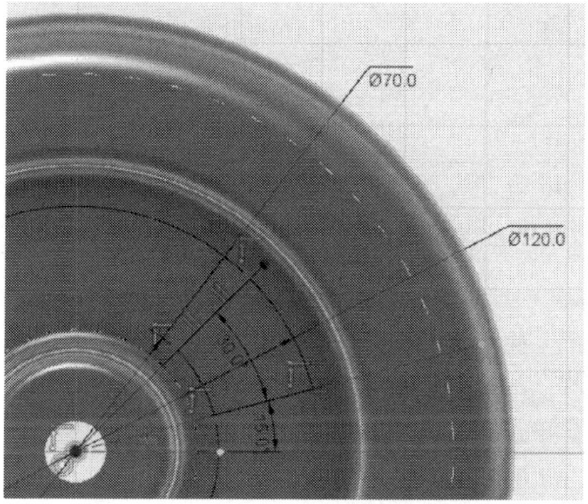

Figure 4-39 *Sketch of the cut feature*

Figure 4-40 *Model after creating the cut feature*

Patterning the Cut Feature

After creating the cut feature, you need to pattern it by using the **Circular Pattern** tool to create other instances of the cut features.

1. Choose **Pattern > Circular Pattern** from the **CREATE** panel; you are prompted to select the faces or features to pattern and also the **CIRCULAR PATTERN** dialog is displayed, as shown in Figure 4-41.

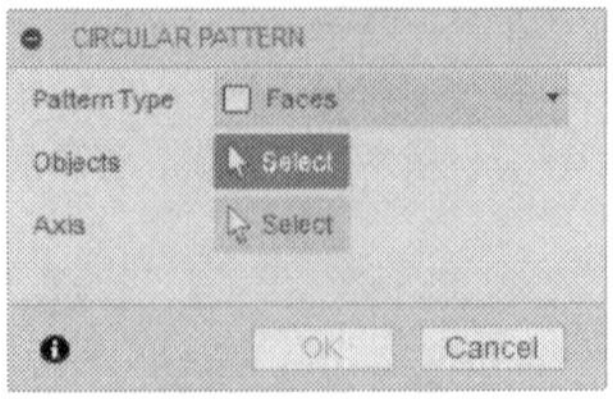

***Figure 4-41** The **CIRCULAR PATTERN** dialog box*

The **Circular Pattern** tool is used to create circular pattern instances of the selected features, faces, and bodies.

2. Select the cut feature created from the timeline in the graphics window.

3. Make sure that the **Features** option is selected in the **Pattern Type** drop-down list.

4. Choose the **Axis Select** button and then select the circular edge of the model; a preview of the circular pattern of the hole feature is displayed in the graphics window. Also, other options will become available such as the **Type** drop-down list, the **Suppress** check box, and the **Quantity** spinner.

5. Select the **Full** option from the **Type** drop-down list and set the value **6** in the **Quantity** spinner.

6. Choose the **OK** button from the **CIRCULAR PATTERN** dialog box; the cut feature is patterned, as shown in Figure 4-42.

***Figure 4-42** Model after patterning the cut feature*

Creating the Extrude Feature

The last feature of the model is an extrude feature which is created by extruding the profile of the Crank Pin.

1. Invoke the **Create Sketch** tool from the **SKETCH** panel in the Ribbon and select the bottom face of the revolved feature; the selected face gets normal to the screen.
2. Create the sketch of Crank Pin of diameter 28 mm, refer to Figure 4-43.
3. Invoke the **EXTRUDE** dialog box and select the created sketch from the graphics window.
4. Select the **One Side** option, if it is not selected by default in the **Direction** drop-down list in the **EXTRUDE** dialog box and enter **-50** in the **Distance** edit box; preview of the feature is displayed in the graphics window. Make sure that the **Join** option is selected in the **Operation** drop-down list.
5. Choose the **OK** button from the **EXTRUDE** dialog box.
6. Similarly, extrude the region between diameter 15 and diameter 6, as shown in Figure 4-44. For extrusion depth, enter **20** in the **Distance** edit box. The final model of the train wheel is shown in Figure 4-45.

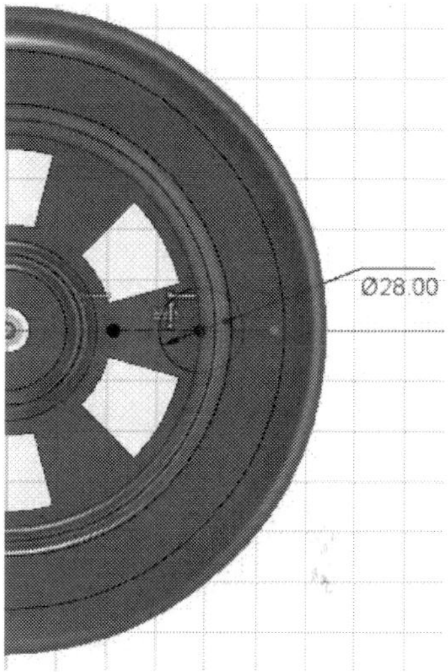

Figure 4-43 *Sketch for the diameter 28 mm*

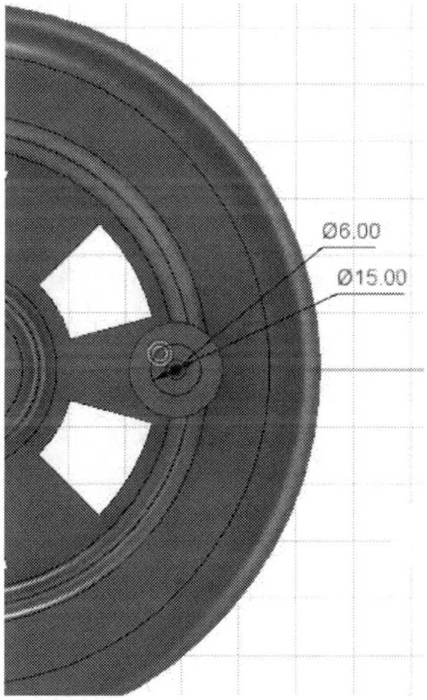

Figure 4-44 *Sketch for the diameter 15 mm and 6 mm*

Figure 4-45 *Final model of the train wheel*

Saving the Model

Next, you need to save the sketch.

1. Choose the **Save** option from the **File** menu or the **Application** menu; the **Save** dialog box is displayed.

2. Enter **c04_Tut_04** in the **Name** edit box.

3. Click on the down arrow on the extreme right side of the **Location** selection box; the **Save** dialog box gets expanded. 

4. Select an existing project from the **PROJECT** column or you can create a new project by choosing the **New Project** button from the **Save** dialog box.

5. Choose the **Save** button to save the sketch and exit the **Save** dialog box.

Self-Evaluation Test

Answer the following questions and then compare them to those given at the end of this chapter:

1. Which of the following tools is used to create another instance of the body with respect to the symmetry plane?

   (a) **Intersect** (b) **Rectangular Pattern**
   (c) **Symmetry** (d) **Mirror**

2. Which entity cannot be used as rotation axis for circular pattern?

   (a) Construction Plane (b) Edge
   (c) Construction Line (d) Cylindrical face

3. __________ is a process of beveling the sharp edges of a model in order to reduce stress concentration.

4. The __________ button from the **Extrude** dialog box is used to reverse the direction of extrusion.

5. If you want to create a chamfer of unequal length, select the __________ option from the **Chamfer Type** drop-down list.

6. The __________ option is selected to revolve the sketch through 360-degree.

7. You can mirror an entire model. (T/F)

8. While creating circular pattern of hole, you can suppress all occurrence except the first one. (T/F)

Review Questions

Answer the following questions:

1. Which of the following construction features can be used as a reference for mirroring features?

   (a) Construction Plane (b) Construction Point
   (c) Construction Line (d) None of these

2. The _________ option is used to terminate the extruded feature upto selected face or plane.

3. The edges that are tangent to the selected edge are selected automatically, if the _________ check box is selected in the **CHAMFER** dialog box.

4. By default, model is displayed in _________ camera type.

5. You cannot preselect the edges or faces for creating a fillet feature. (T/F)

6. Features can be selected only from the timeline. (T/F)

EXERCISES

Exercise 1

Create the model shown in Figure 4-46. The dimensions of the model are shown in Figure 4-47. **(Expected time: 45 min)**

Figure 4-46 *Model for Exercise 1*

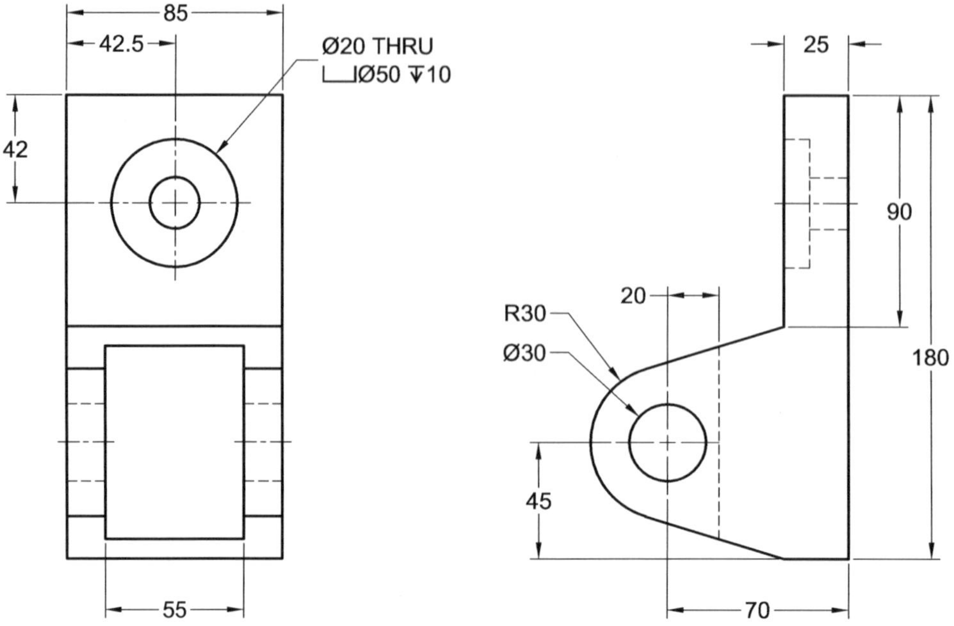

Figure 4-47 *Views and dimensions of the model*

Exercise 2

Create the model shown in Figure 4-48. The views and dimensions of the model are shown in Figure 4-49. **(Expected time: 45 min)**

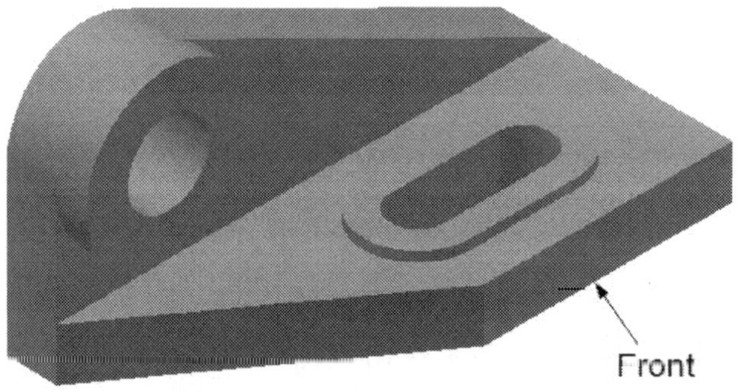

Figure 4-48 *Model for Exercise 2*

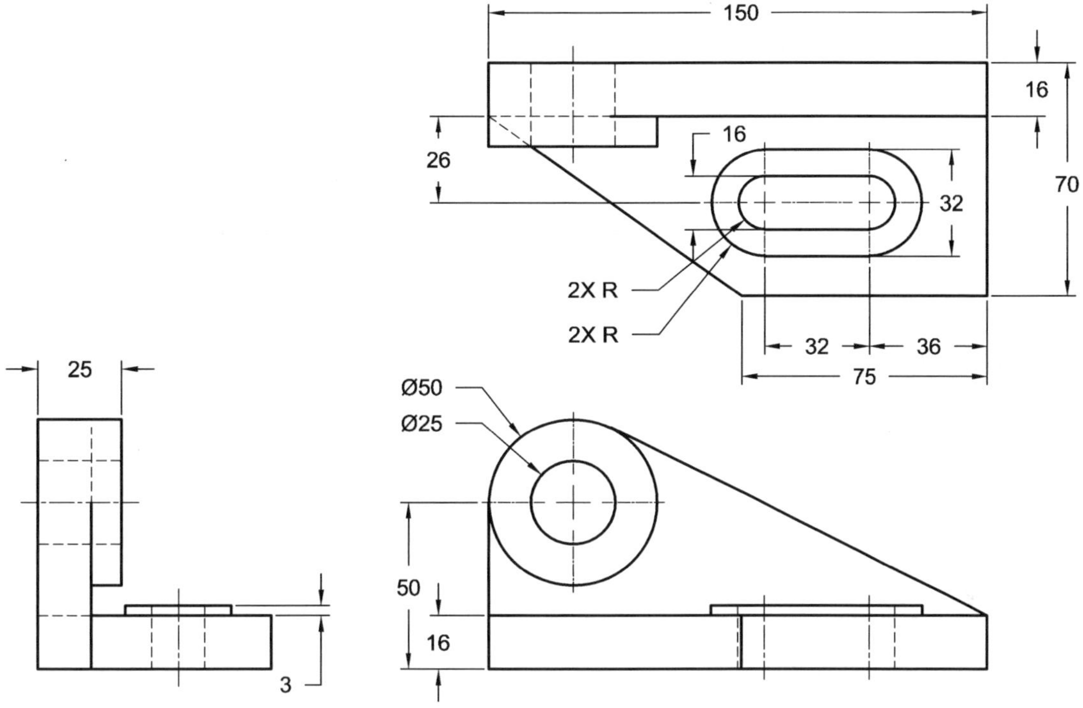

Figure 4-49 The views and dimensions of the model

Exercise 3

In this tutorial, you will create the model shown in Figure 4-50. The dimensions of the model are shown in the same figure. Also, you will create a section view of the model using the **Section View** tool. **(Expected time: 30 min)**

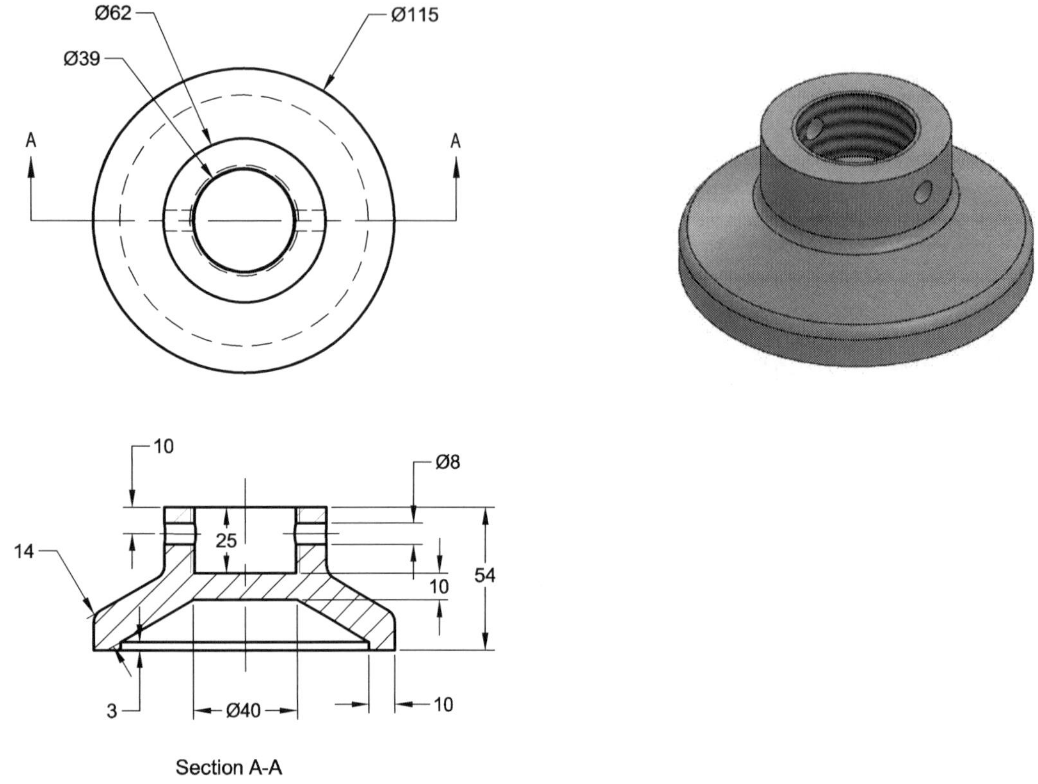

Figure 4-50 Dimensions and views of the model for Exercise 3

Answers to Self-Evaluation Test

1. Mirror, **2.** Construction Plane, **3.** Chamfering, **4. Flip**, **5. Two distances**, **6. Full**, **7.** T, **8.** T

Chapter 5

Creating Reference Geometries

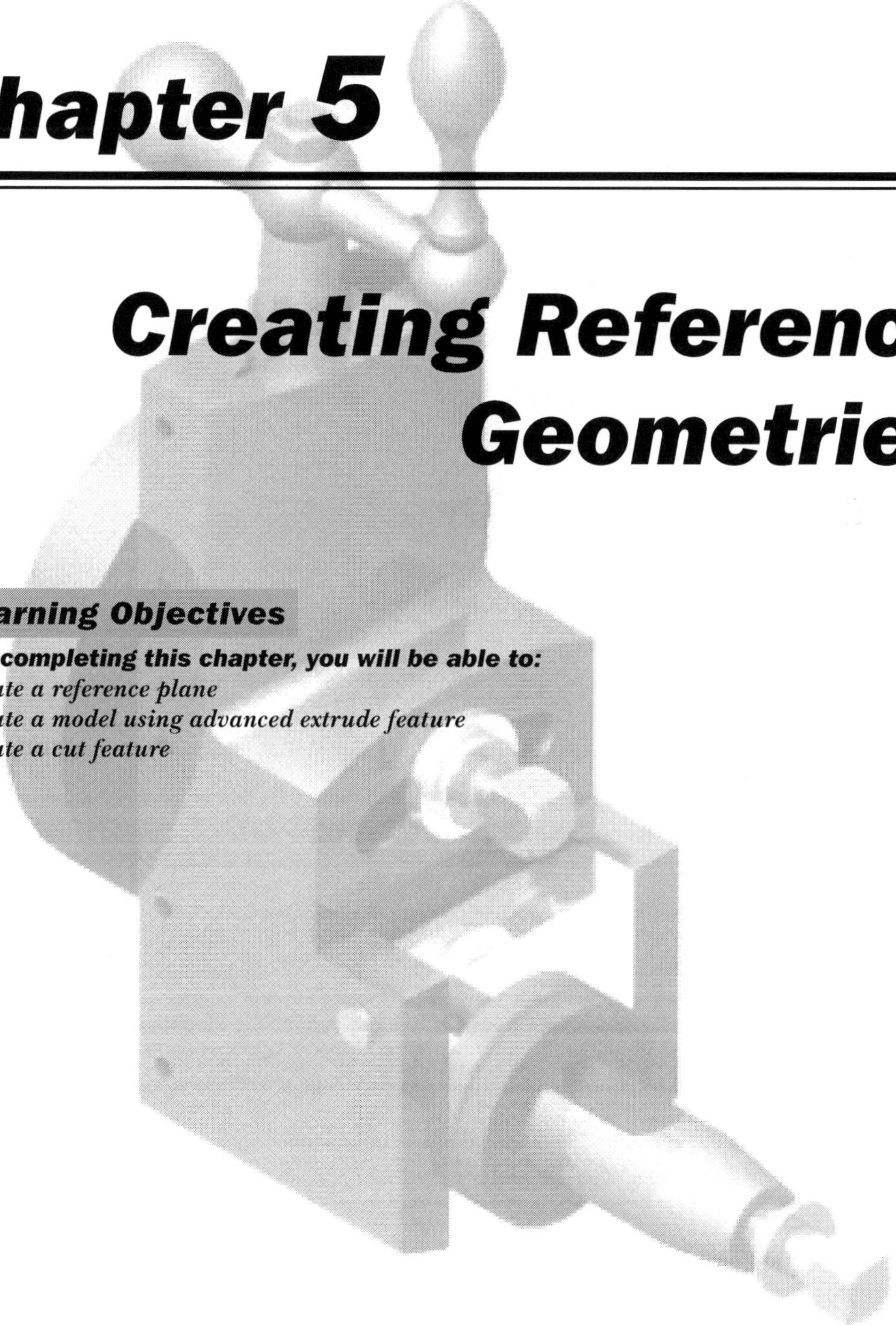

Learning Objectives

After completing this chapter, you will be able to:

- *Create a reference plane*
- *Create a model using advanced extrude feature*
- *Create a cut feature*

IMPORTANCE OF SKETCHING PLANES

In earlier chapters, you created basic models by extruding or revolving the sketches. All those models were created on a single sketching plane. But most mechanical designs consist of multiple sketched features, referenced geometries, and placed features. These features are integrated together to complete a model. Most of these features lie on different planes. When you start a new sketch, you are prompted to select the plane on which you want to draw the sketch. On the basis of design requirements, you can select any plane to create the base feature. To create additional sketched features, you need to select an existing plane or a planar surface, or you need to create a plane that will be used as the sketching plane. For example, consider the model shown in Figure 5-1.

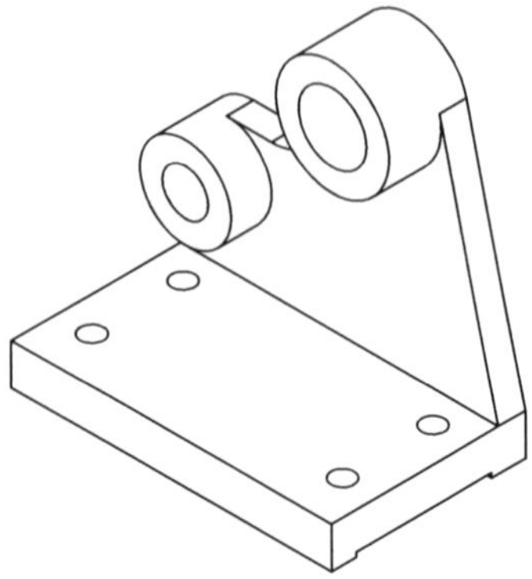

Figure 5-1 *A multifeatured model*

The base feature of this model is shown in Figure 5-2. The sketch for the base feature is drawn on the top plane. After creating the base feature, you need to create other sketched features, placed features, and referenced features, refer to Figure 5-3. The boss features and cut features are the sketched features that require sketching planes where you can draw the sketches of the features.

It is evident from Figure 5-3 that the features added to the base feature are not created on the same plane on which the sketch for the base feature is created. Therefore, to draw the sketches of other sketched features, you need to define other sketching planes.

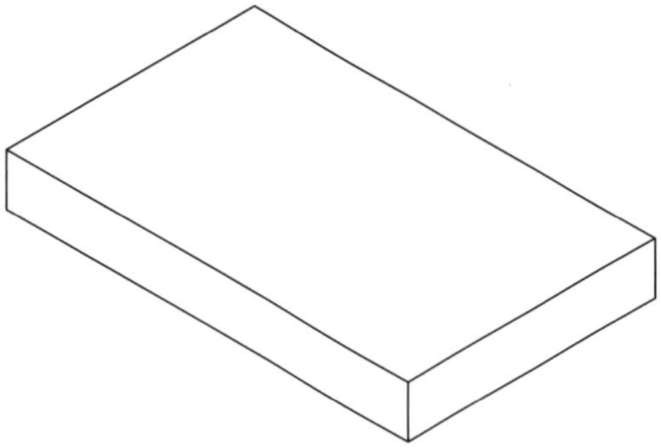

Figure 5-2 *Base feature of the model*

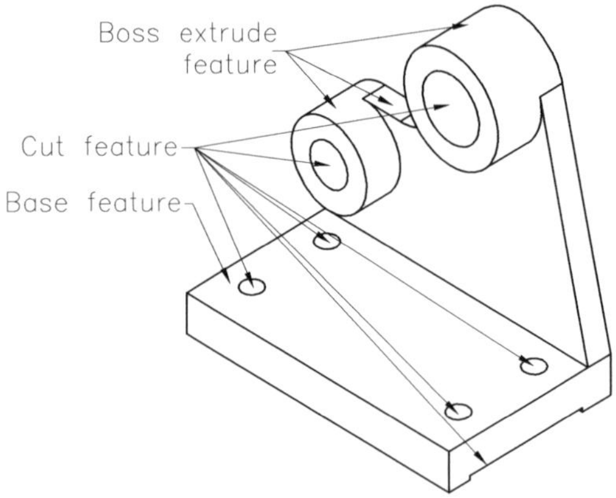

Figure 5-3 *Model after adding other features*

TUTORIALS

Tutorial 1

In this tutorial, you will create the model shown in Figure 5-4. The dimensions of the model are also shown in the same figure. **(Expected time: 30 min)**

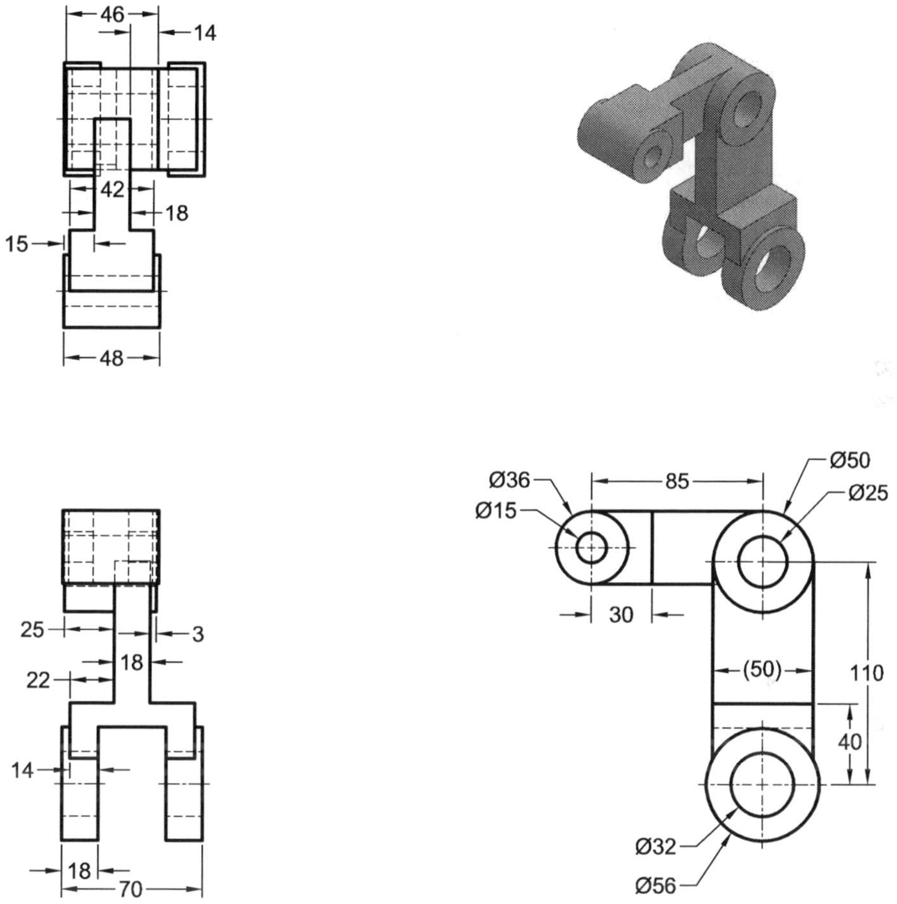

Figure 5-4 *Model and its different views for Tutorial 1*

It is clear from the above figure that the given model is a multi-featured model. It consists of various extrude and cut features. You need to create a separate sketch of each sketched feature for converting them into features.

The following steps are required to complete this tutorial:

a. Start Autodesk Fusion 360 and create the sketch of the base feature on the YZ plane and apply required relations and dimensions to it.

b. Invoke the **Extrude** tool and create the base feature of the model.
c. Create the sketch of the second feature on the YZ plane and apply required relations and dimensions to it.
d. Invoke the **Extrude** tool and extrude the sketch created for the second feature by using the **Midplane** option.
e. Create the third and fourth features of the model by using the Contour Selection method.
f. Create the fifth feature which is an extruded cut feature.
g. Create a reference plane at an offset distance from the XY plane for creating the sketch of the sixth feature.
h. Create the sixth feature of the model which is also an extruded feature.
i. Create a reference plane for creating the sketch of seventh and eighth features.
j. Create the seventh and eighth features of the model.
k. Save the document and then close it.

Starting Autodesk Fusion and Creating the Base Feature

Now, you will create the base feature of the model which is an extrude feature.

1. Choose the **Create Sketch** tool from the **SKETCH** panel of the **MODEL** workspace; the three default planes are displayed in the graphics window.

2. Select the **YZ** plane from the **Origin** node in the **BROWSER** bar and then create the sketch by using the **Circle** tool. Next, apply required dimensions to it, as shown in Figure 5-5.

3. Choose the **Extrude** tool from the **CREATE** panel; isometric view of the sketch is displayed and also the **EXTRUDE** dialog box is displayed.

   Now, you will extrude the feature equally on both sides of the sketching plane.

4. Select the contour between two circle from the graphics window and the **Symmetric** option from the **Direction** drop-down list of the **EXTRUDE** dialog box. On doing so, the **Measurement** area is displayed with two buttons, **Half Length** and **Whole Length**.

   The **Half Length** button is used for creating extrusions of equal length in both directions from the selected axis or plane and the **Whole Length** button is used for dividing the extrusion length equally in both directions from the plane or axis.

5. Choose the **Whole Length** button on the right of the **Measurement** area.

6. Enter **70** in the **Distance** edit box and then choose the **OK** button from the **EXTRUDE** dialog box. The base feature of the model is created, as shown in Figure 5-6.

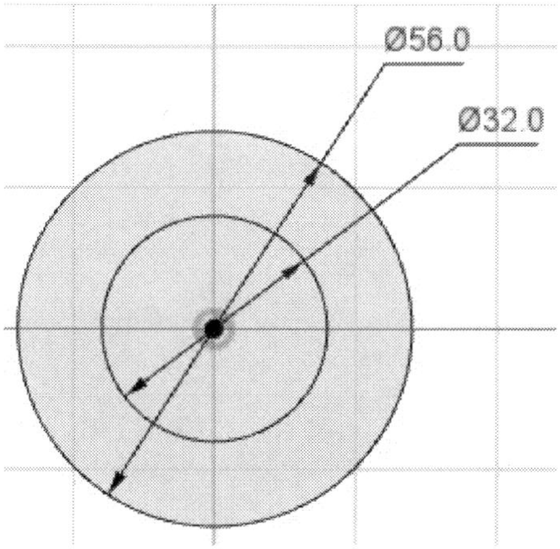

Figure 5-5 Sketch of the base feature

Figure 5-6 The base feature of the model

Creating the Second Feature

Now, you will the create second feature of the model which is also an extrude feature. To create this feature, you need to first create a sketch on the YZ plane and then extrude it by using the **Symmetric** option.

1. Select the **YZ** plane and create the sketch, as shown in Figure 5-7. Make sure the sketch is fully defined.

2. Invoke the **EXTRUDE** dialog box by using the **Extrude** tool and extrude the sketch upto a depth of 62 mm using the **Symmetric** option. Make sure that the **Whole Length** button is chosen on the right of the **Measurement** option. Figure 5-8 shows the model after creating the second feature.

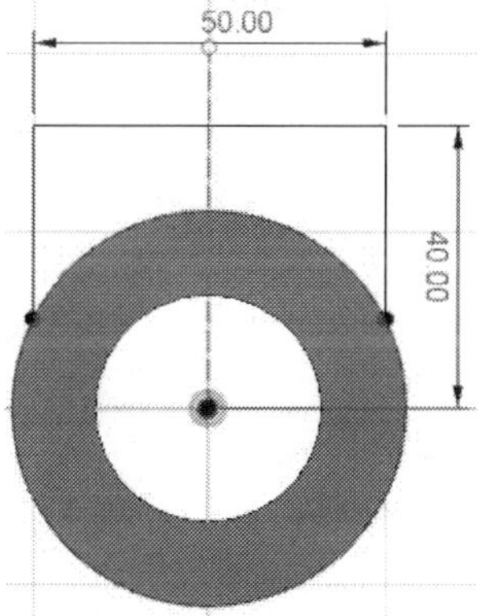

Figure 5-7 Sketch of the second feature

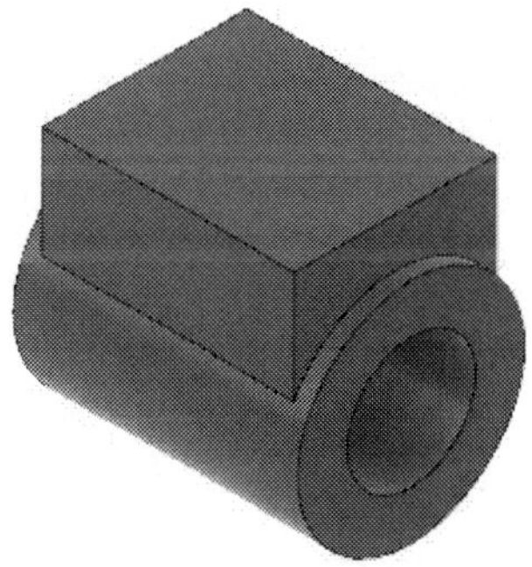

Figure 5-8 Model after creating the second feature

Creating the Third and Fourth Features

Now, you will create third and fourth features of the model. In conventional method, you need to create separate sketches for both these features and then convert them one by one into features. In this tutorial, you will create a sketch and then convert it into features by selecting its contours.

1. Select the **YZ** plane as the sketching plane and then draw the sketch, as shown in Figure 5-9.

   Next, you need to use the contour selection method to create the third and fourth features.

2. Choose the **Extrude** tool from the **CREATE** panel of the Ribbon; the **EXTRUDE** dialog box is displayed in the graphics window and you are prompted to select the sketch profiles or planar faces to extrude.

3. Move the selection cursor toward the bottom and close to the contour of the sketch and then select it when highlighted, refer to Figure 5-10; the preview of the feature is displayed in the graphics window. Also, the **Dimension** edit box is displayed.

4. Select the **Symmetric** option from the **Direction** drop-down list and also choose the **Whole Length** button on the right of the **Measurement** option from the **EXTRUDE** dialog box.

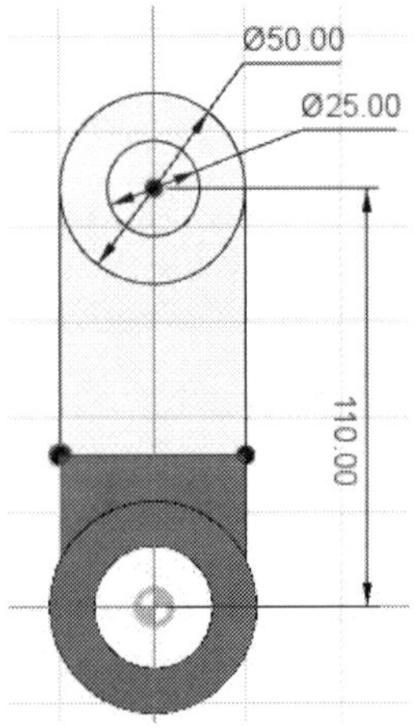

Figure 5-9 *Sketch with multiple contours*

Figure 5-10 *The bottom contour of the sketch to be selected*

5. Enter **18** in the **Distance** edit box and choose the **OK** button. Figure 5-11 shows the model after creating the third feature.

   Now, you will create the fourth feature of the model by using the same sketch used for creating the third extrude feature.

6. Right-click on **Sketch3** (sketch of the third feature) from the expanded **Sketches** sub node in the **BROWSER** bar; a shortcut menu is displayed.

7. Choose the **Show/ Hide** option from the shortcut menu; the selected sketch becomes visible.

8. Invoke the **EXTRUDE** dialog box from the **CREATE** panel. Then, move the cursor toward the closed contour of the sketch formed by two circles and then select it when highlighted, refer to Figure 5-12.

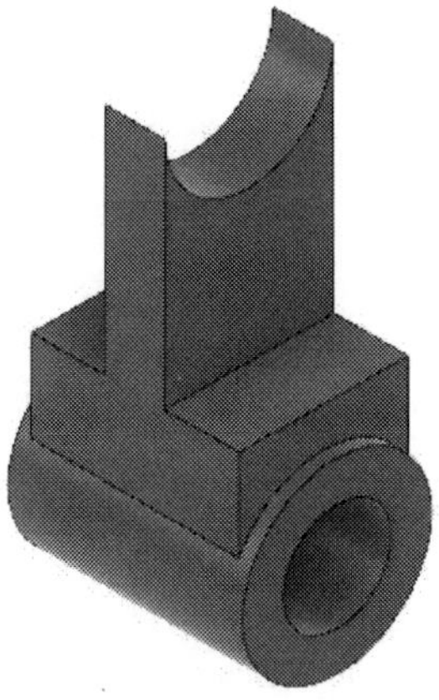

Figure 5-11 *Model after creating the third extrude feature*

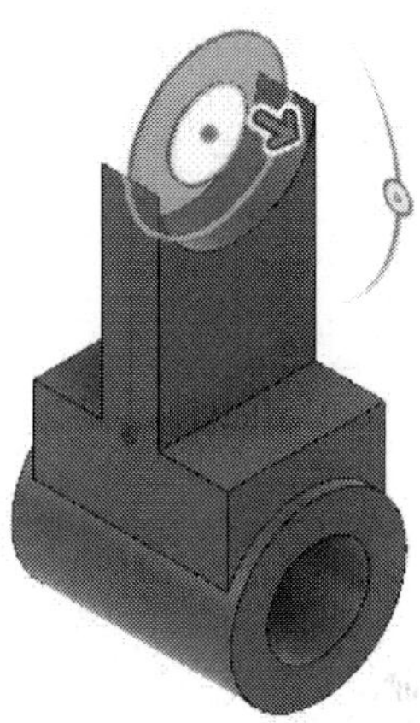

Figure 5-12 *Contour of the sketch to be selected*

9. Select the **Two Sides** option in the **Direction** drop-down list; the expanded **Side 1** and **Side 2** nodes are displayed.

10. Enter **12** in the **Distance** edit box under the expanded **Side 1** node.

11. Similarly, enter **34** in the **Distance** edit box under the expanded **Side 2** node. Next, choose the **OK** button from the **EXTRUDE** dialog box; the extrude feature is created, as shown in Figure 5-13.

Note

To hide/ show an entity, click on the toggle button on the left of the entity in the ***BROWSER*** *bar.*

Figure 5-13 Model after creating the fourth extrude feature

Creating the Fifth Feature

The fifth feature of the model is a cut feature. Now, you will create a cut feature by using the **Extrude** tool.

1. Select the **XY** plane as the sketching plane and then create the sketch, as shown in Figure 5-14.

2. Choose the **Extrude** tool from the **CREATE** panel; the **EXTRUDE** dialog box is displayed and you are prompted to select the sketch profiles or planar faces to extrude.

3. Select the sketch from the graphics window.

   As you have to remove material from the model to create this feature, therefore, you need to perform the cut operation.

4. Select the **Symmetric** option from the **Direction** drop-down list and drag the blue arrow on the model till the cut feature cuts through the entire model; the **Cut** option gets selected in the **Operation** drop-down list.

5. Choose the **OK** button from the dialog box. Figure 5-15 shows the isometric view of the model after creating the cut feature.

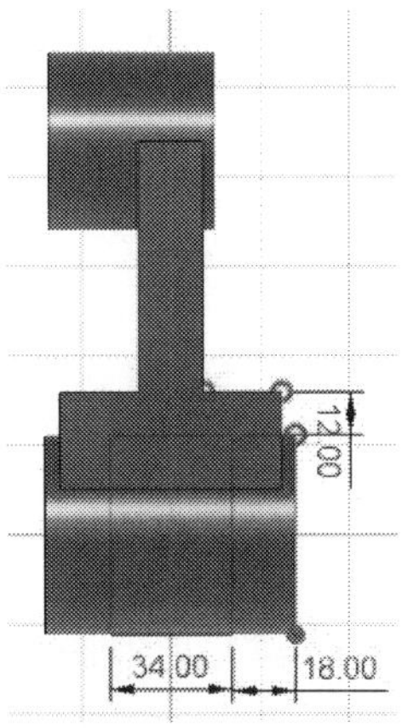

Figure 5-14 Sketch for the cut feature

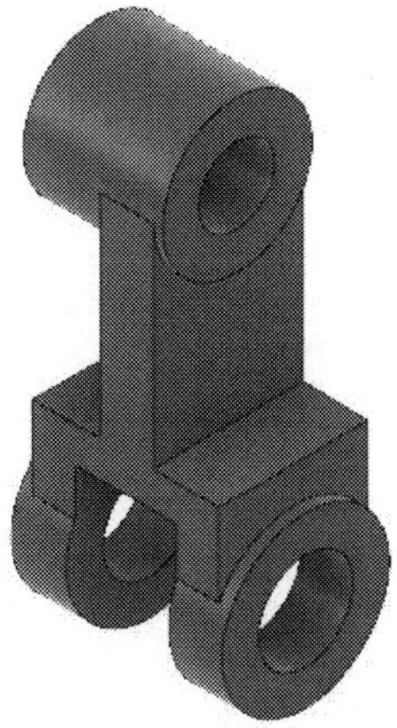

Figure 5-15 Model after creating the cut feature

Creating the Sixth Feature

The sixth feature of the model is an extrude feature. The sketch of this extrude feature is created by using a reference plane created at an offset distance of 55 mm from the XY plane

Generally, all industrial components or designs are multi-featured models. As discussed earlier, all features of a model are not created on the same plane on which the base feature is created. Therefore, you need to select one of the default planes or create a new plane that will be used as the sketching plane for the next feature. It is mentioned above that you can use both the default planes as well as the reference planes created as the sketching planes. The reference plane can be created by choosing different plane options.

- Offset Plane
- Plane at Angle
- Tangent Plane
- Midplane
- Plane Through Two Edges
- Plane Through Three Points
- Plane Tangent to Face at Point
- Plane Along Path

1. Choose **Offset Plane** from the **CONSTRUCT** panel; the **OFFSET PLANE** dialog box is displayed, as shown in Figure 5-16. Also, you are prompted to select the plane, planar face or sketch profile.

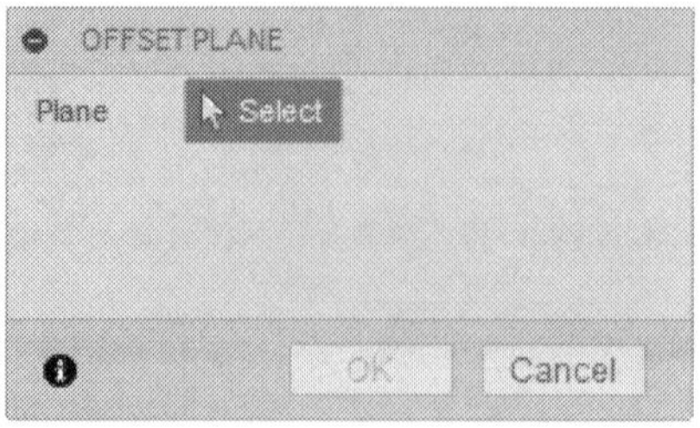

Figure 5-16 *The **OFFSET PLANE** dialog box*

2. Select the **XY** plane from the **BROWSER** bar under the **Origin** node; the **Distance** edit box is displayed in the graphics window and also in the **OFFSET PLANE** dialog box.

3. Enter an offset distance value to **55** in the **Distance** edit box in the dialog box; a reference plane is created at the specified offset distance, as shown in Figure 5-17. Note that size of the plane has been enlarged for better visibility.

4. Choose the **OK** button from this dialog box.

5. Choose the **Create Sketch** tool from the **SKETCH** panel in the Ribbon and select the newly created reference plane from the **BROWSER** bar under the **Construction** node; the selected plane orients normal to the screen.

6. Create the sketch of the sixth feature using **2-Point Rectangle** tool, refer to Figure 5-18.

Figure 5-17 *The plane is created at an offset distance from the XY plane*

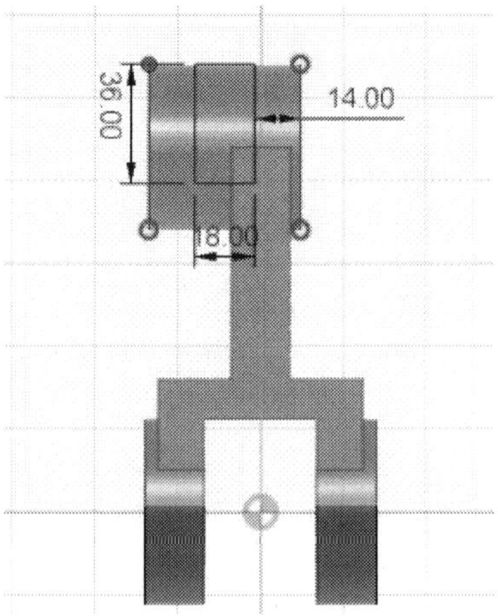

Figure 5-18 *Sketch of the extrude feature*

7. Choose the **Extrude** tool from the **CREATE** panel and then select the sketch from the graphics window; the selected sketch gets highlighted.

8. Select the **To Object** option from the **Extent** drop-down list; you are prompted to select sketch profile, workplane, vertex, faces, or bodies as termination.

9. Select the tangential face (fourth feature) as the termination face of the feature; the preview of the feature is displayed in the graphics window.

10. Choose the **OK** button from the dialog box. The model after creating the sixth feature is shown in Figure 5-19.

Figure 5-19 *Model after creating the sixth feature*

Creating the Seventh and Eighth Features

It is evident from Figure 5-4 that the seventh and eighth features are also extrude features. In this section, you will again use the contour selection method to create seventh and eighth features from a single sketch. Note that the sketch used for these features is created by using the reference plane created at the middle of the planar faces of the fourth feature.

1. Choose the **MidPlane** tool from the **CONSTRUCT** panel in the Ribbon; you are prompted to select two planar faces, planes, or sketch profile.

2. Select both planar faces of the fourth feature one by one as the first and second references, respectively, refer to Figure 5-20; the midplane is created.

3. Choose the **OK** button to exit from the dialog box.

4. Choose the **Create Sketch** tool from the **SKETCH** panel in the Ribbon and select the newly created mid plane; the selected plane gets oriented normal to the screen.

5. Create the sketch of seventh and eighth features with the help of the **Line** and **Center Diameter Circle** tools, as shown in Figure 5-21. For dimensions, refer to Figure 5-4.

Next, you need to extrude the sketch by selecting the contours to create seventh and eighth features.

6. Choose the **Extrude** tool from the **CREATE** panel; the **EXTRUDE** dialog box is displayed in the graphics window and you are prompted to select the sketch profiles or planar faces to extrude.

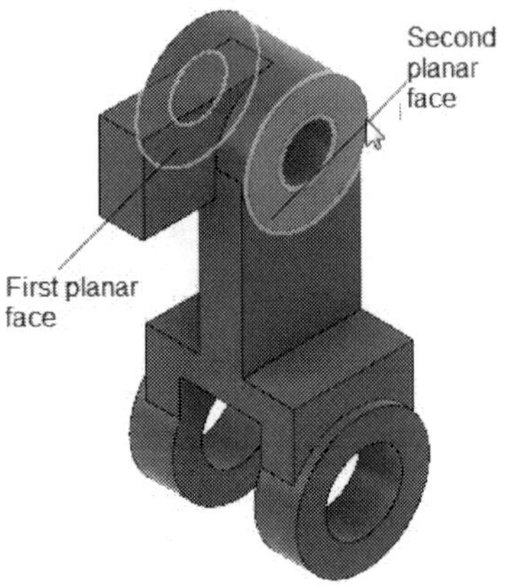

Figure 5-20 The planar faces selected for creating reference plane

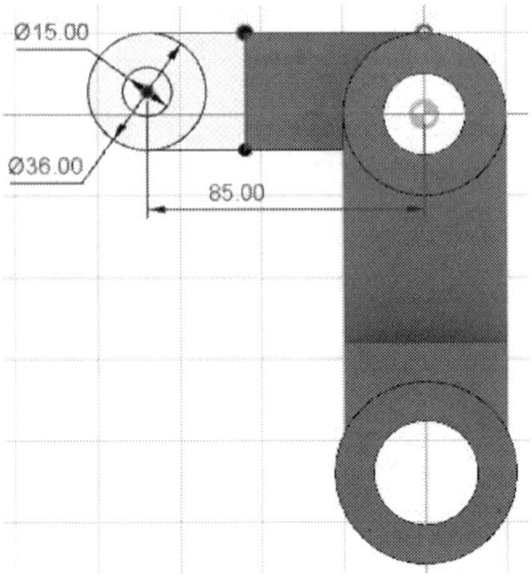

Figure 5-21 Sketch with multiple contours

7. Move the selection cursor toward the closed contour in the sketch and then select it when it is highlighted, refer to Figure 5-22; the preview of the feature is displayed in the graphics window. Also, the **Distance** edit box is displayed.

8. Select the **Symmetric** option from the **Direction** drop-down list and also choose the **Whole Length** button on the right of the **Measurement** area.

9. Enter 42 in the **Distance** edit box and then choose the **OK** button from the **EXTRUDE** dialog box; the selected contour is extruded. Figure 5-23 shows the model after creating the seventh feature.

Figure 5-22 *Contour of the sketch is selected*

Figure 5-23 *The model after creating the seventh feature*

Now, you will create eighth feature of the model by using the same sketch that was used for creating the seventh extrude feature.

10. Right-click on the sketch of the seventh feature in the **BROWSER** bar; a shortcut menu is displayed.

11. Choose the **Show/Hide** option from the shortcut menu; the selected sketch becomes visible.

12. Invoke the **EXTRUDE** dialog box from the **CREATE** panel. Then, move the cursor toward the closed contour formed by two circles in the sketch and then select when it is highlighted, refer to Figure 5-24.

13. Select the **Symmetric** option from the **Direction** drop-down list and also choose the **Whole Length** button on the right of the **Measurement** option.

14. Enter **48** in the **Distance** edit box and then choose the **OK** button from the **EXTRUDE** dialog box; the selected contour is extruded. Figure 5-25 shows the final model after creating the eighth feature.

Figure 5-24 Model after creating the eighth extrude feature

Figure 5-25 Final model

Saving the Model

Next, you need to save the model.

1. Choose the **Save** option from the **File** menu or the **Application** menu; the **Save** dialog box is displayed.

2. Enter **c05_Tut_01** in the **Name** edit box.

3. Click on the down arrow on the extreme right of the **Location** selection box; the **Save** dialog box gets expanded.

4. Select the **CADCIM** project from the **PROJECT** column or you can create a new project by selecting the **New Project** button from the **Save** dialog box.

5. Select the **Save** button to save the sketch and exit the **Save** dialog box.

Tutorial 2

In this tutorial, you will create the model shown in Figure 5-26. The dimensions of the model are given in the same Figure. **(Expected time: 30 min)**

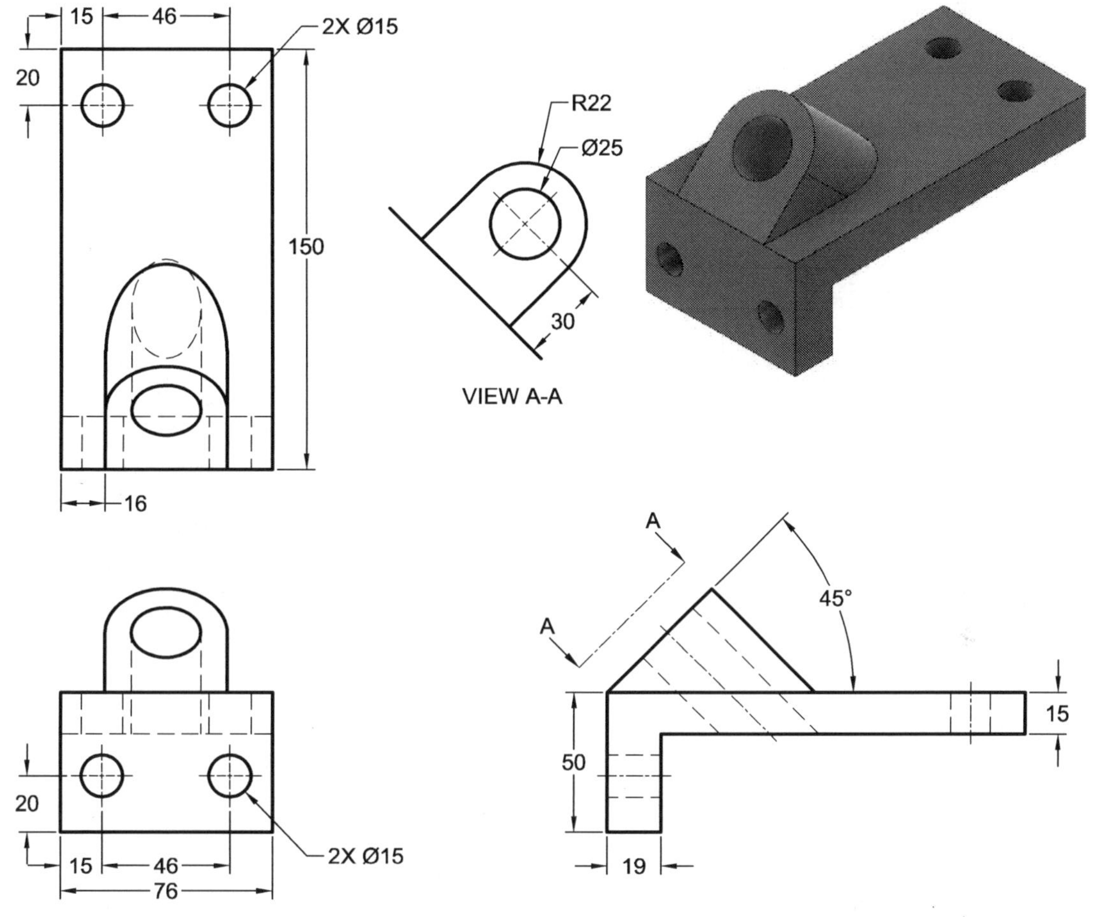

Figure 5-26 *Dimensions for Tutorial 2*

The following steps are required to complete this tutorial:

a. Start Autodesk Fusion 360 and create the sketch of the base feature of the model, refer to Figure 5-27.
b. Invoke the **Extrude** tool and create the base feature of the model, refer to Figure 5-28.
c. Create a reference plane at an angle for creating the second feature, refer to Figure 5-31.
d. Create the second feature of the model, refer to Figures 5-32 through 5-33.
e. Create the fourth feature of the model, refer to Figure 5-36.

Starting Autodesk Fusion and Creating the Base Feature

You will create the base feature of the model which is an extrude feature. The sketch of the base feature is created on the YZ Plane.

1. Start Aotodesk Fusion 360 and choose the **Create Sketch** tool from the **SKETCH** panel of the **MODEL** workspace; the three different default planes are displayed in the graphics window.

2. Select the **YZ** plane from the **BROWSER** bar and then create the sketch by using the **Line** tool. Next, apply the required dimensions to it, as shown in Figure 5-27.

3. Choose the **Extrude** tool from the **CREATE** panel; isometric view of the sketch is displayed and also the **EXTRUDE** dialog box is displayed.

   Now you will extrude the feature equally on both sides of the sketching plane.

4. Select the **Symmetric** option from the **Direction** drop-down list. Also, select the **Whole Length** button from the **Measurement** area.

5. Set the value in the **Distance** edit box to **76** and then choose the **OK** button from the **EXTRUDE** dialog box; base feature of the model is created, as shown in Figure 5-28.

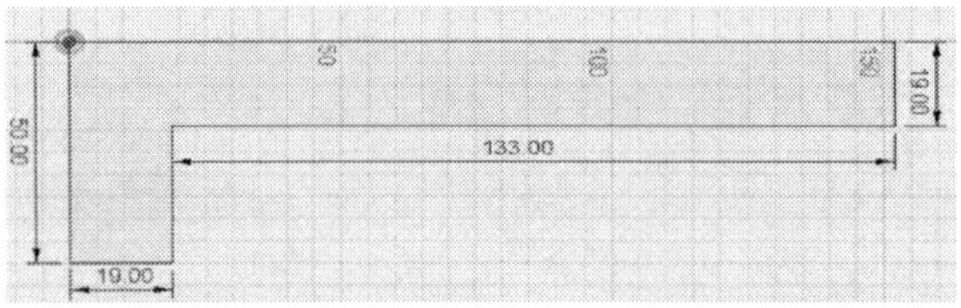

Figure 5-27 *Sketch of the base feature*

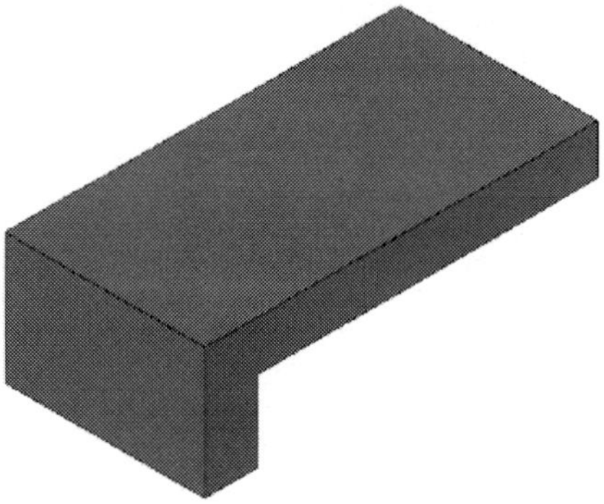

Figure 5-28 *The base feature of the model*

Creating the Second Feature

The second feature of the model is also an extrude feature. The sketch of this extruded feature is created on a reference plane which is at an angle of 135 degrees on the left edge of the model.

1. Choose the **Plane at Angle** tool from the **CONSTRUCT** panel; the **PLANE AT ANGLE** dialog box is displayed, as shown in Figure 5-29 and you are prompted to select line edge, sketch line, or axis.

2. Select the left edge of the top face, refer to Figure 5-30; an **Angle** edit box is displayed in the dialog box.

3. Enter **135** in the **Angle** edit box; preview of the reference plane is displayed in the graphics window.

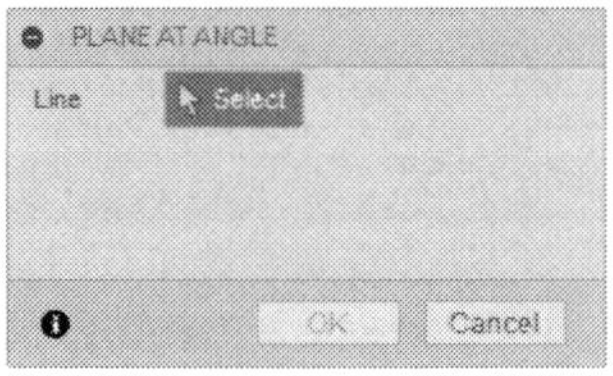

Figure 5-29 *The **PLANE AT ANGLE** dialog box*

4. Choose the **OK** button from the dialog box; the reference plane is created in the graphics window, as shown in Figure 5-31.

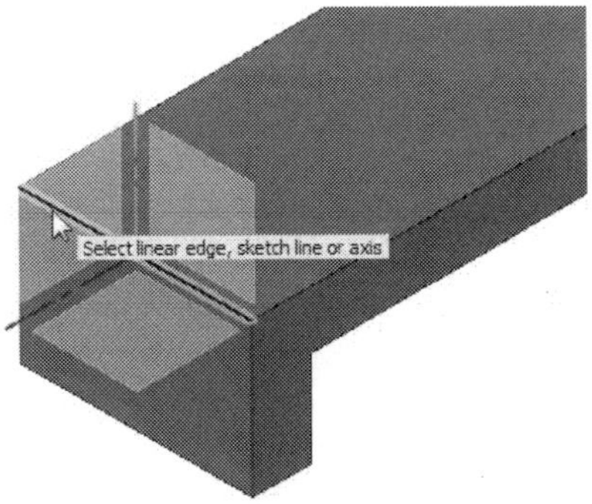

Figure 5-30 *The edge to be selected*

Figure 5-31 *The reference plane created*

Now, you will create a sketch for the second feature by selecting the newly created plane as the sketching plane.

5. Choose the **Create Sketch** tool from the **CREATE** panel and create sketch for the second feature, as shown in Figure 5-32.

6. Extrude the sketch of the model by using the **To Object** option from the **Extent** drop-down list. The model after creating the second feature is shown in Figure 5-33.

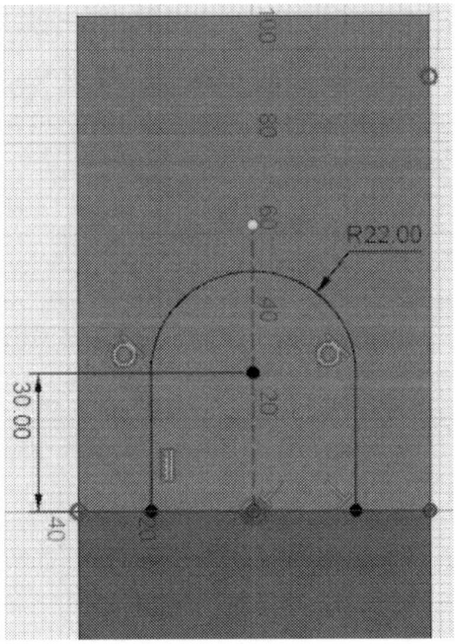

Figure 5-32 *Sketch of the second feature*

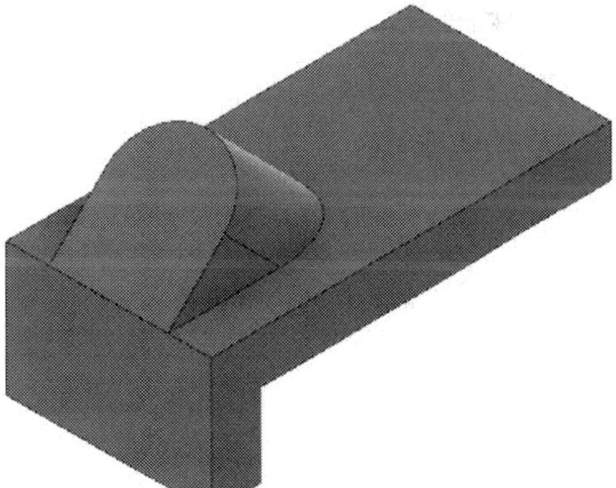

Figure 5-33 *Model after creating the second feature*

Creating the Third Feature

The third feature of the model is a cut feature. It will be created by using a sketch drawn on the second feature and base feature.

1. Choose the **Create Sketch** tool from the **SKETCH** panel of the **Model** workspace and select the planar face of the second feature; the selected face orients normal to the screen.

2. Create sketch for the third feature, as shown in Figure 5-34.

3. Choose the **Extrude** tool from the **CREATE** panel; isometric view of the sketch is displayed and also the **EXTRUDE** dialog box is displayed.

   Now you will create a through cut in the body.

4. Select the circle created and the **All** option from the **Extent** drop-down list and if required, select the **Flip** button to flip the direction of extrusion.

5. Make sure that the **Cut** option is selected from the **Operation** drop-down list.

6. Similarly, create the other cut features of the model. For dimension refer to Figure 5-26. The isometric view of the final model is shown in Figure 5-35.

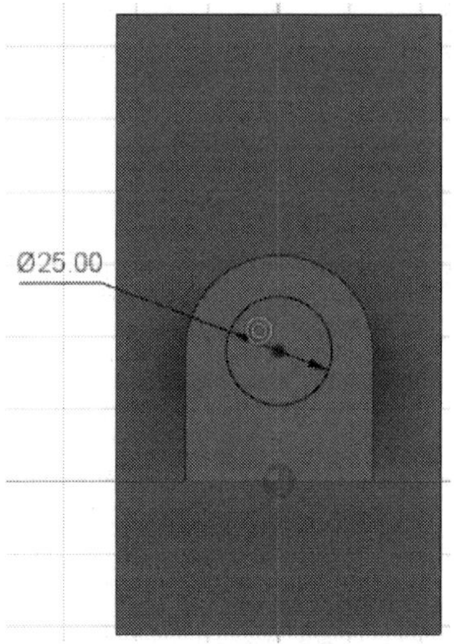

Figure 5-34 Sketch of the cut feature

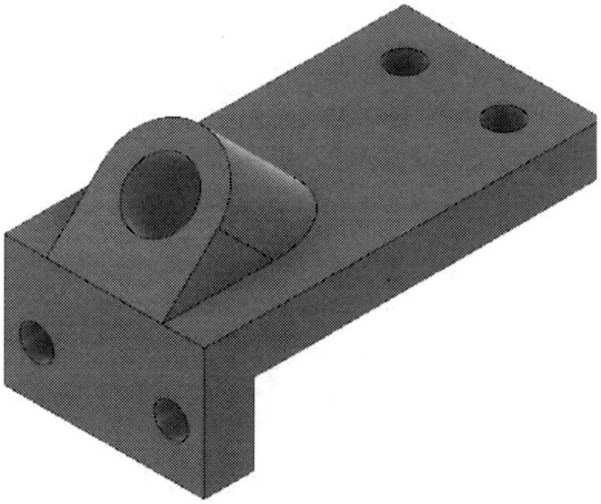

Figure 5-35 The final model

Saving the Model

Next, you need to save the model.

1. Choose the **Save** option from the **File** menu or the **Application** menu; the **Save** dialog box is displayed.

2. Enter **c05_Tut_02** in the **Name** edit box.

3. Click on the down arrow on the extreme right of the **Location** selection box; the **Save** dialog box gets expanded.

4. Select the **CADCIM** project from the **PROJECT** column or create a new project by selecting the **New Project** button from the **Save** dialog box.

5. Select the **Save** button to save the sketch and exit the **Save** dialog box.

Self-Evaluation Test

Answer the following questions and then compare them to those given at the end of this chapter:

1. The __________ tool is used to create a construction plane through an edge, axis, or line at a specified angle.

2. The __________ option in the **Extent** drop-down list is used to specify a body where the extruded feature will terminate.

3. You can use the __________ option to create a reference axis that passes through the center point of a cylindrical or conical surface.

4. A construction axis can be used as a reference when creating a construction plane. (T/F)

5. When you choose **Create Sketch** from the **SKETCH** panel, Autodesk Fusion 360 provides you with two default planes. (T/F)

6. A construction point must be located on a face or construction plane. (T/F)

Review Questions

Answer the following questions:

1. You can create construction plane tangent to a face and aligned to a point using __________ tool.

2. Choose the **Mid Plane** tool from the **CONSTRUCT** panel in the Ribbon to create a plane between two planar face. (T/F)

3. You can create a plane at an offset distance from the selected planar face by using the tool available in the **CONSTRUCT** panel. (T/F)

4. When you create a circular feature, a temporary axis is displayed. (T/F)

5. You can hide the construction planes in the graphics window. (T/F)

EXERCISES

Exercise 1

Create the solid model shown in Figure 5-36. The dimensions of the model are given in the same figure. **(Expected time: 30 min)**

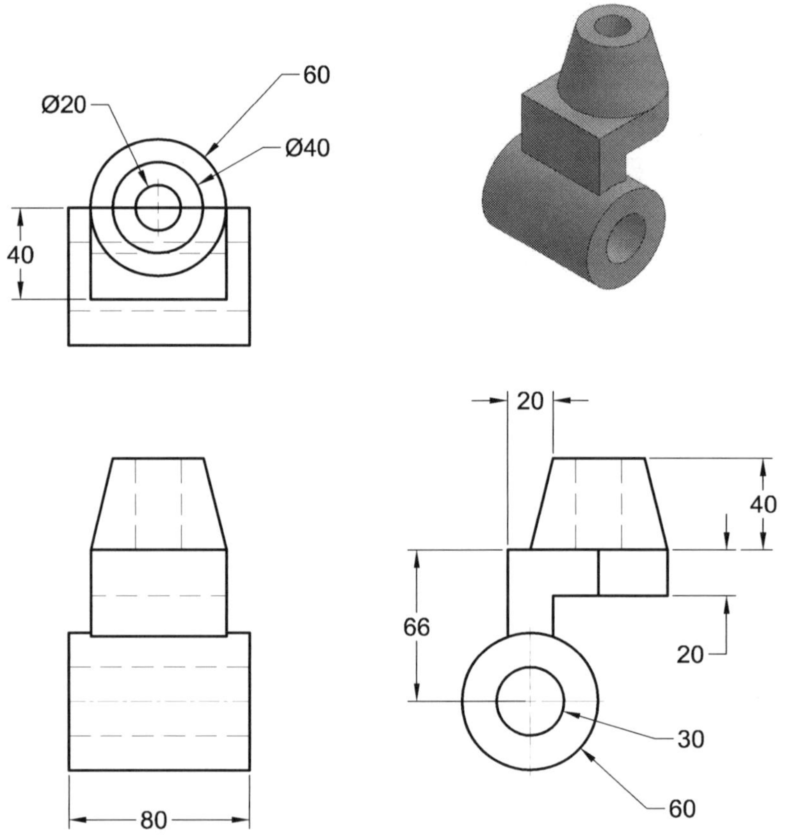

Figure 5-36 *Solid model and its dimensions for Exercise 1*

Exercise 2

Create the model shown in Figure 5-37. The dimensions of the model are given in Figure 5-38. **(Expected time: 30 min)**

Figure 5-37 *Solid model for Exercise 2*

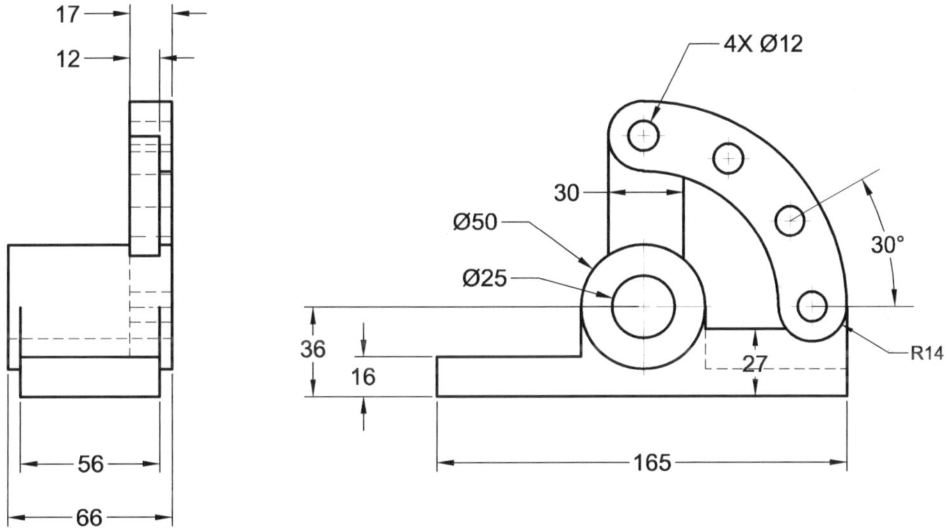

Figure 5-38 *Dimensions for the solid model*

Exercise 3

Create the model shown in Figure 5-39. The dimensions of the model are given in the same figure. **(Expected time: 30 min)**

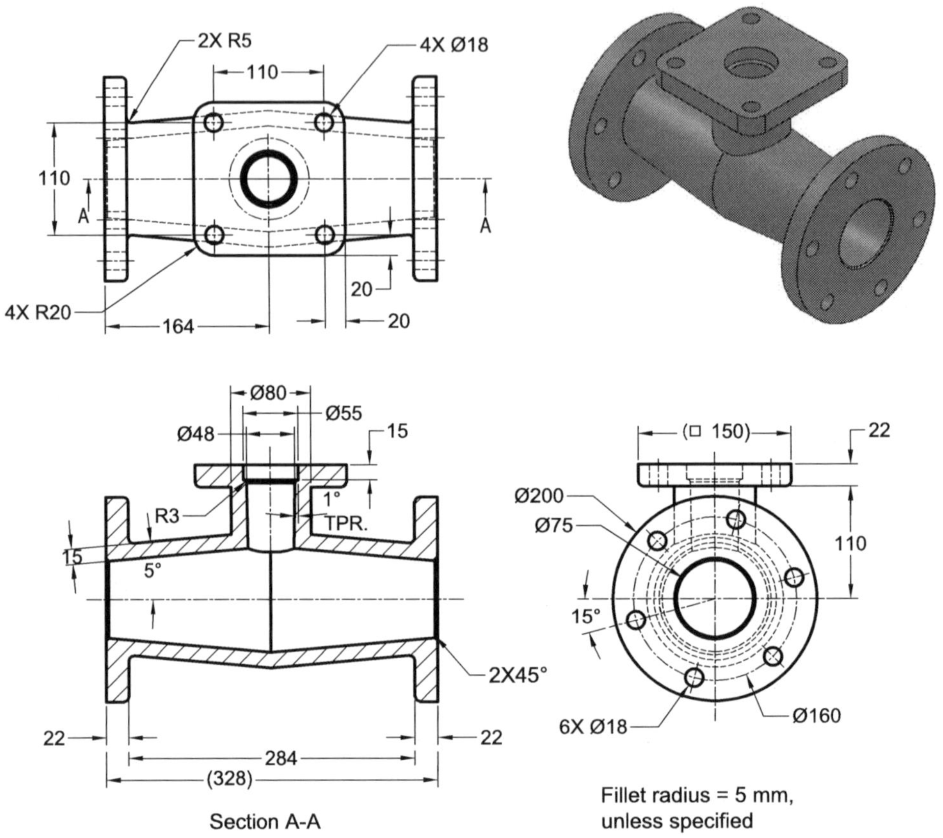

Figure 5-39 The model and its dimensions for Exercise 3

Answers to Self-Evaluation Test

1. Plane at Angle, **2. To Object**, **3. Axis Through Cylinder/Cone/Torus**, **4.** T, **5.** F, **6.** T

Chapter 6

Advanced Modeling-II

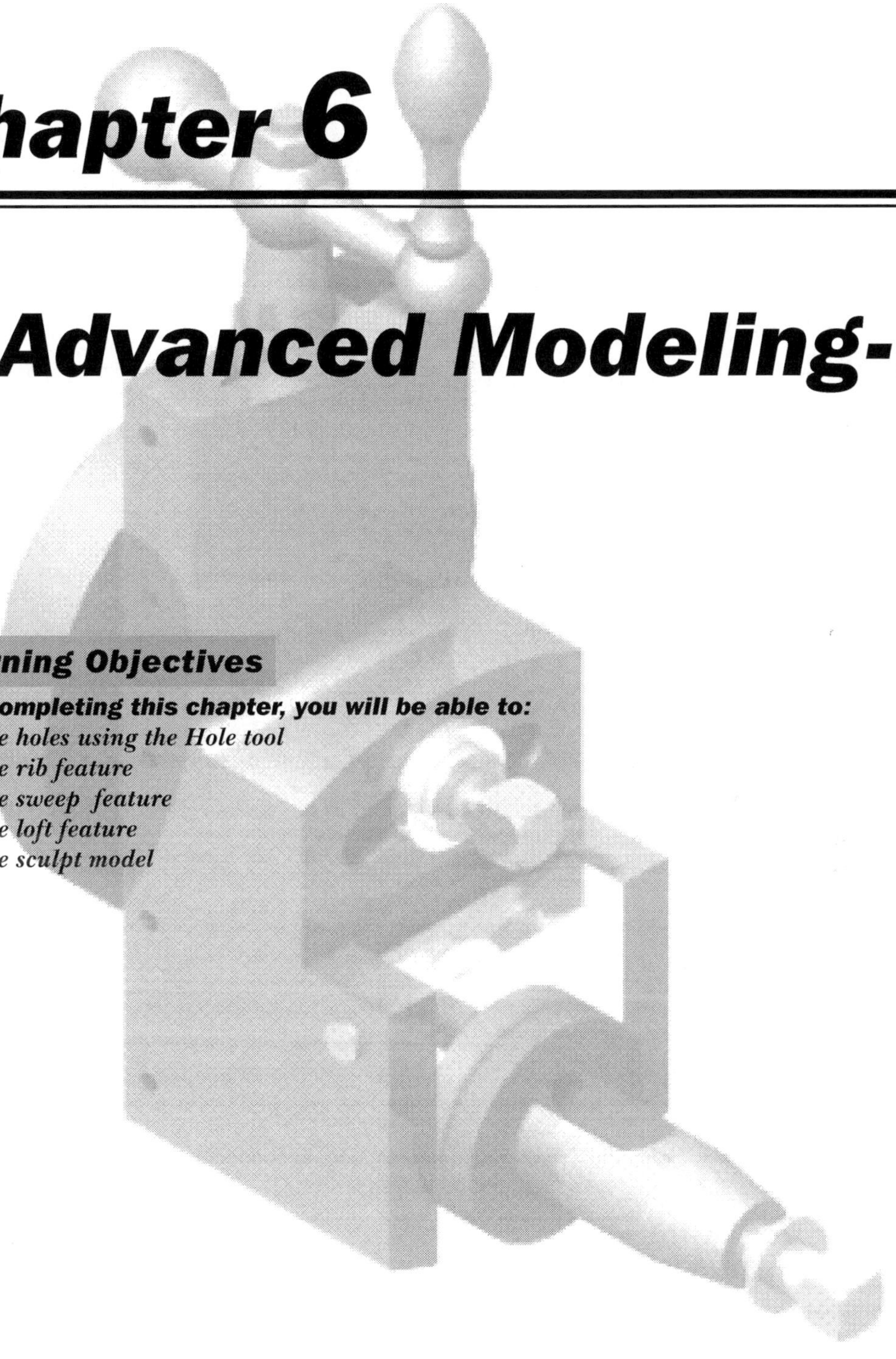

Learning Objectives

After completing this chapter, you will be able to:

- *Create holes using the Hole tool*
- *Create rib feature*
- *Create sweep feature*
- *Create loft feature*
- *Create sculpt model*

INTRODUCTION

This chapter discusses various advanced modeling tools available in Autodesk Fusion 360 that assist you in creating an accurate design by capturing the design intent of a model. In previous chapters, you have learned to create holes using the **Extrude** tool with cut operation. In this chapter, you will learn to create holes using the **Hole** tool. You will also learn about some more advanced modelling tools that will help you in creating complex features of the model. The tools discussed in this chapter are **Rib**, **Sweep**, **Loft** and also the tools available in the Sculpt workspace. Sculpting in Autodesk Fusion 360 allows you to create freeform solid bodies and surfaces. In the Sculpt workspace, you can rapidly create different forms by simply pressing and pulling on subdivided surfaces. Sculpted forms can be easily converted into solid bodies and can be used in conjunction with the solid modelling tools of the Autodesk Fusion 360.

TUTORIALS

Tutorial 1

In this tutorial, you will create the model shown in Figure 6-1. The dimensions and views of the model are shown in the same figure. **(Expected time: 30 min)**

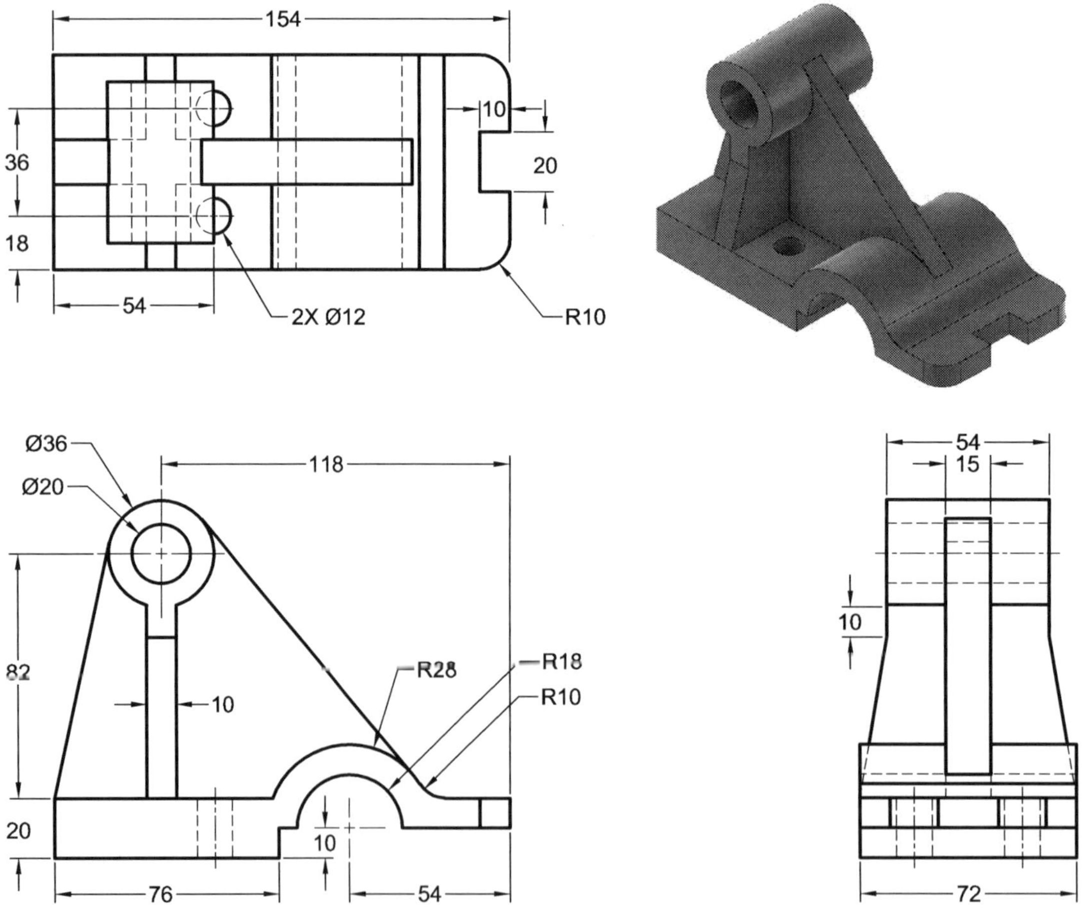

Figure 6-1 *Dimensions and views for Tutorial 1*

The following steps are required to complete this tutorial:

a. Start Autodesk Fusion 360 and create the base feature of the model on the **xz** plane.
b. Create the second feature, which is an extrude feature, on the mid plane or on the **yz** plane of the base feature.
c. Create the feature on the top planar face of the second feature.
d. Create the extrude feature on the top planar face of the third feature.
e. Create the extrude cut feature on the top planar face of the third feature.
f. Create the rib feature using the **Rib** tool.
g. Create a extrude feature on the base feature.
h. Add a simple hole of diameter 10 mm using the **Simple Hole** tool.
i. Create a linear pattern of hole of 10 mm diameter.
j. Save the model.

Creating the Base Feature

1. Start a Autodesk Fusion 360 and open new design using the **New Design** option from the **File** menu or **Application** menu.

2. Choose the **Create Sketch** tool from the **SKETCH** panel. Next select the **XY** plane from the **BROWSER** bar as the sketching plane; the selected plane gets oriented normal to the screen.

3. Create a sketch of the base feature of the model, as shown in Figure 6-2.

4. Invoke the **EXTRUDE** dialog box using the **Extrude** tool and select the created sketch.

5. Select the **Symmetric** option from the **Direction** drop-down list and choose the **Whole Length** button from the **Measurement** area.

6. Enter **72** as the value in the **Distance** edit box and then choose the **OK** button from the dialog box; base feature of the model is created. The isometric view of the model is shown in Figure 6-3.

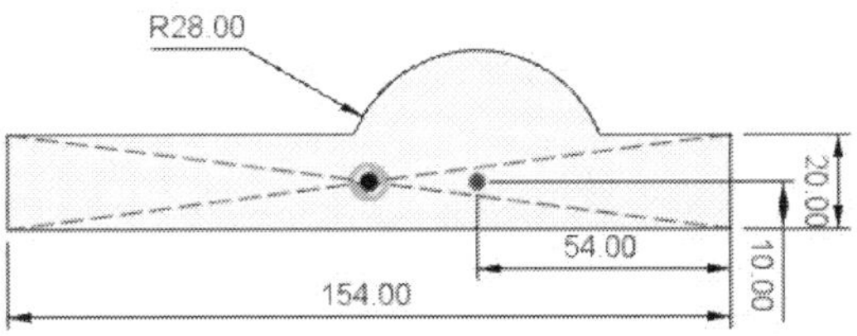

Figure 6-2 *Sketch of the base feature*

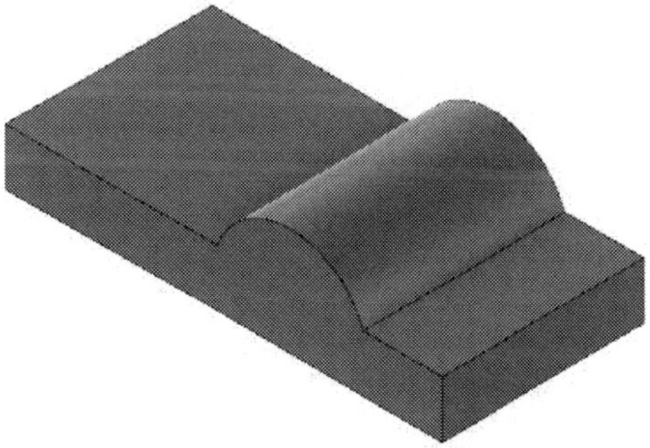

Figure 6-3 *Model after creating the base feature*

Creating the Extrude Feature

The second feature of the model is also an extrude feature. The sketch of this extrude feature will be drawn on the xy plane.

1. Choose the **Create Sketch** tool from the **SKETCH** panel. Next select the **XY** plane from the **BROWSER** bar as the sketching plane; the selected plane gets oriented normal to the screen.

2. Create the sketch for the second feature of the model, as shown in Figure 6-4.

3. Invoke the **EXTRUDE** dialog box using the **Extrude** tool and select the profile to be extruded.

4. Select the **Symmetric** option from the **Direction** drop-down list of the **EXTRUDE** dialog box and choose the **Whole Length** button from the **Measurement** option.

5. Enter **54** in the **Distance** edit box and then choose the **OK** button from the dialog box; second feature of the model is created. The isometric view of the model is shown in Figure 6-5.

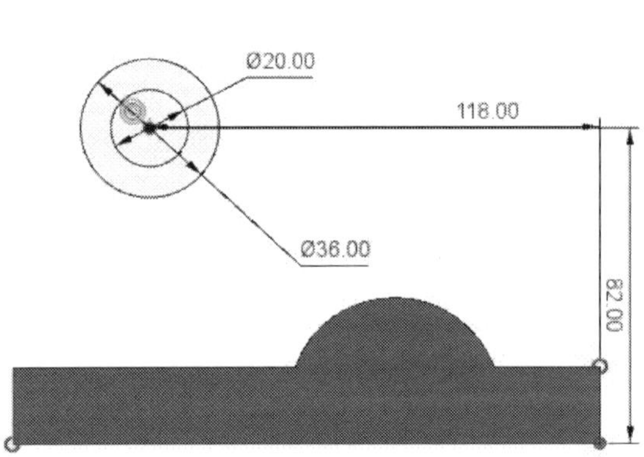

Figure 6-4 *Sketch of the second feature*

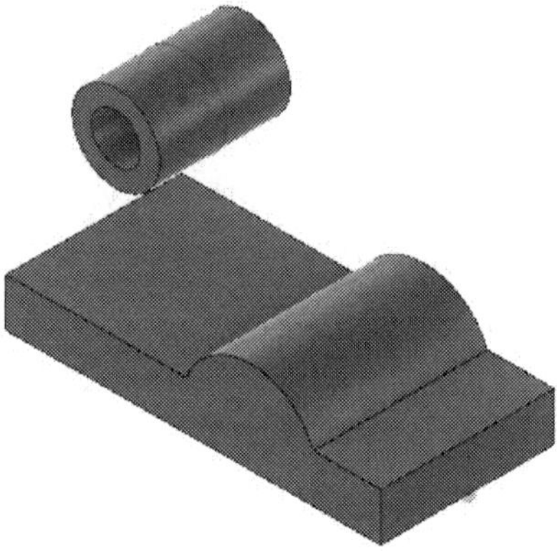

Figure 6-5 *Model after creating the second feature*

Creating the Third Feature

The third feature of the model is an extrude feature. The sketch of the extrude feature will be drawn on the xy plane. This sketch will be extruded using the **Symmetric** option to create the extrude feature.

1. Choose the **Create Sketch** tool from the **SKETCH** panel and orient the model normal to the sketching plane which is **XY** plane.

2. Select the **Slice** check box from the **SKETCH PALETTE** window gets sliced the model with respect to the current sketching plane, as shown in Figure 6-6.

3. Choose **SKETCH>Project/Include>Intersect** from the Ribbon; you are prompted to select the objects to intersect and also the **INTERSECT** dialog box is displayed, as shown in Figure 6-7.

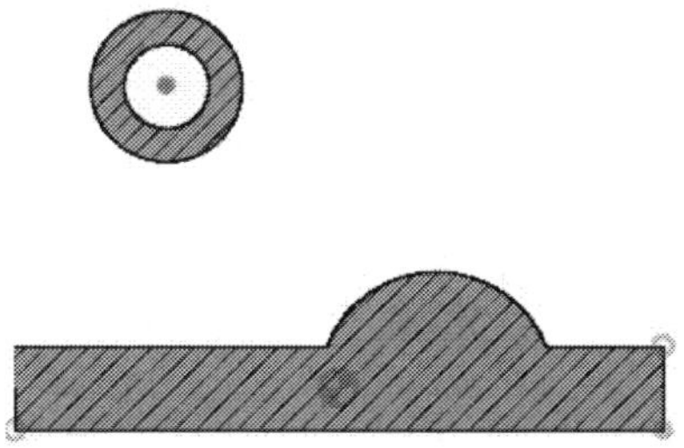

Figure 6-6 *Sliced view of the model*

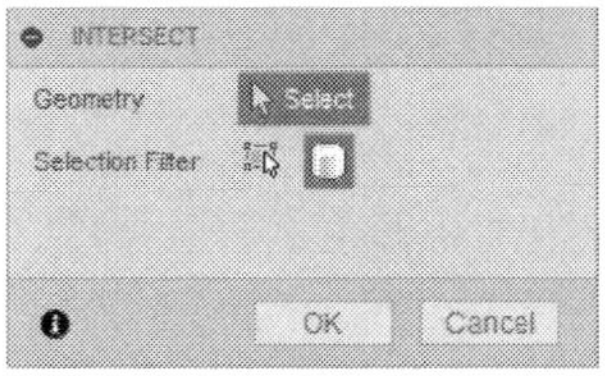

Figure 6-7 *The* ***Intersect*** *dialog box*

4. Select the sliced object; the edge of the sliced object is projected on the sketching plane. This makes creating the sketch more easier.

   The **Intersect** tool is used to project points, model edges and geometries that intersect the selected sketch plane. You can select the bodies or entities of the bodies by specifying selection filter option. The **Specified entities** button in the **INTERSECT** dialog box is selected to project the entities of model on the selected plane whereas the **Body** button is selected to project the model on the sketching plane.

5. Choose the **Line** tool from the **SKETCH** panel in the Ribbon, create the sketch, and apply the required constraint, as shown in Figure 6-8.

6. Invoke the **EXTRUDE** dialog box by using the **Extrude** tool and select the created sketch.

7. Select the **Symmetric** option from the **Direction** drop-down list and also choose the **Whole Length** button from the **Measurement** option.

8. Enter **15** in the **Distance** edit box and then choose the **OK** button from the dialog box; third feature of the model is created. The isometric view of the model is shown in Figure 6-9.

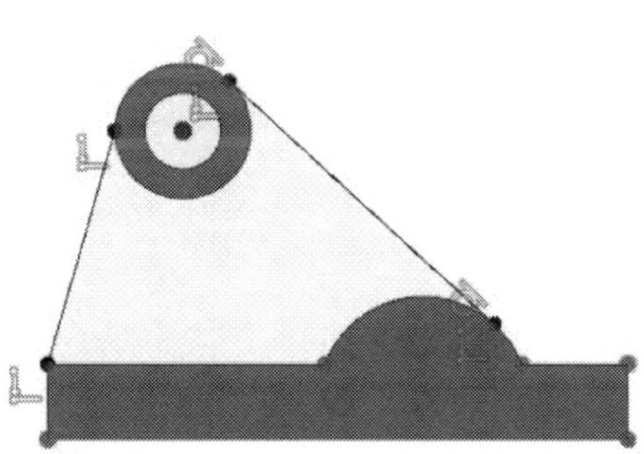

Figure 6-8 *Sketch of the third feature*

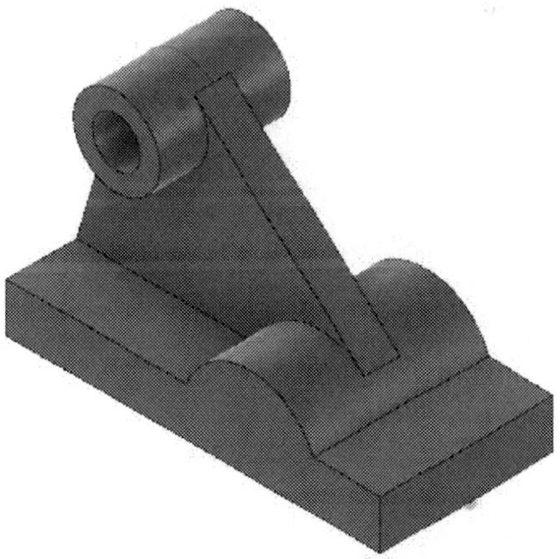

Figure 6-9 *Model after creating the third feature*

Creating the Rib Feature

Fourth feature of the model is a rib feature. The sketch of this rib feature will be created on the Offset Plane.

1. Choose the **Offset Plane** tool from the **CONSTRUCT** panel in the Ribbon; you are prompted to select a plane, planar face or sketch profile. Also, the **OFFSET PLANE** dialog box is displayed.

2. Select the right face of the model and enter **-118** in the **Distance** edit box

3. Choose **OK** from the **OFFSET PLANE** dialog box; an offset plane is created.

4. Choose the **Create Sketch** tool from the **SKETCH** panel in the Ribbon and select the created plane from the **BROWSER** bar under the **Construction** node; the plane is oriented normal to the screen.

5. Select the **Slice** check box from the **Options** node in the **SKETCH PALETTE** window. Also, choose the **SKETCH >Project/Include > Intersect** from the Ribbon and select the sliced object; the edge of the sliced object gets projected on the sketching plane.

6. Create the sketch of the **Rib** feature using the **Line** tool, as shown in Figure 6-10.

7. Choose the **Rib** tool from the **CREATE** panel in the Ribbon; the **RIB** dialog box is displayed, as shown in Figure 6-11. Also, you are prompted to select the sketch to create the rib. 

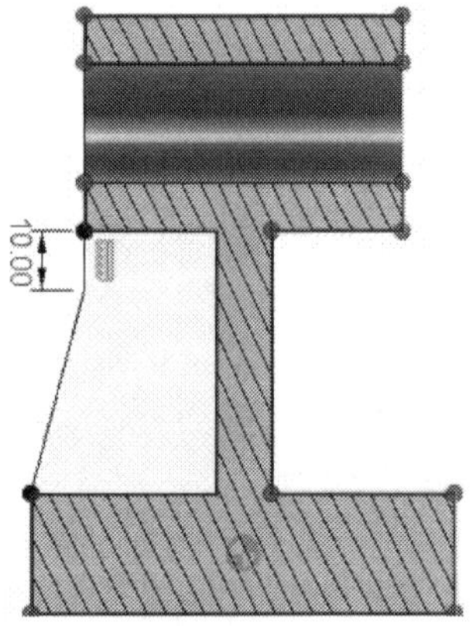

Figure 6-10 *Sketch of the **Rib** feature*

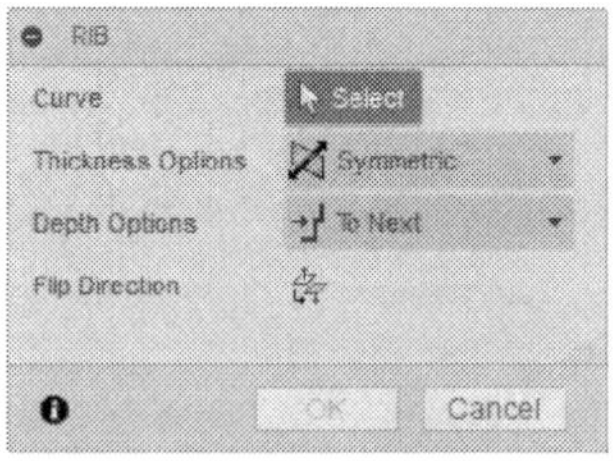

Figure 6-11 *The **RIB** dialog box*

The **Rib** tool is used to create a rib feature. Ribs are thin-walled structures that are used to increase the strength of the entire structure of a component so as to protect it from failing under an increased load.

8. Select the sketch from the graphics window and enter the value of thickness as **10** in the **Thickness** edit box.

The **Thickness** edit box is used to specify the thickness of the rib feature.

9. Make sure that the **Symmetric** option is selected from the **Thickness Options** drop-down list of the **RIB** dialog box.

Note

*By default, the **Symmetric** option is selected in the **Thickness Options** drop-down list. As a result, the rib will be created on both sides of the sketch. You can also select the **One Direction** option from this drop-down list to create ribs on either sides of the sketch by accordingly entering negative or positive value in the **Thickness** edit box.*

10. Also make sure that the **To Next** option is selected in the **Depth Options** drop-down list

    The **Depth Options** drop-down list is used to add material to the model. The **To Next** option is used to add the material upto the next intersecting entity in the model and the **Depth** option adds material upto the specified depth.

11. Choose the **OK** button from the **RIB** dialog box; a rib feature is created, as shown in Figure 6-12.

Figure 6-12 *Model after creating the rib feature*

Creating the Mirror Feature

Now you need to create a mirror copy of the last created rib feature by using the **Mirror** tool.

1. Choose the **Mirror** tool from the **CREATE** panel; the **MIRROR** dialog box is displayed. Also, you are prompted to select the faces or feature to mirror.

2. Select the **Features** option from the **Pattern Type** drop-down list in the **MIRROR** dialog box.

3. Select the rib feature from the timeline. Also, select the **Mirror Plane** option from the dialog box; you are prompted to select the plane or planar face.

4. Select the **XY** plane from as a mirror plane; preview of the mirrored rib feature is displayed in the graphics window.

5. Choose the **OK** button from the dialog box; the mirrored feature is created. The rotated view of the model after creating the mirrored feature is shown in the Figure 6-13.

Figure 6-13 *Model after creating the mirror feature*

Creating the Hole Feature

Now, you will create hole features on the model. You can create a hole using the **Hole** tool. However, in this tutorial, you will create hole features arbitrarily on the top planar face of the base feature by using the **Hole** tool and position them later.

Note

*In the previous chapters, you learned to create holes by extruding circle. Now, you will learn how to create hole feature using the **Hole** tool. If you use this tool, you do not need to draw the sketch of the hole. The holes created using this tool act as placed features.*

1. Choose the **Hole** tool from the **CREATE** panel of the Ribbon; you are prompted to select face, plane, or sketch point to place hole and also the **HOLE** dialog box is displayed, as shown in Figure 6-14.

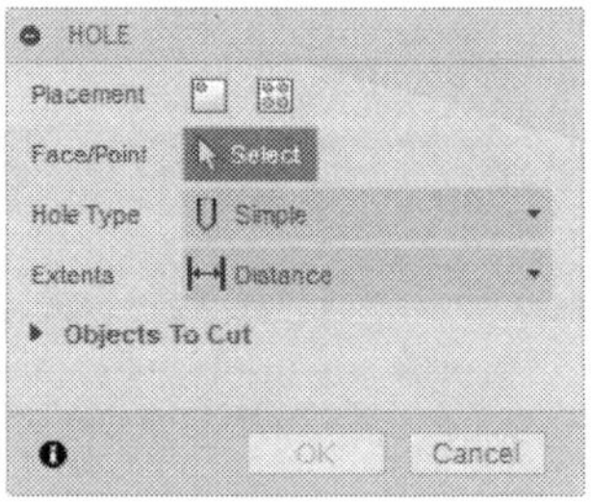

Figure 6-14 *The HOLE dialog box*

2. Select the top planar face of the base feature as the placement plane for the hole feature; preview of the placed hole is displayed in the graphics window.

3. Select the edge labelled 1 on the top face of the base feature, refer to Figure 6-15.

4. Enter **100** to th**e Dimension** edit box displayed in the graphics window.

5. Similarly, select the edge labelled 2 on the top face of the base feature and enter **18** in the **Dimension** edit box, refer to Figure 6-15.

6. Set the value of the diameter of the hole in the preview window to 12 and select the **All** option from the **Extents** drop-down list.

7. Choose the **OK** button from the dialog box.

8. Mirror the created hole feature along the xy plane. The rotated model after creating the hole feature and mirror of the hole feature is shown in Figure 6-16.

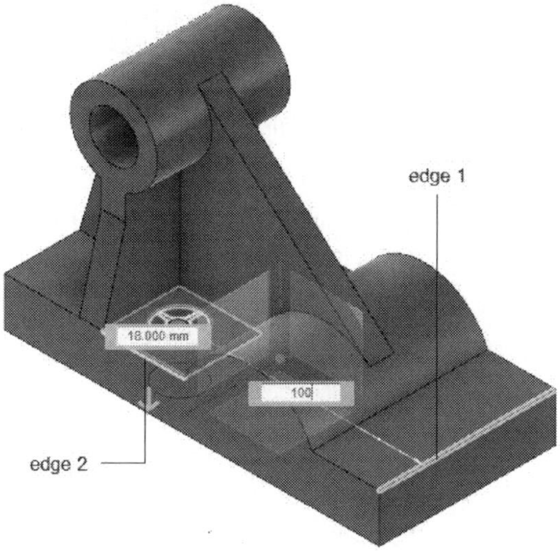

Figure 6-15 *Placement for the hole feature*

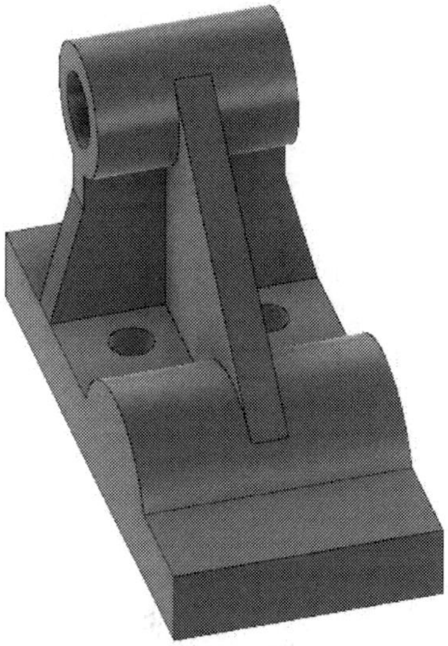

Figure 6-16 *Rotated view of a model*

Creating the Cut Feature

1. Choose the **Create Sketch** from the **SKETCH** panel. Next, select the **XY** plane from the **BROWSER** bar as the sketching plane. Also, select the **Slice** check box from the **SKETCH PALLETE** window.

2. Create sketch of the cut feature on the model and apply required constraints, as shown in Figure 6-17.

3. Invoke the **EXTRUDE** tool and then select the **Two Sides** option from the **Direction** drop-down list and also select the **All** option from the **Extent** drop-down list under both the **Side1** and **Side2** nodes.

4. Make sure that the **Cut** option is selected in the **Operation** drop-down list.

5. Keep rest of the settings as it is and choose the **OK** button from the dialog box.

6. Similarly, create another cut feature on the base feature, as shown in Figure 6-18. Note that for clarity, applied constraints are hidden.

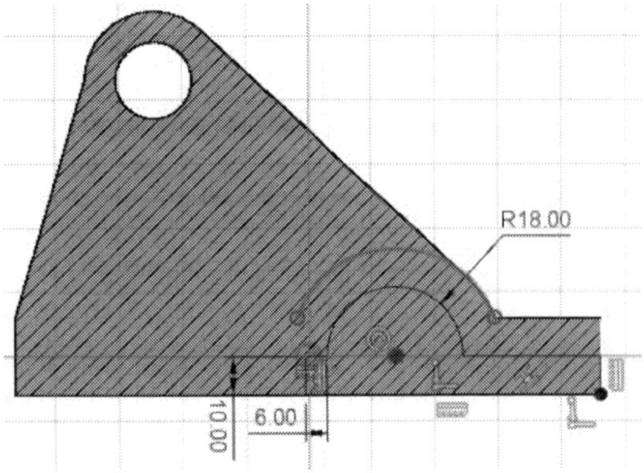

Figure 6-17 *Sketch of the cut feature on the XY plane*

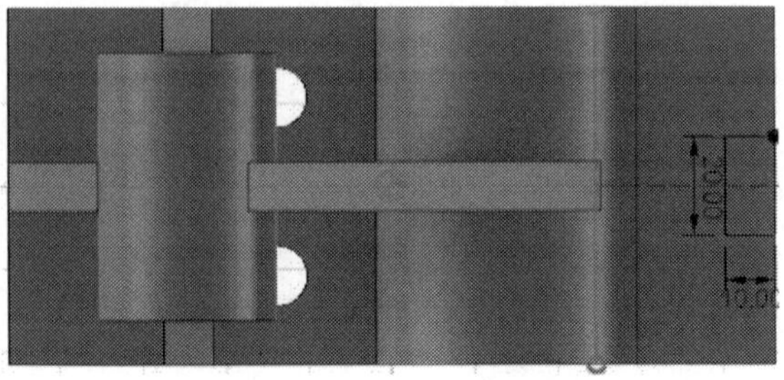

Figure 6-18 *Sketch of the cut feature on base feature*

7. Apply fillets on the created model. For dimensions, refer to Figure 6-1. Final model after applying fillets is shown in Figure 6-19.

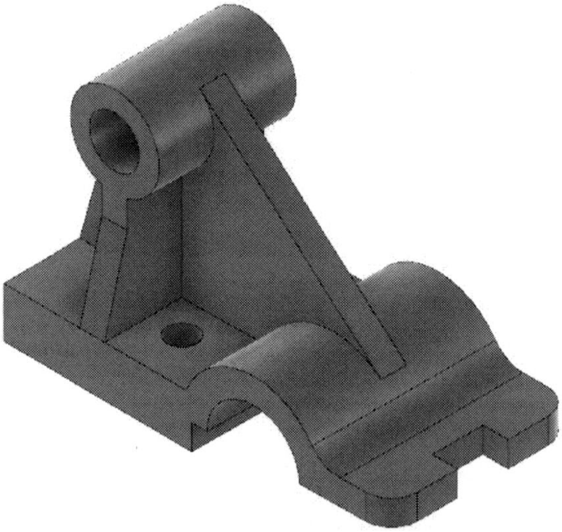

Figure 6-19 *The final model*

Saving the Model

Next, you need to save the model.

1. Choose the **Save** option from the **File** menu or the **Application** menu; the **Save** dialog box is displayed.

2. Enter **c06_Tut_01** in the **Name** edit box.

3. Click on the down arrow on the extreme right of the **Location** selection box; the **Save** dialog box gets expanded. 

4. Select the **CADCIM** project from the **PROJECT** column or create a new project by choosing the **New Project** button from the **Save** dialog box.

5. Choose the **Save** button to save the sketch and exit the **Save** dialog box.

Tutorial 2

In this tutorial, you will create the model shown in Figure 6-20. The dimensions of the model are shown in the same figure. Also, you will create a section view of the model using the **Section View** tool. **(Expected time: 45 min)**

The following steps are required to complete this tutorial:

a. Start Autodesk Fusion 360 and create the base feature of the model by revolving the sketch.
b. Create the sweep feature.
c. Add a fillet to the base feature.
d. Display section view of the model.
e. Save the model.

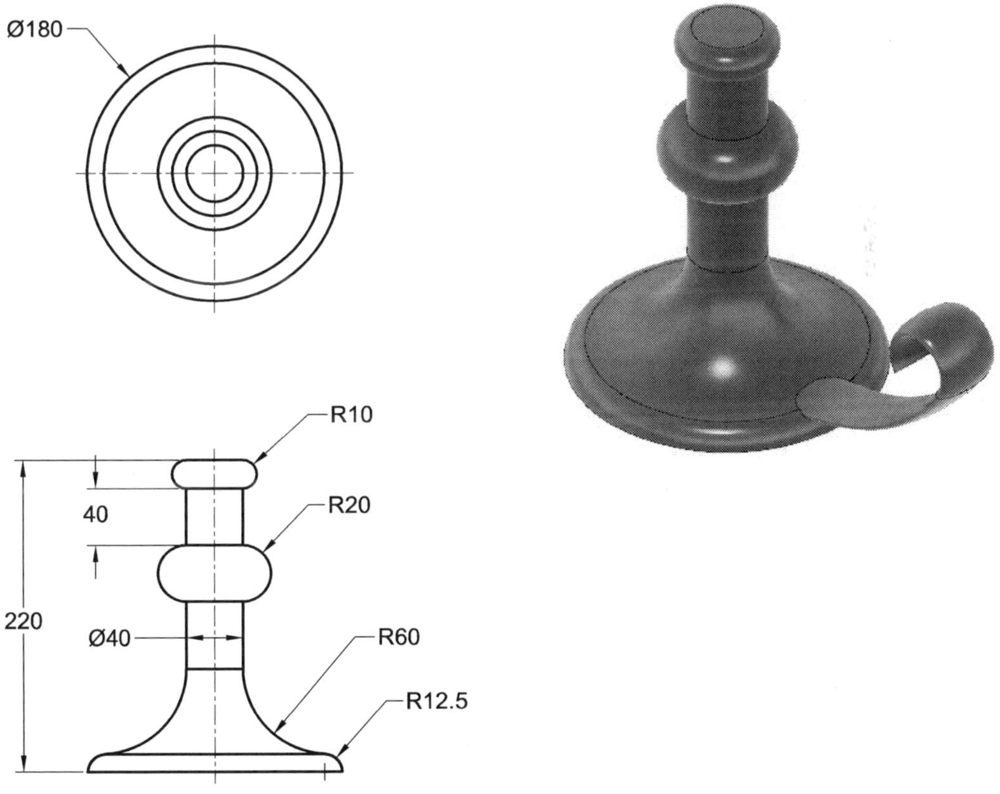

***Figure 6-20** Dimensions and views for Tutorial 2*

Starting Autodesk Fusion 360 and creating the Base Feature

The base feature of the model is a revolve feature.

1. Start Autodesk Fusion 360 and then open new design using the **New Design** option from the **File** menu or from the **Application** menu.

2. Select the **Create Sketch** tool from the **SKETCH** panel. Next select the **XY** plane from the **BROWSER** bar as the sketching plane; the selected plane orients normal to the screen.

3. Create sketch of the base feature of the model, as shown in Figure 6-21.

4. Invoke the **Revolve** tool and revolve the sketch by selecting its vertical centerline as the axis of revolution. The isometric view of the model after creating the revolved feature is shown in Figure 6-22.

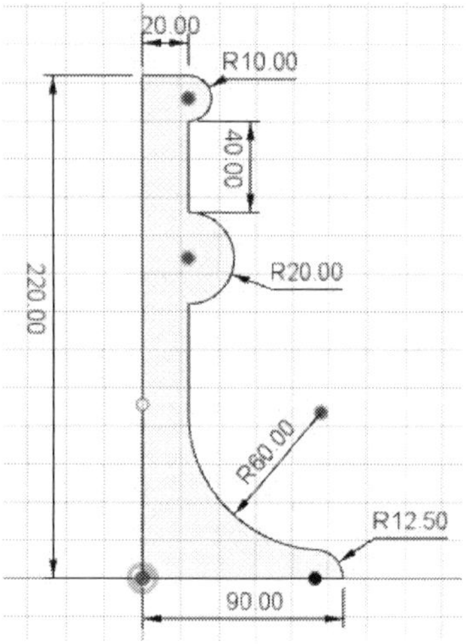

Figure 6-21 *Sketch of the base feature*

Figure 6-22 *Model after creating the base feature*

Creating Sweep Feature

To create the sweep feature, you first need to create its path and profile. A profile is a section for the sweep feature and a path is the route taken by the profile while creating the sweep feature.

1. Choose **Create Sketch** from the **SKETCH** panel and orient the model normal to the sketching plane which is **XY** plane.

2. Select the **Slice** check box from the **SKETCH PALETTE** window and create the sketch, as shown in Figure 6-23.

3. Choose **STOP SKETCH** from the Ribbon; the isometric view the sketch is displayed.

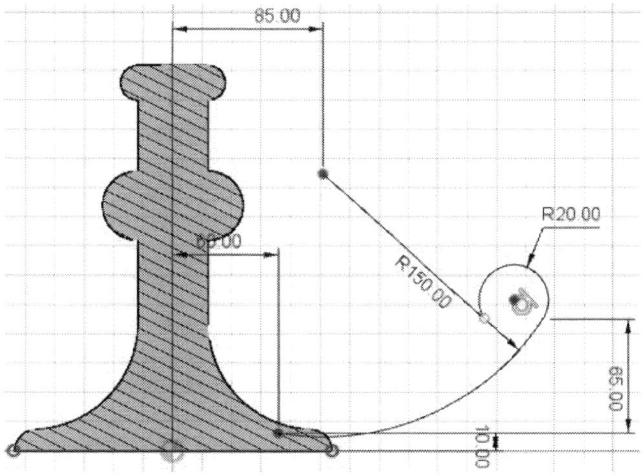

Figure 6-23 *Sketch of the path*

After creating the path for the sweep feature, you need to create its profile. To create the profile, first you need to create a reference plane normal to the path and then select it as the sketching plane.

4. Choose the **Plane Along Path** tool from the **CONSTRUCT** panel; the **PLANE ALONG PATH** dialog box is displayed, as shown in Figure 6-24, and you are prompted to select the edges or sketch curves.

5. Select the previously created sketch as the path from the graphics window; an arrow on the curve is displayed and also the **Distance** edit becomes available in the dialog box.

6. Enter **1** in the **Distance** edit box and choose **OK** from the **PLANE ALONG PATH** dialog box.

7. Select the newly created reference plane and create the sketch of the profile, as shown in Figure 6-25.

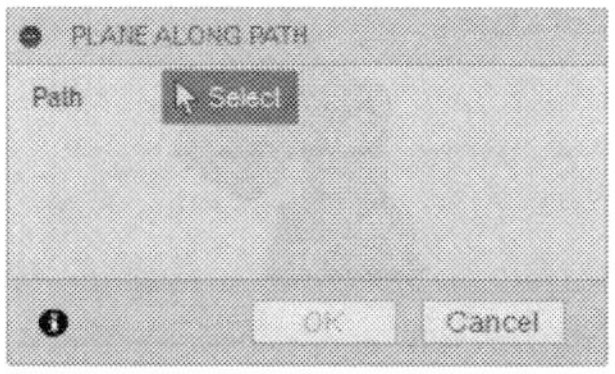

Figure 6-24 *The* ***PLANE ALONG PATH*** *dialog box*

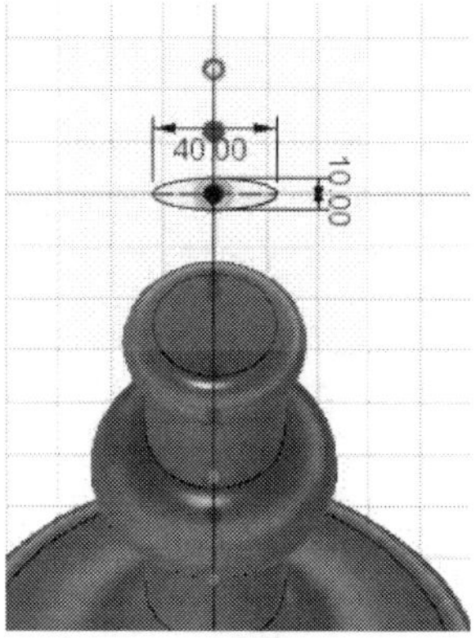

Figure 6-25 *Sketch of the profile*

8. Choose the **Sweep** tool from the **CREATE** panel in the Ribbon; the **SWEEP** dialog box is displayed, as shown in Figure 6-26. Also you are prompted to select the sketch profiles or planar faces to sweep.

   The **Sweep** tool is one of the most important advanced modeling tools. This tool is used to extrude a closed profile along an open or a closed path. Therefore, you need a profile and a path to create a sweep feature.

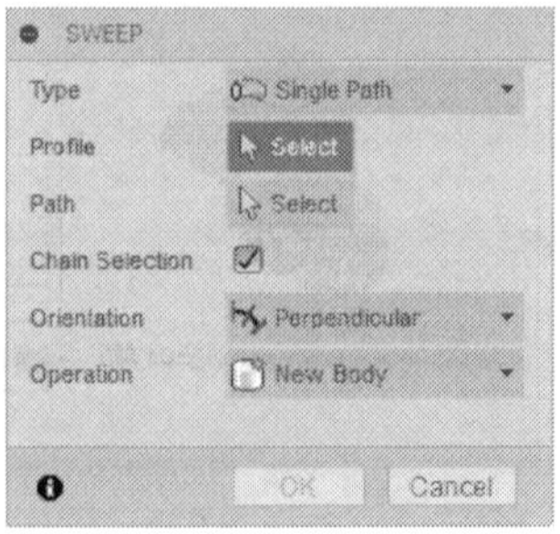

Figure 6-26 *The* ***SWEEP*** *dialog box*

9. Select the sketch profile from the graphics window for the sweep feature. Next, select the **Path** option from the **SWEEP** dialog box; you are prompted to select the sketch curve or edges as a path.

10. Select the path for the sweep feature; the preview of the sweep feature is displayed in the graphics window.

11. Choose the **Join** option from the **Operation** drop-down list.

12. Retain rest of the settings in the **SWEEP** dialog box and then choose the **OK** button; the final model is created, as shown in Figure 6-27.

Figure 6-27 *Model after creating the sweep feature*

Displaying Section View of the Model

Next, you need to display section view of the model. Section view of the model will be created using the **Section Analysis** tool.

The **Section Analysis** tool is used to display the section view of the model by cutting it using a plane or a face. You can also save the section view with a name to generate the section view directly on the drawing sheet in the drawing environment.

1. Orient the model in isometric view.

2. Choose the **Section Analysis** tool from the **INSPECT** panel; the **SECTION ANALYSIS** dialog box is displayed, as shown in Figure 6-28 and you are prompted to select a plane or a planar face to define the section view.

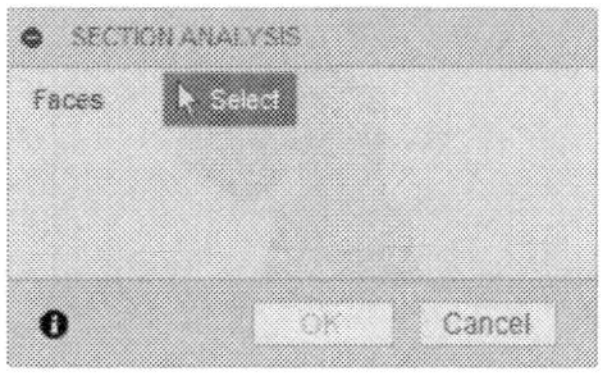

Figure 6-28 *The **SECTION ANALYSIS** dialog box*

3. Select the **XY** plane from the **BROWSER** bar; the preview of the section view is displayed in the graphics window and the **SECTION ANALYSIS** dialog box gets modified with the addition of three edit boxes namely **Y Distance**, **X Angle**, and **Z Angle**.

   The **Y Distance** edit box is used to specify the distance of the sectioning plane from the origin. The **X Angle** and **Z Angle** edit boxes are used to specify the angles for the sectioning plane.

4. Accept the default values and choose the **OK** button from the dialog box. The resultant section view of the model is shown in the Figure 6-29.

Figure 6-29 *Resultant section view of the model*

5. Click on toggle button located before of the **Analysis** folder in the **BROWSER** bar to show the full model.

Saving the Model

Next, you need to save the model.

1. Choose the **Save** option from the **File** menu or the **Application** menu; the **Save** dialog box is displayed.

2. Enter **c06_Tut_02** in the **Name** edit box.

3. Click on the down arrow on the extreme right of the **Location** selection box, the **Save** dialog box gets expanded.

4. Select the **CADCIM** project from the **PROJECT** column or create a new project by choosing the **New Project** button from the **Save** dialog box.

5. Select the **Save** button to save the model and exit the **Save** dialog box.

Tutorial 3

In this tutorial, you will create the model shown in Figure 6-30. The dimensions of the model are shown in the Figure 6-31. **(Expected time: 45 min)**

The following steps are required to complete this tutorial:

a. Start Autodesk Fusion 360 and create a revolve feature as the base feature of the model.
b. Create a loft feature between two sections along a specified path.
c. Create a circular pattern of the loft feature.
d. Save the model.

Figure 6-30 *Model for Tutorial 3*

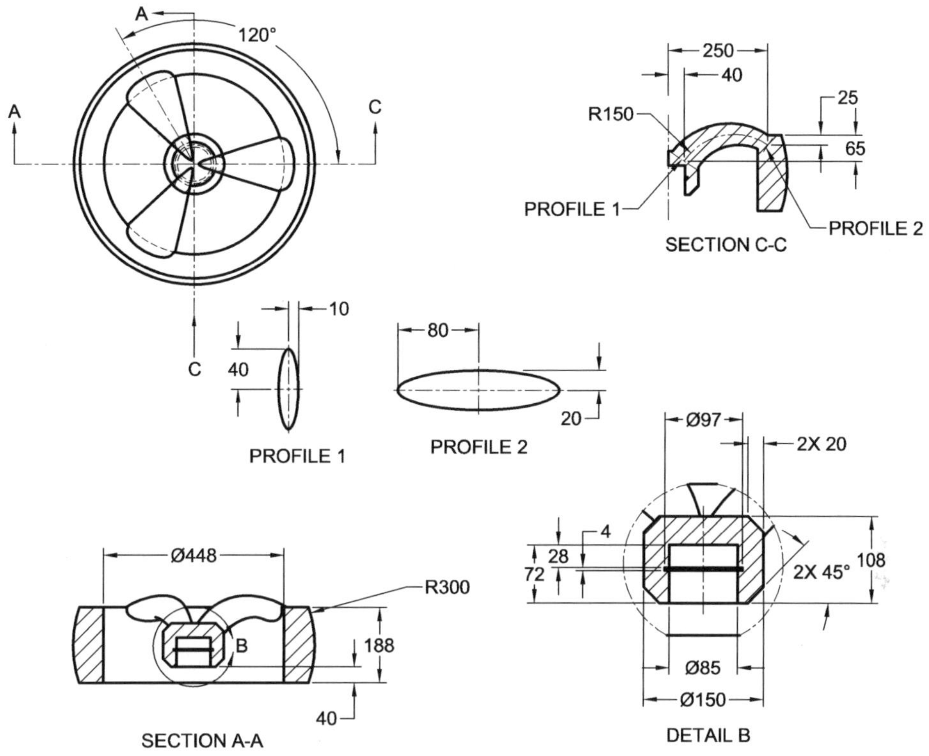

Figure 6-31 Dimensions and views for Tutorial 3

Creating the Base Feature

You will create the base feature of the model which is a revolve feature.

1. Start Autodesk Fusion 360 and open a new design file using the **New Design** option from the **File** menu or the **Application** menu.

2. Select the **Create Sketch** tool from the **SKETCH** panel. Next select the **XY** plane from the **BROWSER** bar as the sketching plane.

3. Create sketch of the revolve feature of the model, as shown in Figure 6-32.

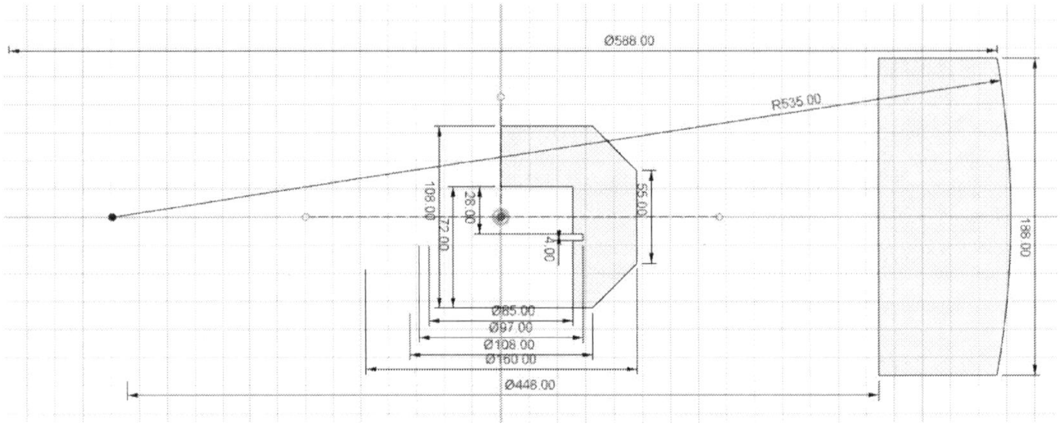

Figure 6-32 *Sketch of the revolve feature*

4. Choose the **Revolve** tool from the **CREATE** panel in the Ribbon; you are prompted to select the sketch profiles or planar faces to revolve. Also, the **REVOLVE** dialog box is displayed.

5. Select the sketch as the profile and vertical axis from the graphics window; the preview of the revolve feature is displayed in the graphics window.

6. Make sure that the value of the angle is set to **360** degrees in the **Angle** edit box.

7. Choose the **OK** button from the **REVOLVE** dialog box; the revolve feature is created, as shown in Figure 6-33.

Figure 6-33 *Model after creating revolve feature*

Creating the Loft Feature

The second feature of the model is a loft feature which can be created by using the **Loft** tool. The loft features are created by blending more than one similar or dissimilar profiles together to get a free-form shape. These similar or dissimilar profiles may or may not be parallel to each other. In this tutorial, you will create a loft feature by blending two sections along specified path. Therefore, first you will create a path and then sections for creating the loft feature.

1. Choose the **Create Sketch** tool from the **SKETCH** panel and select **XY** plane as the sketching plane for creating the path.

2. Change the current display style of the model to hidden visible line by choosing **Display Settings > Visual Style > Wireframe with Hidden Edges**.

3. Create the sketch of the path, as shown in Figure 6-34.

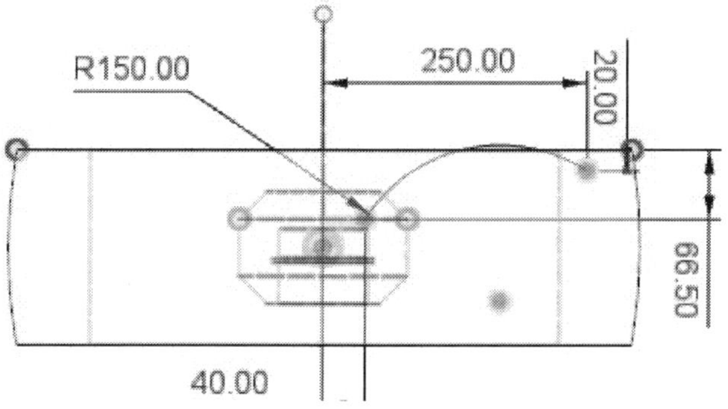

Figure 6-34 *Sketch of the path*

After creating the sketch of the path, you need to create section for creating the loft feature. In this tutorial, you need to create reference planes normal to the path at its both ends for creating the loft sections.

4. Change the current display style of the model to shaded with visible edges.

5. Select **Plane Along Path** from the **CONSTRUCT** panel of the Ribbon; the **PLANE ALONG PATH** dialog box is displayed and you are prompted to select the edges or sketch curves as a path.

6. Select the previously created sketch as the path from the graphics window and drag the arrow on the model to the end of the path; preview of the reference plane normal to sketch is displayed at one end. Next, choose the **OK** button from the **PLANE ALONG PATH** dialog box.

7. Similarly, create another reference plane normal to the path at the other endpoint of the path. Figure 6-35 shows the isometric view of the model after creating the references plane.

After creating the planes, you need to create loft sections on it.

8. Select the reference plane to create the sketch of the first section, refer to Figure 6-36, and select the **Create Sketch** tool from the **SKETCH** panel. Then create the elliptical section, as shown in Figure 6-37.

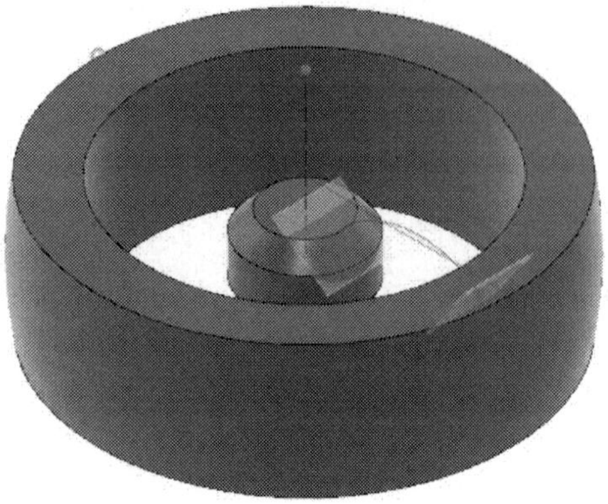

Figure 6-35 *Isometric view of the model after creating the reference plane*

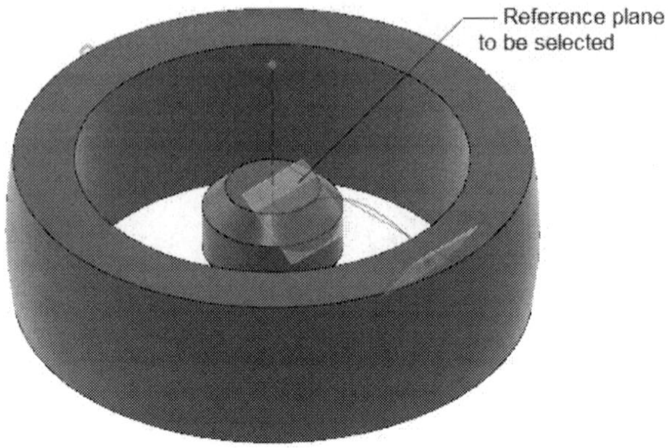

Figure 6-36 *The reference plane to be selected*

9. Similarly, create the second section of the loft feature, as shown in Figure 6-38 by selecting the second reference plane created earlier previously as the sketching plane.

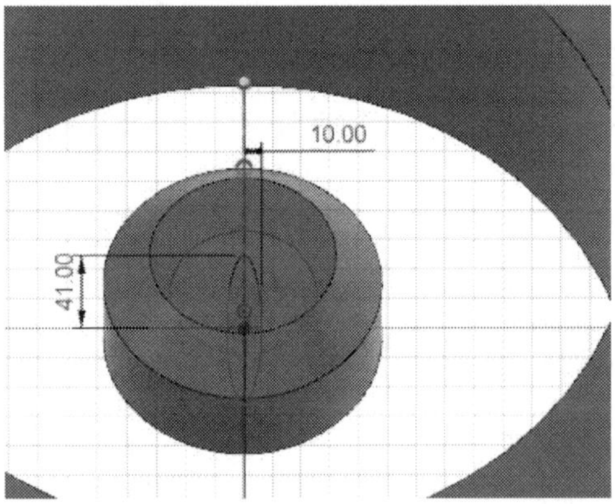

Figure 6-37 *Model after creating reference plane*

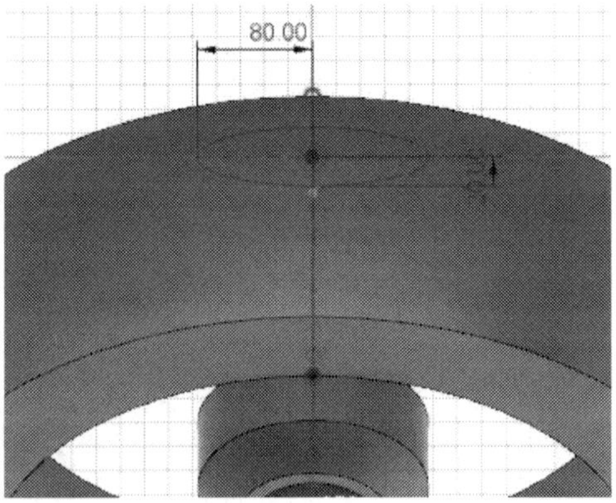

Figure 6-38 *The second section created for the loft feature*

Now, using the **Loft** tool you will create the loft feature by blending the previously created sections along the path.

10. Choose the **Loft** tool from the **CREATE** panel in the Ribbon; the **LOFT** dialog box is displayed in the graphics window. Also, you are prompted to select loft inputs.

    As discussed earlier, the **Loft** tool is used to create a loft feature by blending more than one similar or dissimilar sections together to get a free-form shape. These similar or dissimilar sections may or may not be parallel to each other. Note that the sections for the solid loft should be a closed sketch.

11. Select both the sections from the graphics window one by one; preview of the loft feature is displayed. Also names of the selected sections or profiles are displayed in the **Profiles** area.

12. Choose the **Centerline** button from the **Guide Type** area of the **LOFT** dialog box and then select path created earlier from the graphics window; preview of the loft feature is displayed in the graphics window.

13. Make sure that the **Join** option should be selected in the **Operation** drop-down list.

14. Retain the rest of the settings and the choose the **OK** button from the dialog box. Figure 6-39 shows the model after creating loft feature.

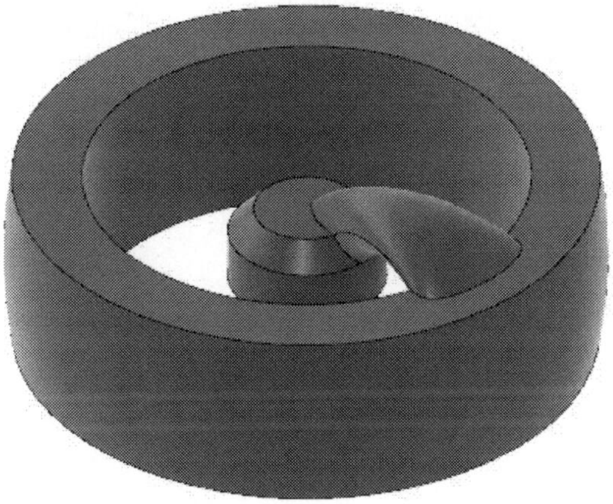

Figure 6-39 *Model after creating loft feature*

Patterning the Loft Feature

After creating the loft feature, you need to pattern it using the **Circular Pattern** tool.

1. Choose the **Circular Pattern** tool from the **CREATE > Pattern**; the **CIRCULAR PATTERN** dialog is displayed.

2. Select the loft feature from the timeline in the graphics window as the feature to pattern.

3. Make sure that the **Features** option is selected from the **Pattern Type** drop-down list.

4. Click on the **Axis** selection box in the **CIRCULAR PATTERN** dialog box and then select the circular edge of the model; preview of the circular pattern of the loft feature is displayed. Also, the options like **Type** drop-down list, the **Suppress** check box, the **Quantity** spinner becomes available.

5. Select the **Full** option from the **Type** drop-down list and set the value of 3 in the **Quantity** spinner.

6. Choose the **OK** button from the **CIRCULAR PATTERN** dialog box; the loft feature is patterned. Final model after loft features is patterned is shown in Figure 6-40.

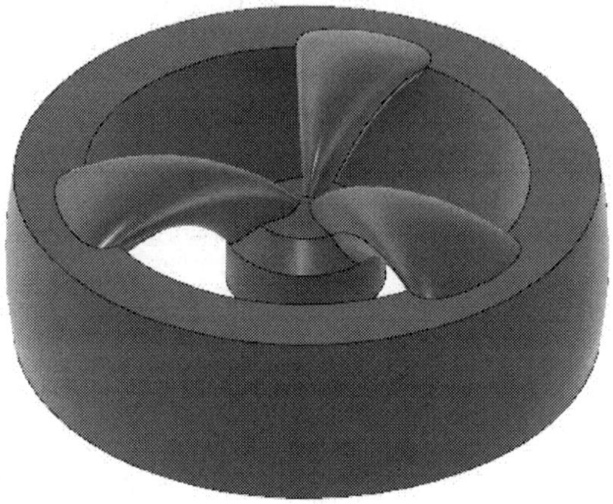

Figure 6-40 *Final model of the Tutorial 3*

Saving the Model

Next, you need to save the model.

1. Choose the **Save** option from the **File** menu or the **Application** menu; the **Save** dialog box is displayed.

2. Enter **c06_Tut_03** in the **Name** edit box.

3. Click on the down arrow on the extreme right side of the **Location** selection box, the **Save** dialog box gets expanded. 

4. Select the **CADCIM** project from the **PROJECT** column or create a new project by selecting the **New Project** button from the **Save** dialog box.

5. Select the **Save** button to save the sketch and exit the **Save** dialog box.

Tutorial 4

In this tutorial, you will create the model shown in Figure 6-41.

(Expected time: 45 min)

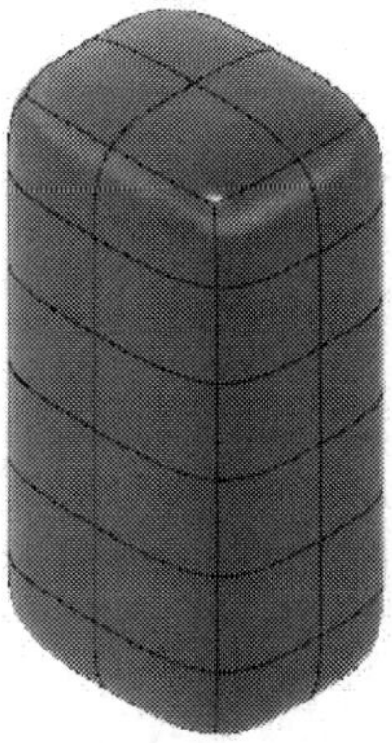

Figure 6-41 *The model after creating box*

The following steps are required to complete this tutorial:

a. Start Autodesk Fusion 360 and invoke the **Create Form** tool.
b. Create a box at the center using box primitives.
c. Deform the box by using edges.
d. Save the model.

Starting Autodesk Fusion 360 and creating Design in Sculpt workspace

To create a new design in the sculpt workspace, you need to invoke the **Create Form** tool.

1. Choose the **Create Form** tool from the **CREATE** panel in the Ribbon; the **SCULPT** workspace is invoked and the **Sculpt Environment** message box is displayed, as shown in Figure 6-42.

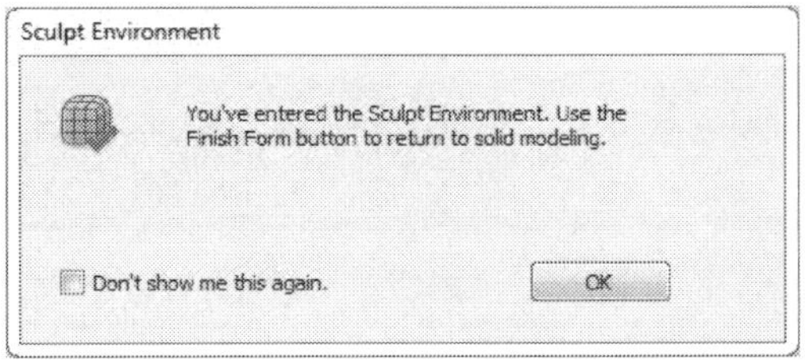

Figure 6-42 *The* ***Sculpt Environment*** *message box*

2. Choose the **OK** button from the message box.

Creating the Box

1. Choose the **Box** tool from the **CREATE** panel in the Ribbon; the **BOX** dialog box is displayed, as shown in Figure 6-43. Also you are prompted to select a plane or a planar face.

2. Select **XZ Plane** from the **BROWSER** bar; you are prompted to specify the center point of the box.

3. Select the origin to specify the center point of the box in the graphics window; you are prompted to specify the size of the rectangle.

4. Enter **50** in the length input box in the graphics window. Next, press the TAB key and enter **40** in the width input box.

5. Press the TAB key again to lock the dimensions; the **BOX** dialog gets modified and the **Height** edit box is displayed in the graphics window.

   In the **BOX** dialog box, you will observe that other options are also available. You can specify the number of faces with respect to the width of the box in the **Width Faces** edit box and number of faces with respect to the height of the box in the **Height Faces** edit box. Similarly, you can specify the number of faces with respect to the length of the box in the **Length Faces** edit box. The direction of extrusion of the primitives can be defined using the **Direction** drop-down list.

6. Enter **100** in the **Height** edit box in the graphics window or in the **Height** edit box in the **BOX** dialog box.

7. Retain rest of the settings as a default and the choose the **OK** button from the dialog box. Figure 6-44 shows the model after creating box.

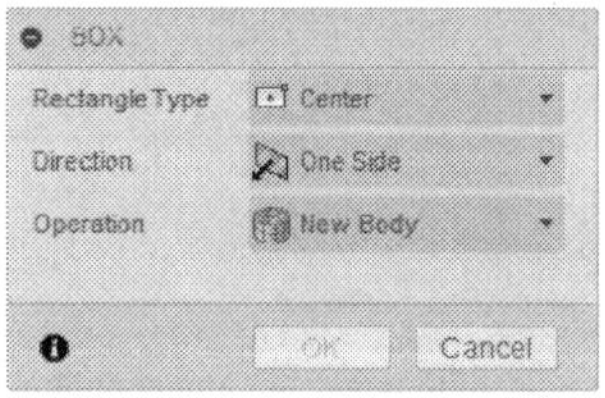

Figure 6-43 *The **BOX** dialog box*

Figure 6-44 *The model after creating box*

Creating the Mirror

Now you need to create mirror symmetry with the elements of the T-Spline body.

1. Choose the **Mirror - Internal** option from the **SYMMETRY** panel in the Ribbon; the **MIRROR - INTERNAL** dialog box is displayed. Also, you are prompted to select the face on the master side.

2. Select the front face of the box, as shown in Figure 6-45; you are prompted to select the face on the mirror side.

3 Select the back face of the box; a green symmetric line is displayed on the box in the graphics window, as shown in Figure 6-46.

4. Choose the **OK** button from the dialog box.

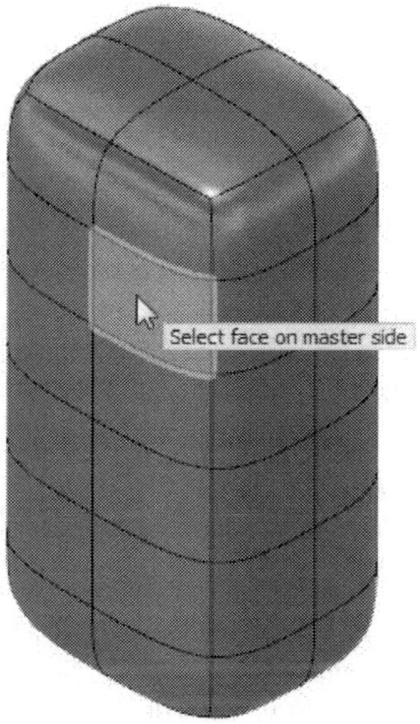

Figure 6-45 *Selecting front face of the box*

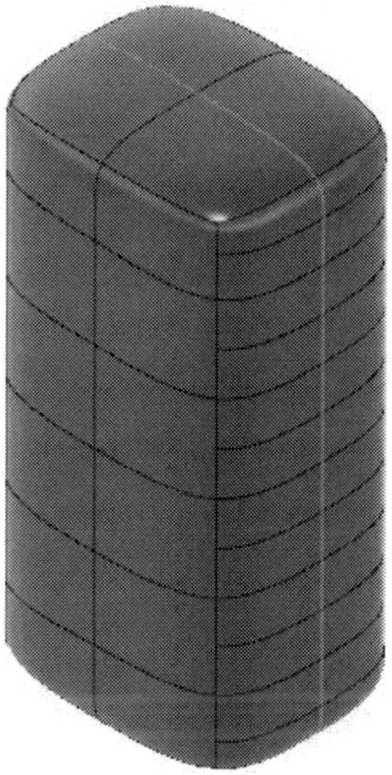

Figure 6-46 *Symmetric line created*

Inserting Edges

You need to insert edges along original edges in the model.

1. Choose the **Insert Edge** tool from the **MODIFY** panel in the Ribbon; the **INSERT EDGE** dialog box is displayed and you are prompted to select the edges.

2. Select edge of the box in the graphics window, as shown in Figure 6-47.

3. Make sure that the **Both** option is selected in the **Insertion Side** drop-down list of the **INSERT EDGE** dialog box.

4. Retaining rest of the values, choose the **OK** button from the dialog box; the edges are created at both sides of the selected T-Spline edge.

5. Similarly insert other edges on both sides of the symmetric line.

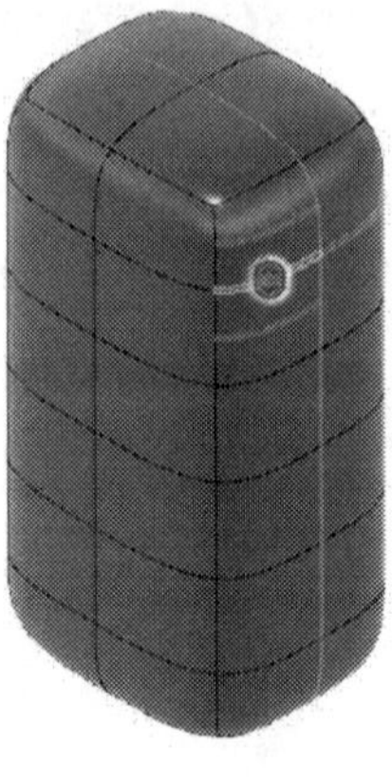

Figure 6-47 *Inserting edges along selected T-Spline edge*

Creasing and Editing the edges

1. Choose the **Crease** tool from the **MODIFY** panel of the Ribbon; the **CREASE** dialog box is displayed in the graphics window and you are prompted to select the edges and vertices.

2. Select the edges created in the last step and choose the **OK** button from the dialog box. The model after only creasing the edges is shown in Figure 6-48.

   Note that you have to select edges on one side of the symmetric line. The crease will be automatically mirrored to the other side of the symmetric line after choosing the **OK** button from the **CREASE** dialog box.

3. Choose the **Edit Form** tool from the **MODIFY** panel in the Ribbon; the **EDIT FORM** dialog box is displayed and you are prompted to select edges, vertices, or faces.

4. Select all the creasing edges by pressing CTRL key; the triad and the **X Distance** edit box get displayed in the graphics window.

5. Enter **10** in the **X Distance** edit box or dynamically drag the arrow of the triad in the X direction.

6. Choose the **OK** button from the **EDIT FORM** dialog box. The final model is shown in Figure 6-49.

7. Choose the **FINISH FORM** panel from the Ribbon to exit the **SCULPT** workspace.

8. Add fillet of 2mm to the creased edges.

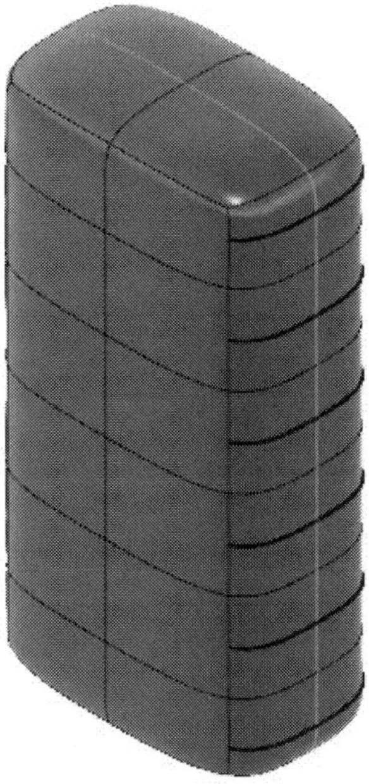

Figure 6-48 The model with creased edges

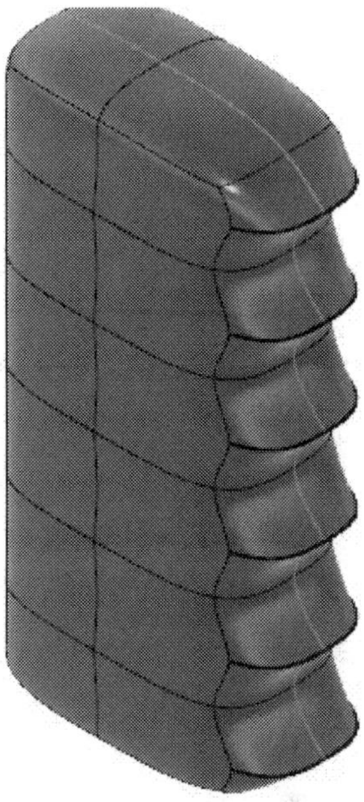

Figure 6-49 The final model

Saving the Model

Next, you need to save the model.

1. Choose the **Save** option from the **File** menu or the **Application** menu; the **Save** dialog box is displayed.

2. Enter **c06_Tut_04** in the **Name** edit box.

3. Click on the down arrow on the extreme right of the **Location** selection box; the **Save** dialog box gets expanded. 

4. Select project from the **PROJECT** column or create a new project by choosing the **New Project** button from the **Save** dialog box.

5. Select the **Save** button to save the sketch and exit the **Save** dialog box.

Self-Evaluation Test

Answer the following questions and then compare them to those given at the end of this chapter:

1. Which of the following tools is used to add an additional edge to a T-Spline?

(a) **Bevel Edge** (b) **Merge Edge**
(c) **Crease Edge** (d) **Insert Edge**

2. Which of the following elements must be there in a design before creating a sweep feature?

   (a) Profile (b) Path
   (c) Section (d) Start Point

3. You can view the section view of a model by using the ___________ check box.

4. __________ are thin wall-like structures used to bind joints together.

5. The __________ tool can directly manipulate faces, edges, and vertices.

6. A sweep feature is the one whose geometry is swept along one defined path. (T/F)

7. You can create simple, counterbore, or countersink holes using the **Hole Type** drop-down list. (T/F)

8. The sketches for the solid loft features should be closed profiles. (T/F)

9. The swept features are created by blending more than one dissimilar geometries. (T/F)

10. The first reference for a hole feature must be a sketch plane. (T/F)

Review Questions

Answer the following questions:

1. You need to select the __________ panel in the Ribbon to exit the Sculpt workspace,

2. The __________ tool is used to create a loft feature by blending more than one similar or dissimilar sections together to get a free-form shape.

3. The__________ tool is used to display the section view of a model by cutting it using a plane or a face.

4. The __________ tool is used to create sharp edges between the selected faces.

5. For creating a solid sweep feature, the profile must be a closed sketch. (T/F)

6. Sculpting in Autodesk Fusion 360 allows you to create freeform solid bodies and surfaces. (T/F)

EXERCISES

Exercise 1

Create a solid model for Exercise 1, as shown in Figure 6-50. The dimensions to be used for creating the model are given in the same Figure. **(Expected time: 45 min)**

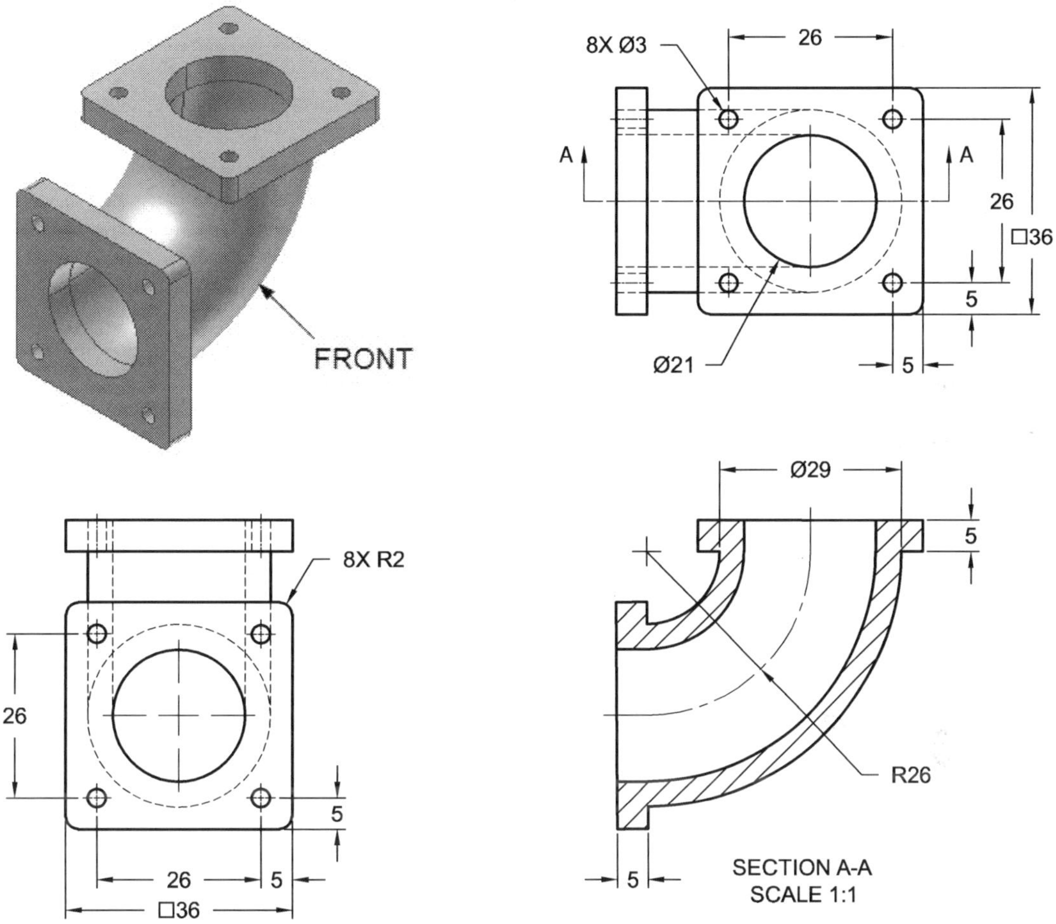

Figure 6-50 *Views and dimensions of the model*

Exercise 2

Create the model shown in Figure 6-51. The views and dimensions of the model are given in Figure 6-52. **(Expected time: 1 hr)**

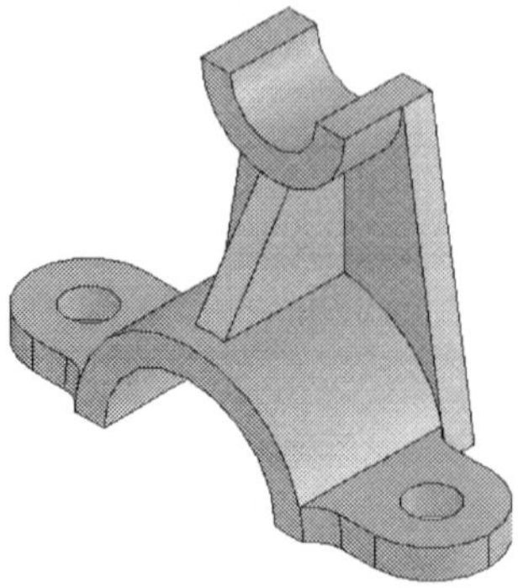

Figure 6-51 *Model for Exercise 2*

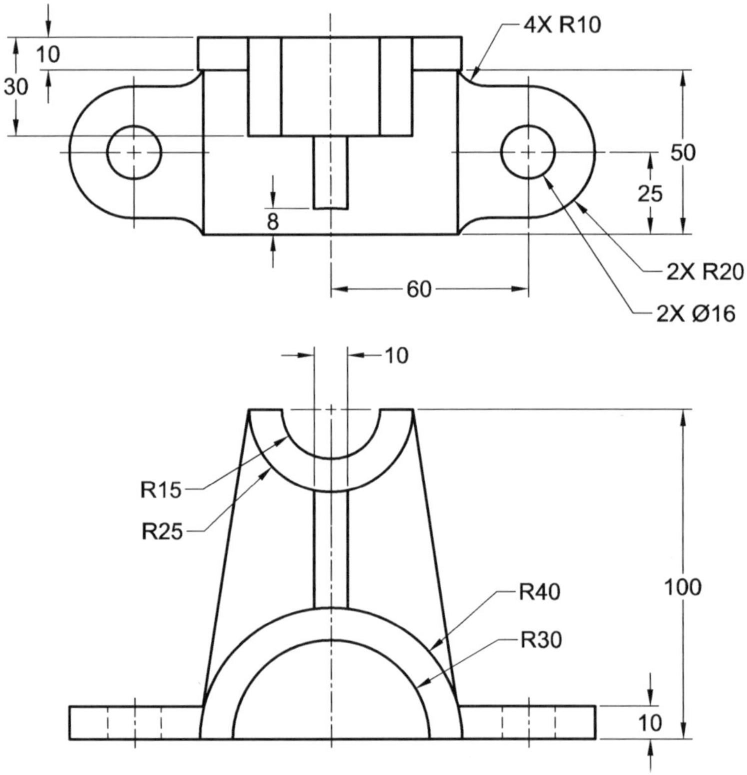

Figure 6-52 *Views and dimensions of the model for Exercise 2*

Answers to Self-Evaluation Test

1. d, **2**. a, **3**. **Slice**, **4**. **Ribs**, **5**. **Edit Form**, **6**. T, **7**. T, **8**. T, **9**. F, **10**. T

Chapter 7

Assembling Components

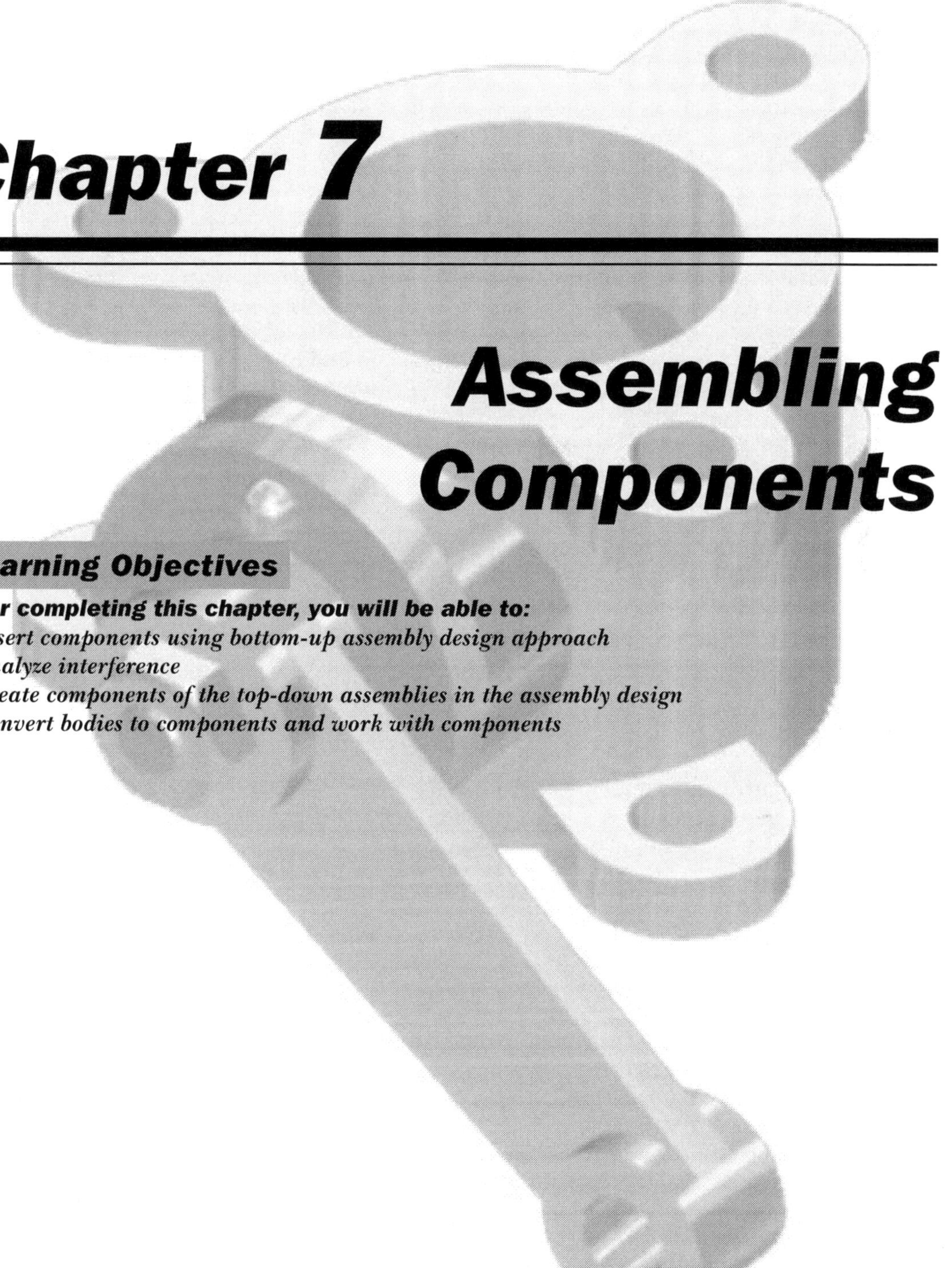

Learning Objectives

After completing this chapter, you will be able to:

- *Insert components using bottom-up assembly design approach*
- *Analyze interference*
- *Create components of the top-down assemblies in the assembly design*
- *Convert bodies to components and work with components*

ASSEMBLY DESIGN

Assembly design is a process of assembling two or more components at their respective work positions by applying joints. These joints are applied in a way such that the components can perform as required. The assembly joints allow you to restrict the degrees of freedom of the components. Autodesk Fusion 360 allows the models to be fully designed in a single workspace, starting from base sketch to a full-scale assembly.

Top-down Assembly Design Approach

In the top-down assembly design approach, components are created within the same design. Therefore, there is no need to create separate sketch for the components. Here, you can create a sketch of the given assembly as a master sketch which also includes all the sketches of the components of that assembly. These sketches can be defined in a relation with each other by applying constraints and they can be also extruded separately.

By adopting the top-down assembly design approach, the user gets distinctive advantage of using the geometry of one component to define the geometry of another component. Here, the construction and assembling of components take place simultaneously. As a result, the user can view the development of the product in real time. This design approach is highly preferred where the reference of previously created components is required to develop a new component.

Bottom-up Assembly Design Approach

The bottom-up assembly design approach is the most preferred approach, for creating assembly models. In this approach, the components are created individually and then inserted into the assembly design.

By adopting the bottom-up assembly design approach, the user gets an opportunity to pay more attention to the details of the components while they are being designed individually. This approach is preferred for creating large assemblies, especially for those which have intricate individual components.

Bodies and Components

Body is feature or set of features which form a hybrid design. You can convert a body into a component to achieve the final design of the assembly. For example, you may create a coffee cup by creating a body for the cup and another body for its handle, and then you can join them together.

Component is an independent part of the assembly. It does not have position as well as kinematic relationship with other components until joints are applied. Its motion and relationship with other components in the assembly is defined by using joints so that the design is functionally accurate. For example, a door hinge consists of three components: one is on the left, and other is on the right, and the third one is the pin that holds the two together. In this case, all these three components are independent of each other.

TUTORIALS

Tutorial 1

In this tutorial, you will create components of a Bench Vice and then assemble them, as shown in Figure 7-1. You need to create all the components as separate files. The exploded view of the assembly is shown in Figure 7-2. The views and dimensions of the components are shown in Figure 7-3 through 7-6. **(Expected time: 3 hrs 30 min)**

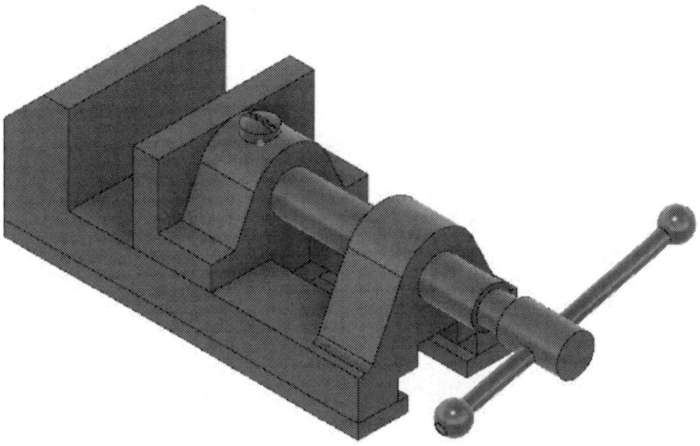

Figure 7-1 *The Bench Vice Assembly*

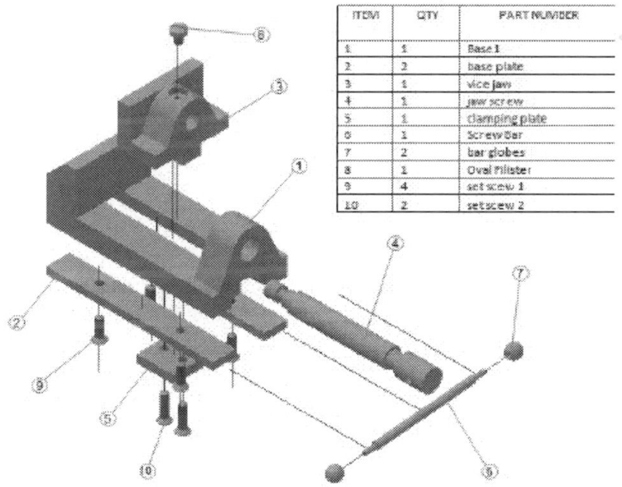

ITEM	QTY	PART NUMBER
1	1	Base 1
2	2	base plate
3	1	vice jaw
4	1	jaw screw
5	1	clamping plate
6	1	Screw Bar
7	2	bar globes
8	1	Oval Fillister
9	4	set scew 1
10	2	setscew 2

Figure 7-2 *Exploded view of the Bench Vice Assembly*

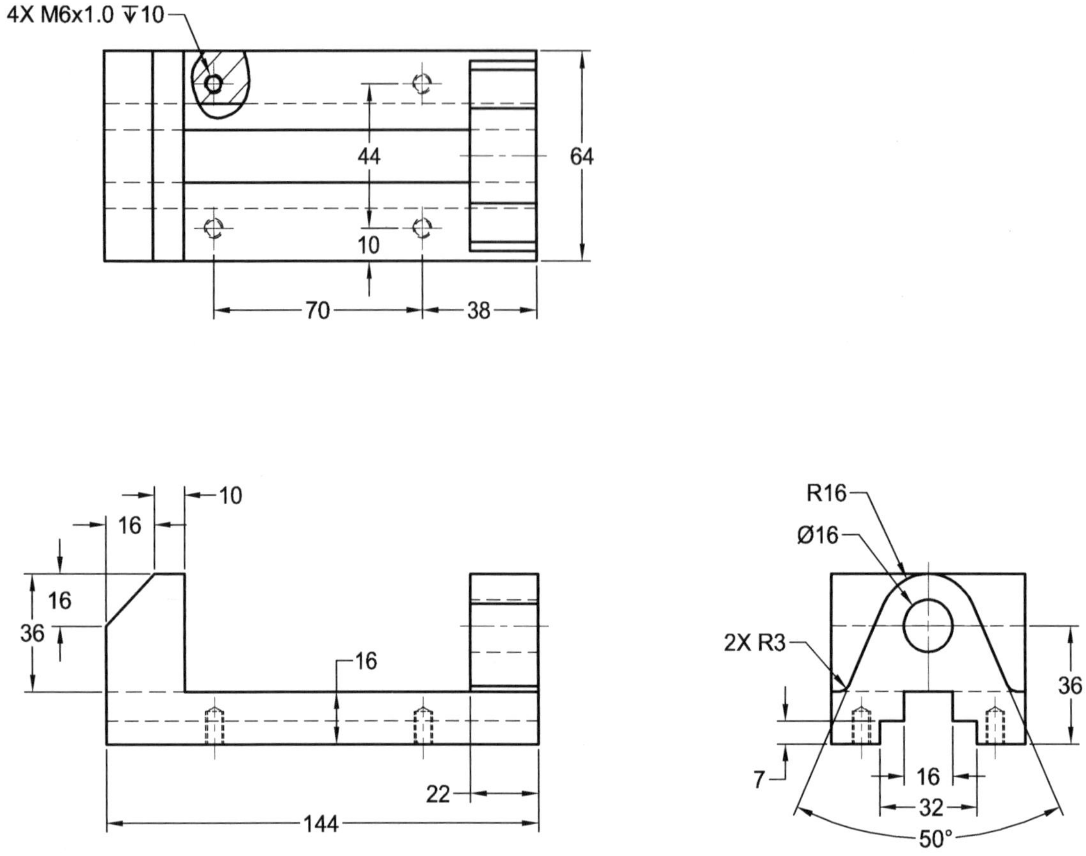

Figure 7-3 Orthographic views and dimensions of the Base

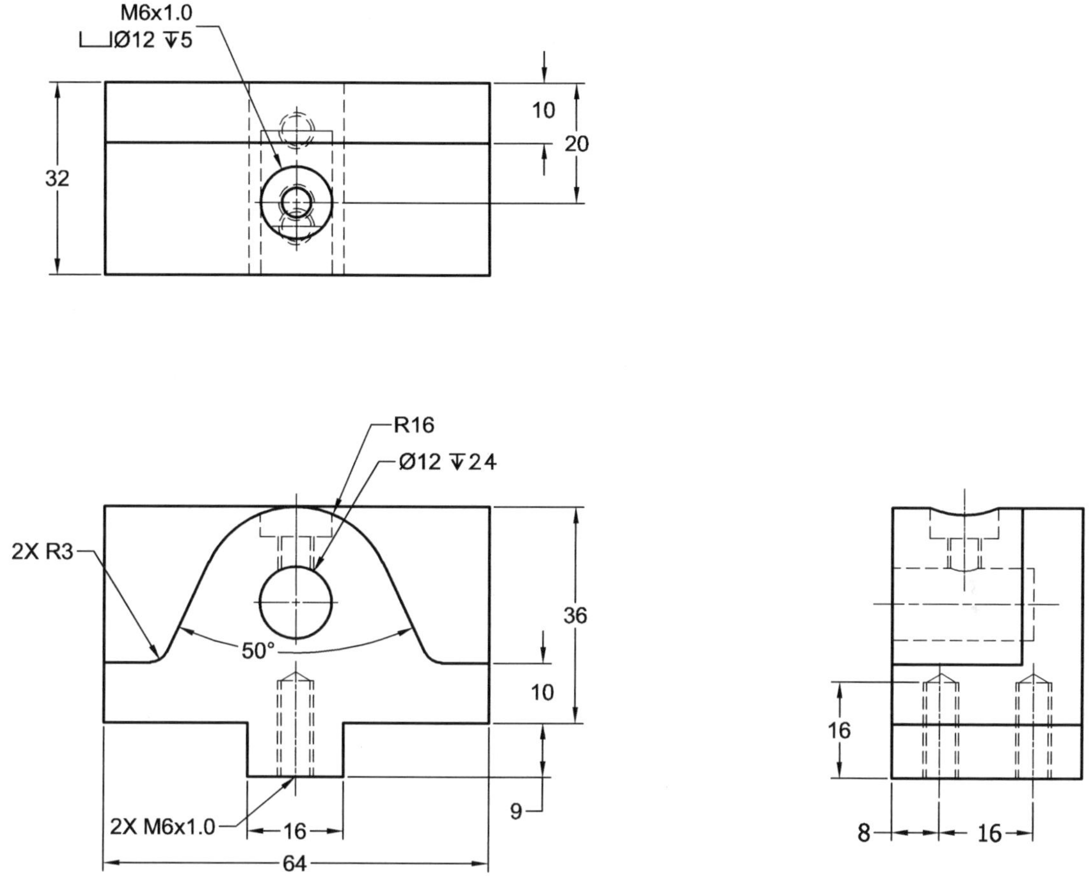

Figure 7-4 *Orthographic views and dimensions of the Vice Jaw*

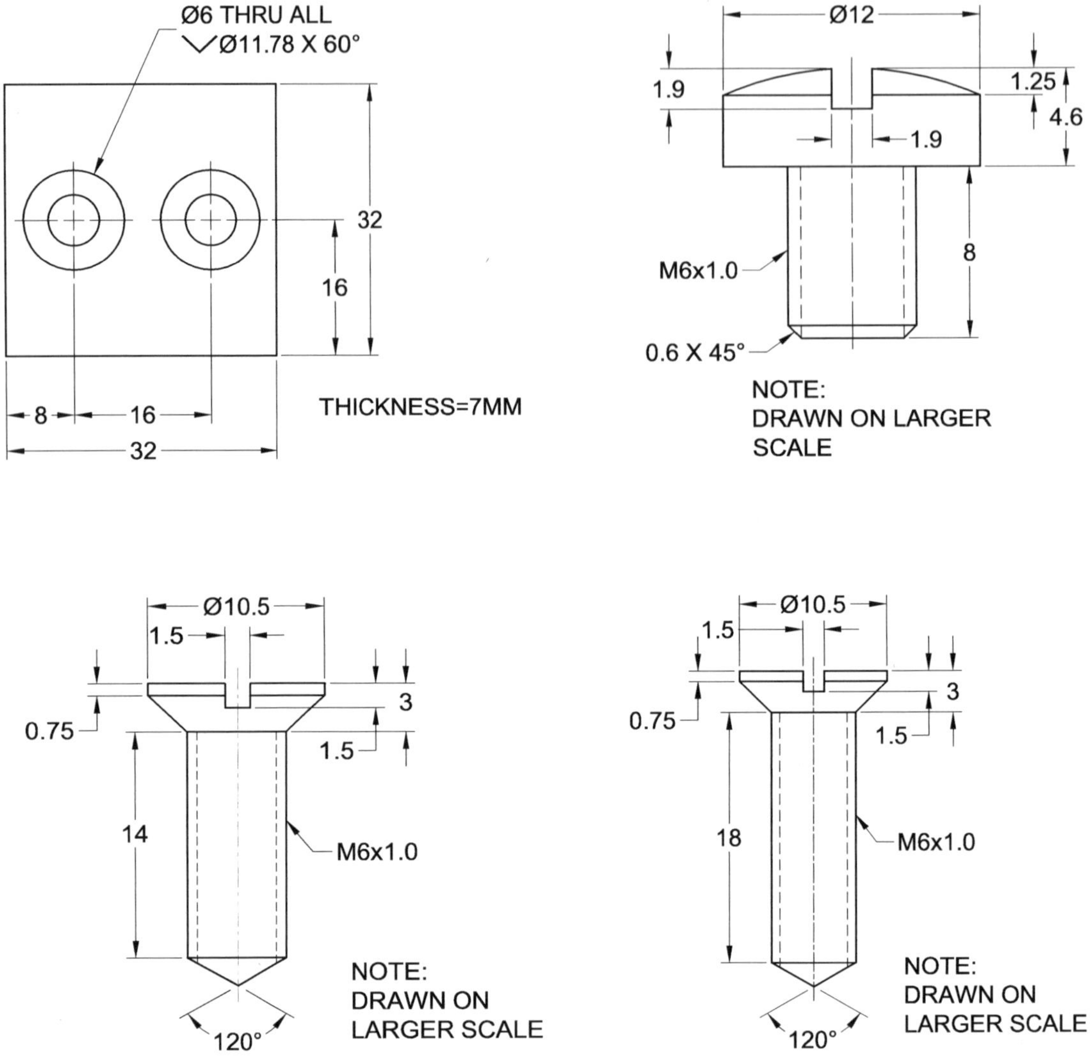

Figure 7-5 *Orthographic views and dimensions of various components of the Bench Vice assembly*

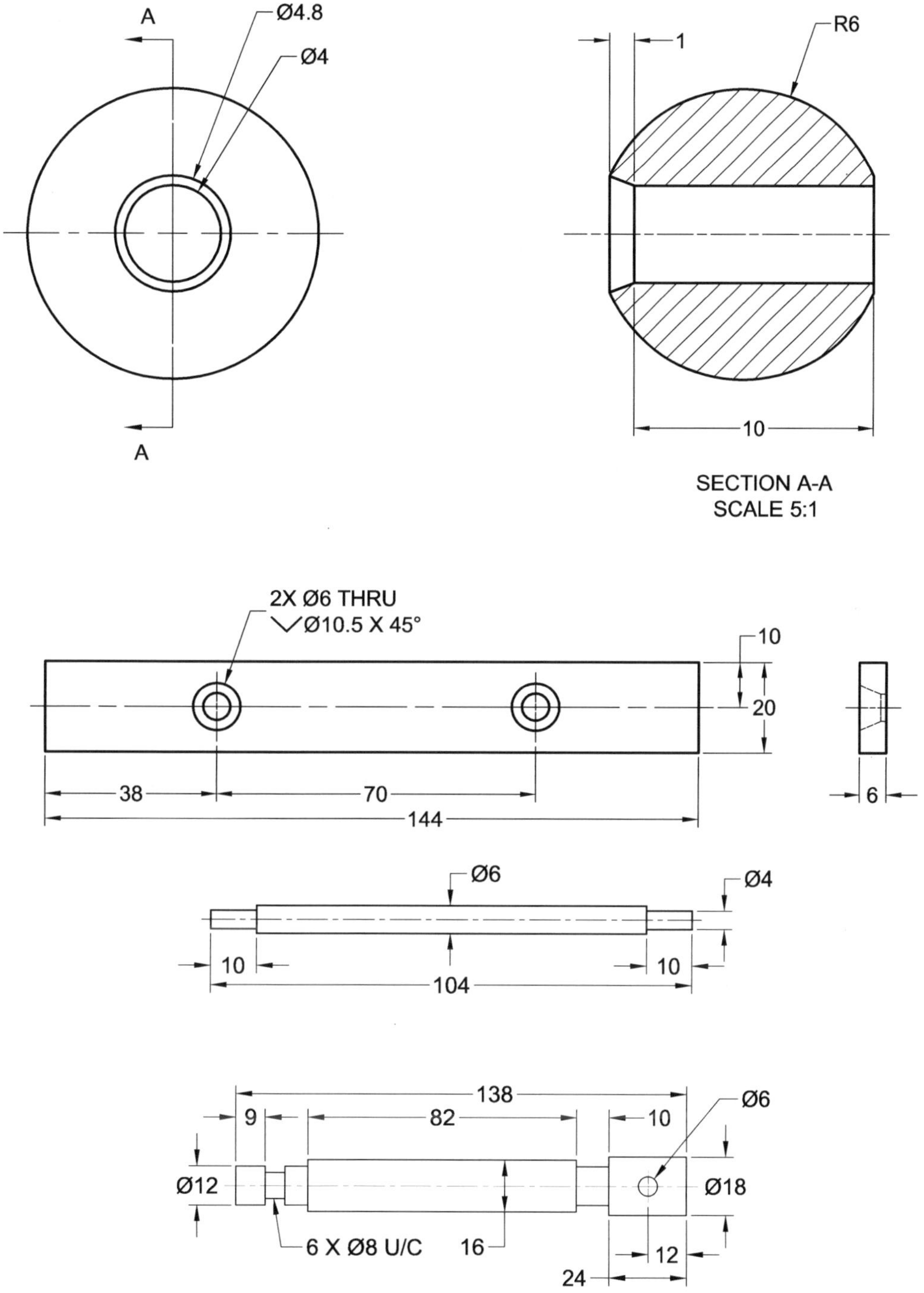

Figure 7-6 *Orthographic views and dimensions of Bar Globe, Base, Screw Bar, and Jaw Screw*

The following steps are required to complete this tutorial:

a. Create all components of the assembly as separate bodies.
b. Insert components in the design.
c. Assemble the components using Joints.
d. Create contact sets.
e. Analyze the interference.
f. Save the assembly.

Creating the Folder

Before creating the components for the assembly, create a folder with the name *Bench Vice Assembly* in the **Data Panel**.

1. Close all Autodesk Fusion 360 files and then choose the **New Folder** button from the **Data Panel** at the top-left side of Application bar; the **New Folder** is created in the **Data Panel**.

2. Rename the created folder with the Bench Vice Assembly.

Creating the Components

The Bench-Vice assembly will be created with the bottom-up assembly design approach. You need to create all components of this assembly separately and then save them with their respective names.

First create the Base part of the assembly.

1. Choose the **Create Sketch** tool from the **SKETCH** panel of the **MODEL** workspace; the default planes are displayed in the graphics window.

2. Create Base of the Bench Vice assembly on the desired plane using the dimensions given in Figures 7-3.

3. Choose the **Save** option from the **Application** menu; the **Save** dialog box is displayed.

4. Enter **Base** in the **Name** edit box.

5. Click on the down arrow on the extreme right side of the **Location** selection box; the **Save** dialog box gets expanded.

6. Choose the **CADCIM** option from the **PROJECT** area; all the projects in this area will be displayed. Next, choose **Bench Vice Assembly** from the **CADCIM >master** area.

7. Choose the **Save** button to save the sketch and exit the **Save** dialog box.

8. Similarly, create all other components of the Bench Vice Assembly and save them on the specific location.

Inserting Components in Assembly Design

1. Choose the **Show Data Panel** button from the Application bar, if the **Data Panel** is not displayed by default.

2. Double-click on the **Bench Vice Assembly** folder in the **Data Panel**; the Bench Vice folder is opened, as shown in Figure 7-7.

3. Right-click on the Base in the **Data Panel**; a shortcut menu is displayed.

   Note that if you have not saved the design earlier then a warning message will be displayed prompting you to save this design before inserting components.

4. Choose the **Insert into Current Design** option from the shortcut menu displayed; the Base is inserted in the design and the **MOVE/COPY** dialog box is displayed in the graphics window, as shown in Figure 7-8.

5. Keep the inserted component at its default location and choose the **OK** button to close the dialog box.

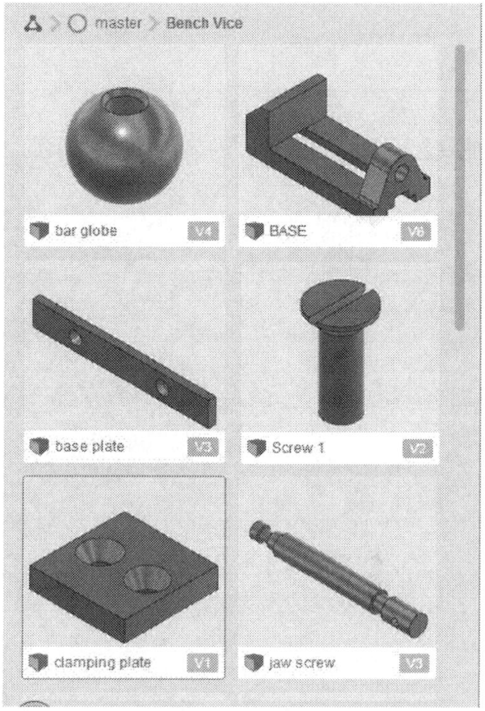

***Figure 7-7** Partial view of the Data Panel displaying components*

Tip
*You can insert the components in an assembly design by dragging and dropping them from the **Data Panel**.*

Note
*You can sort the selection of the components by using the options available in the **Settings** flyout which is invoked by clicking on the **Settings** button.*

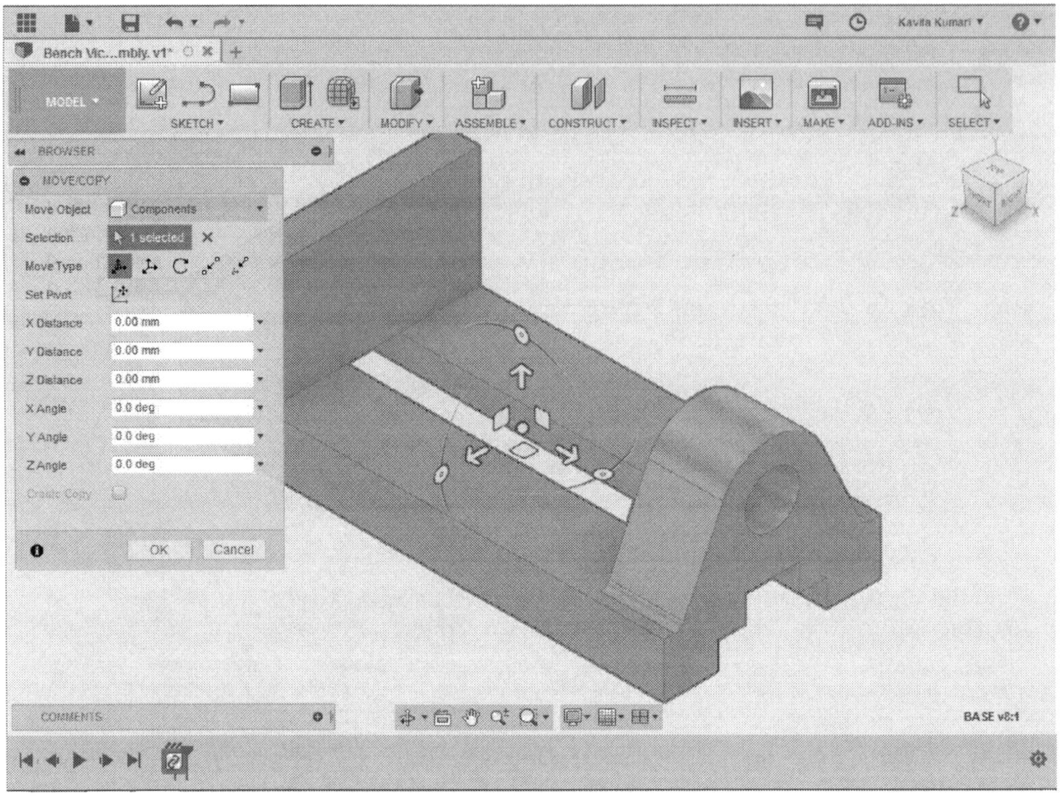

Figure 7-8 Component inserted into the design

6. Right-click on the Base in the **BROWSER** bar; a shortcut menu is displayed.

7. Select the **Ground** option from the shortcut menu; the selected component gets grounded in the assembly design. By doing this, you have restricted all the Degrees of Freedom of the component.

   You will notice that after the Base component gets grounded, a push pin icon is added next to the component. This icon indicates that the component is grounded and now it cannot be moved or rotated.

Inserting and Assembling the Vice Jaw

Next, you need to insert the Vice Jaw in the assembly design.

1. Right-click on the Vice Jaw in the Data Panel and choose the **Insert into Current Design** option from the shortcut menu displayed; the selected component is inserted in the assembly design and the **MOVE/ COPY** dialog box is also displayed along with the inserted component. If the location of the component is not the same as shown in Figure 7-9, then you need to reposition it using triad and the **MOVE/ COPY** dialog box.

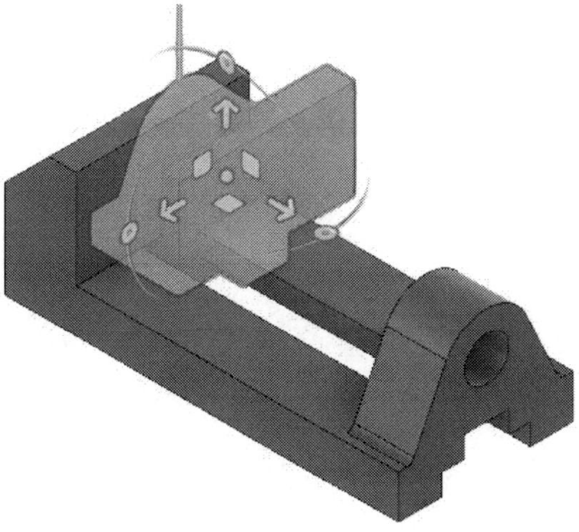

Figure 7-9 *Vice Jaw inserted in the assembly design*

2. Rotate the required manipulator by dragging it or enter **180** in the **Y Angle** edit box of the **MOVE/ COPY** dialog box.

3. Choose the **OK** button from the **MOVE/ COPY** dialog box.

   Now, you need to assemble the inserted components into the assembly by using the Joint tools in the **ASSEMBLE** panel. Joint tools are used to connect two components or bodies and allow the motion between the connected components or the sub-assembly.

4. Choose the **Joint** tool from the **ASSEMBLE** panel of the **MODEL** workspace; a message box is displayed, as shown in Figure 7-10.

5. Choose the **Capture Position** button from this message box; the **JOINT** dialog box is displayed with the **Component1** option selected by default, as shown in Figure 7-11.

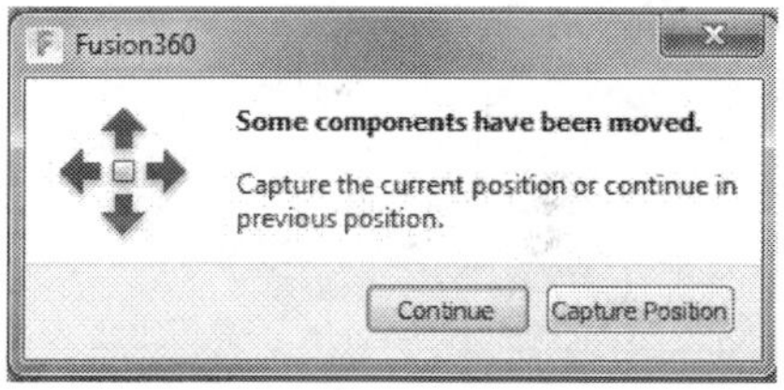

Figure 7-10 *A message box*

6. Move the cursor in the assembly design; you are prompted to select the location on the component to move.

7. Select the location on the component or the Vice Jaw, as shown in Figure 7-12. On doing so, the **Component 2** option gets selected and you are prompted to select the location on the second component.

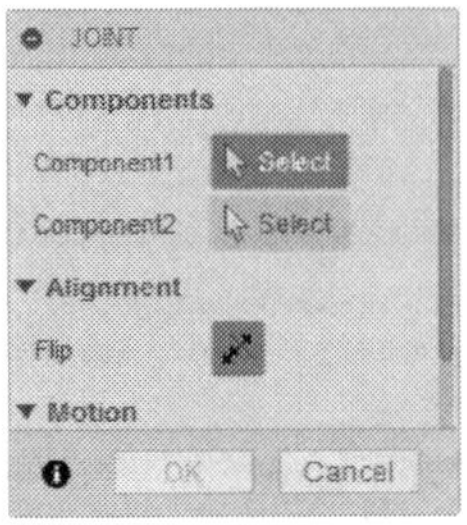

*Figure 7-11 The **JOINT** dialog box*

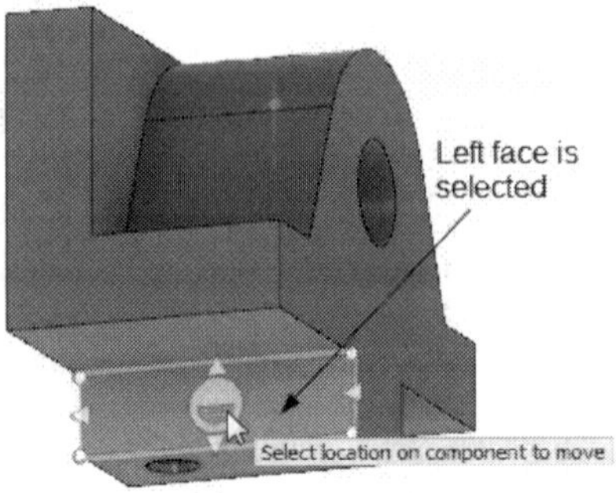

Figure 7-12 Selecting location on the Vice Jaw

8. Select the location on the Base, as shown in Figure 7-13; the Vice Jaw gets assembled with the Base and the **JOINT** dialog box gets modified, as shown in Figure 7-14.

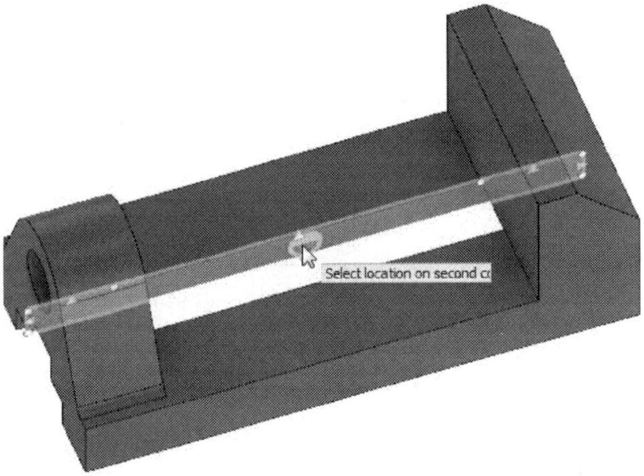

Figure 7-13 Selecting location on the Base

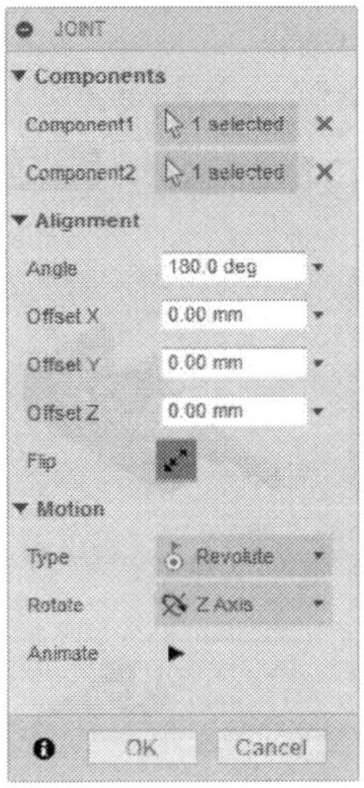

*Figure 7-14 The modified **JOINT** dialog box*

You can modify the angle and the offset distance in the X, Y, Z direction under the **Alignment** area by using their respective edit boxes on the basis of the type of joints selected.

Note

*If two or more components are there in an assembly design then after invoking the **JOINT** dialog box, the grounded component gets highlighted indicating that the component cannot be selected as **Component 1** while applying joints.*

9. Select the **Slider** option from the **Type** drop-down list under the **Motion** area; the **Axis** drop-down list displayed under the **Mot**ion area.

10. Select the **X Axis** option in the **Axis** drop-down list to slide the Vice Jaw on the Base.

11. Choose the **OK** button from this dialog box. The assembly after applying the Slider joint between the Base and the Vice Jaw is shown in Figure 7-15.

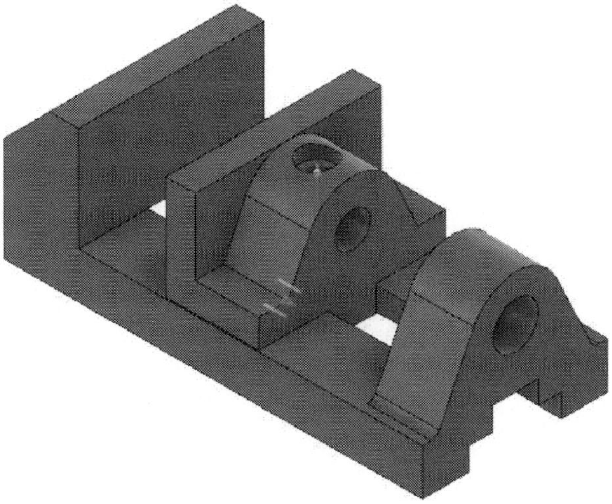

Figure 7-15 *The assembly after assembling the Vice Jaw*

Creating Contact Set

Next, you need to create contact set between Base and Vice Jaw. The Contact sets constrain components so that they cannot pass through each other, and thus behave as they would in the real world.

1. Choose the **Enable All Contact** tool from the **ASSEMBLE** panel; the **New Contact Set** tool is added to the **ASSEMBLE** panel and the **Contact: all** node is added to the **BROWSER** bar.

2. Right-click on the **Contact: all** node in the **BROWSER** bar; a shortcut menu is displayed.

3. Choose the **New Contact Set** option from the shortcut menu; the **NEW CONTACT SET** dialog box is displayed and you are prompted to select two or more components to create a contact set between them. You can also invoke this dialog box by choosing the **New Contact Set** tool from the **ASSEMBLE** panel.

4. Select Base and Vice Jaw from the graphics window to create a contact between them.

5. Choose the **OK** button in the **JOINT** dialog box; the contact is applied between the components.

Assembling Jaw Screw

Next, you need to assemble the Jaw Screw with the assembly.

1. Right click on the Jaw Screw in the **Data Panel**; a shortcut menu is displayed. Choose the **Insert into Current Design** option; the Jaw Screw is inserted in the assembly and the **MOVE/ COPY** dialog box is displayed.

   By default, Jaw Screw is positioned in such a way that it overlaps the existing components of the assembly. Therefore, you need to use the traid or the **MOVE/ COPY** dialog box to reposition it in the open space.

2. Choose the **OK** button from the **MOVE/ COPY** dialog box after repositioning the Jaw Screw.

3. Choose the **Joint** tool from the **ASSEMBLE** panel of the **Model** Workspace; a message box is displayed, refer to Figure 7-10.

4. Choose **Capture Position** from this message box; the **JOINT** dialog box is displayed with the **Component 1** option selected by default. Move the cursor in the assembly design; you are prompted to select the location on the component to move.

5. Select location on the Jaw Screw, as shown in Figure 7-16. On doing so, the **Component 2** option gets selected in the dialog box and you are prompted to select the location on the second component.

6. Select location on the Vice Jaw, as shown in Figure 7-17; the Jaw Screw gets assembled with the Vice Jaw and the **JOINT** dialog box gets modified.

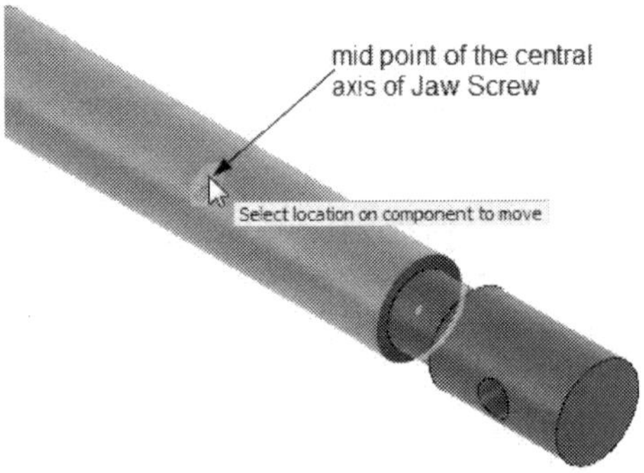

Figure 7-16 *Selecting location on the Jaw Screw*

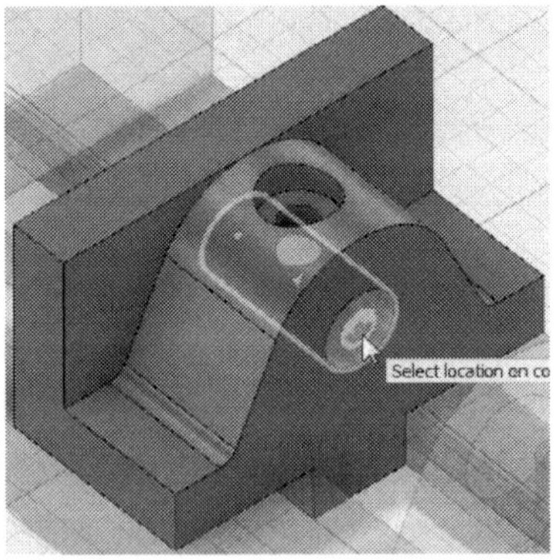

Figure 7-17 *Selecting location on the Vice Jaw*

7. Select the **Cylindrical** option from the **Type** drop-down list under the **Motion** area; the **Axis** drop-down list gets added under the **Motion** area.

8. Select **Z Axis** from the **Axis** drop-down list.

9. Choose the **OK** button from the **JOINT** dialog box; the Jaw Screw gets aligned with the central axis of the Vice Jaw.

   You need to fix the Jaw Screw linearly.

10. Invoke the **JOINT** dialog box again and select the face of the Jaw Screw, as shown in Figure 7-18; you are prompted to select the location on the second component.

11. Select the location on the Vice Jaw, as shown in Figure 7-19; a message box is displayed, as shown in Figure 7-20.

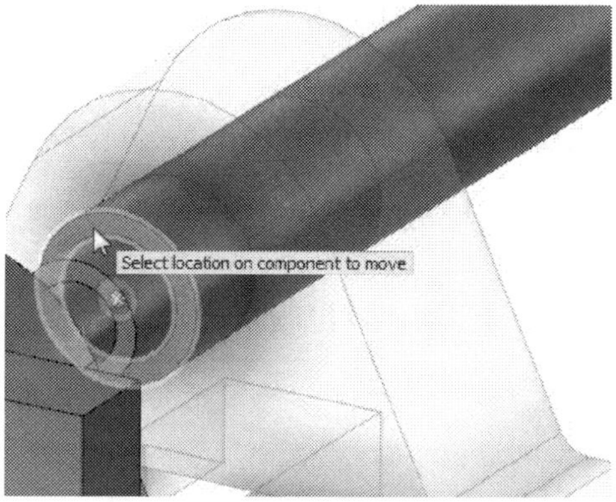

Figure 7-18 *Selecting location on the Jaw Screw*

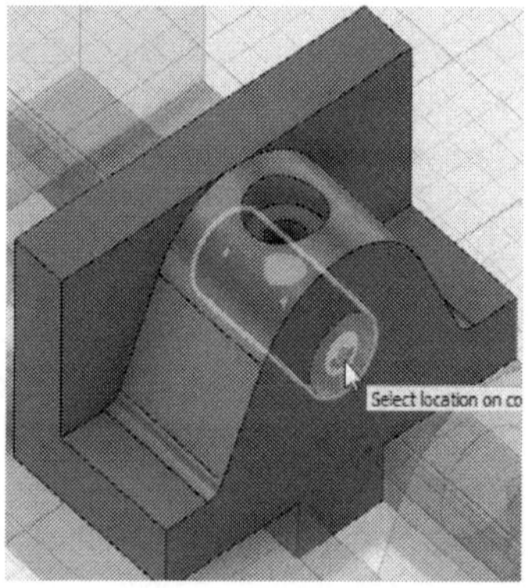

Figure 7-19 *Selecting location on the Vice Jaw*

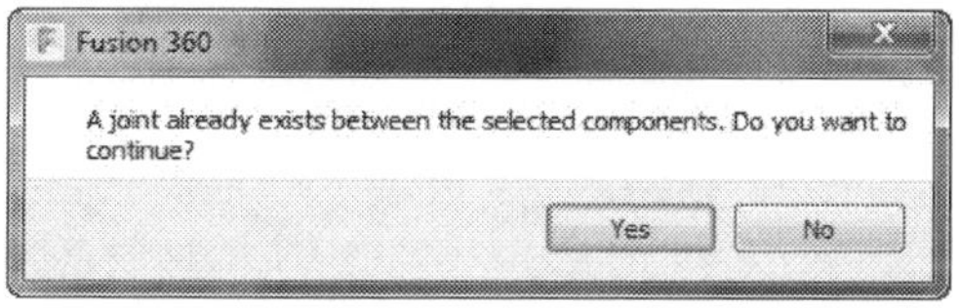

Figure 7-20 *A message dialog box*

12. Choose the **Yes** button from this message dialog box.

13. Select the **Rigid** option from the **Type** drop-down list and choose the **OK** button from this dialog box; the Jaw Screw gets assembled with the Vice Jaw. The assembly after assembling the Jaw Screw to the Vice Jaw is shown in Figure 7-21.

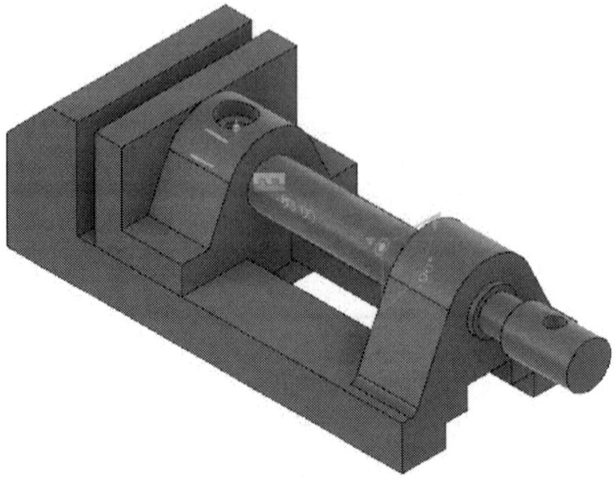

Figure 7-21 *The assembly after assembling Jaw Screw*

Assembling the Oval Fillister

Next, you need to assemble the Oval Fillister with the assembly.

1. Insert the Oval Fillister in the assembly and reposition it in such a manner that there is no overlapping between the Oval Fillister and the existing components of the assembly.
2. Invoke the **JOINT** dialog box.
3. Select the face of the Oval Fillister as **Component1**, as shown in Figure 7-22, and the inner face of Vice Jaw as **Component2**, as shown in Figure 7-23.
4. Select the **Rigid** joint from the **Type** drop-down list and choose the **OK** button from this dialog box; the Oval Fillister gets assembled with the Vice Jaw.

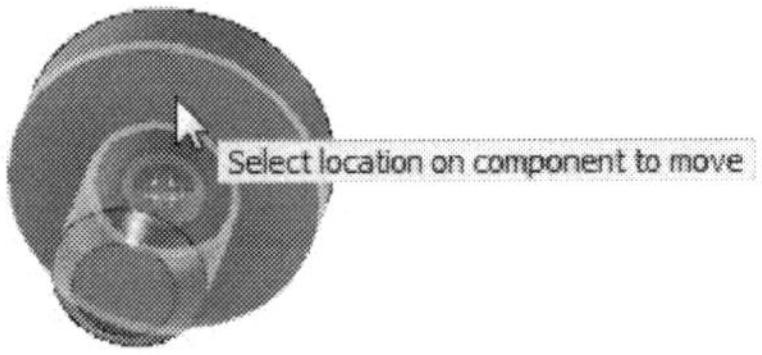

Figure 7-22 *Face to be selected*

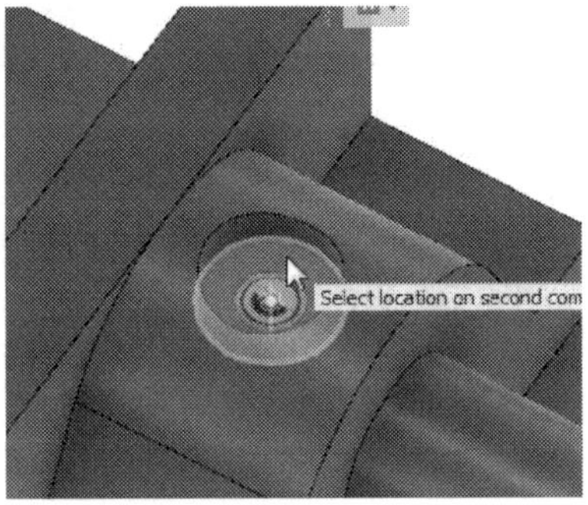

Figure 7-23 *Selecting location on the Vice Jaw*

> **Tip**
> *After inserting a component in the assembly design if you want to hide/show its origin or its sketches then you need to open that component and then choose the **Hide/Show** button from the **BROWSER** bar.*

Assembling the Screw Bar and Bar Globes

1. Insert the Screw Bar in the assembly and reposition it, if required.

2. Invoke the **JOINT** dialog box and select the **Slider** joint from the **Type** drop-down list.

3. Select the face of the Screw Bar, as shown in Figure 7-24, and select the cylindrical face of the hole of the Jaw Screw, as shown in Figure 7-25.

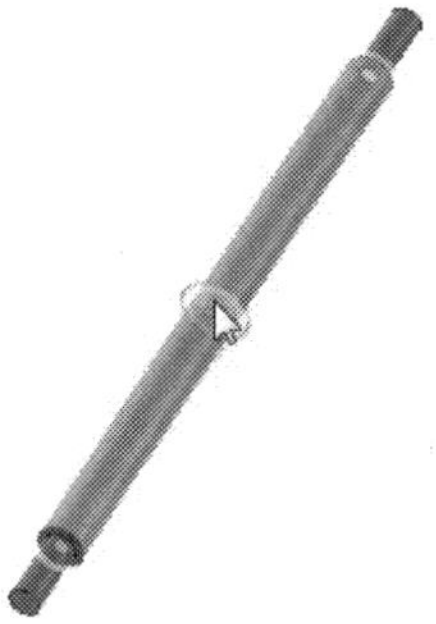

Figure 7-24 *Selecting location on the Screw Bar*

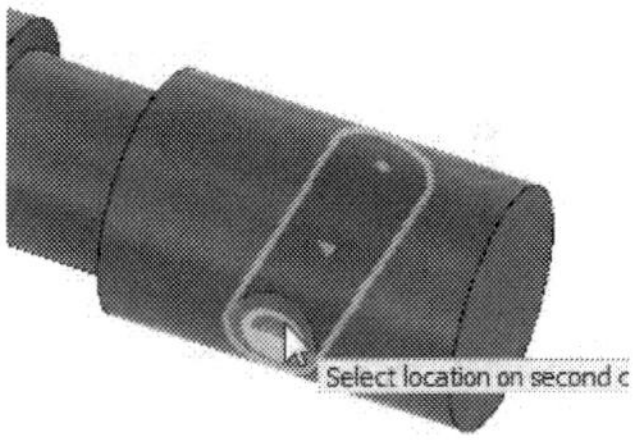

Figure 7-25 *Selecting location on the Jaw Screw*

4. Choose the **OK** button from the **JOINT** dialog box.

5. Insert two instances of Bar Globes in the assembly design from the **Data Panel** and reposition them so that its planar surface is toward the Screw Bar where it will be assembled.

> **Tip**
> *You can create a copy of the inserted instance of the component by using the **MOVE/COPY** dialog box. To create the copy, select the component from the **BROWSER** bar and then select the **Create Copy** check box in the dialog box. Next, move the component to place the copied instance and choose the **OK** button.*

6. Invoke the **JOINT** dialog box from the **ASSEMBLE** panel.

7. Select the cylindrical surface of Bar Globe, as shown in Figure 7-26 and the circular edge of the Screw Bar, as shown in Figure 7-27.

8. Select the **Rigid** option from the **Type** drop-down list under the **Motion** area and then choose the **OK** button in the **JOINT** dialog box.

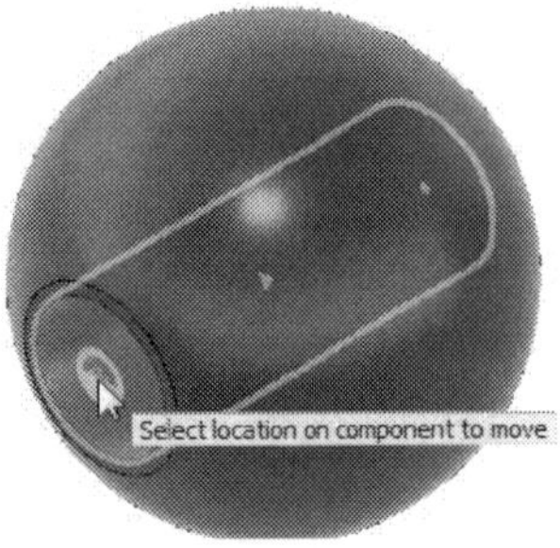

Figure 7-26 *Selecting cylindrical surface*

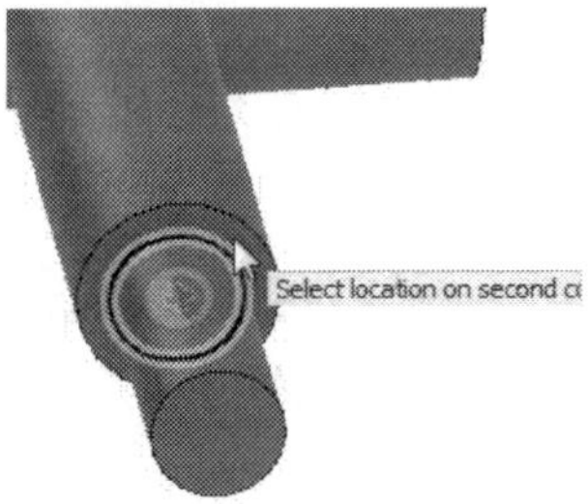

Figure 7-27 *Selecting circular edge*

9. Similarly, apply Rigid joint between the second Bar globe and the other end of the Screw bar. The assembly after assembling above components is shown in Figure 7-28.

10. Invoke the **NEW CONTACT SET** dialog box and select the Screw Bar, Jaw Screw, and both Bar Globes from the **BROWSER** bar to create contact sets between them.

11. Choose the **OK** button from the dialog box.

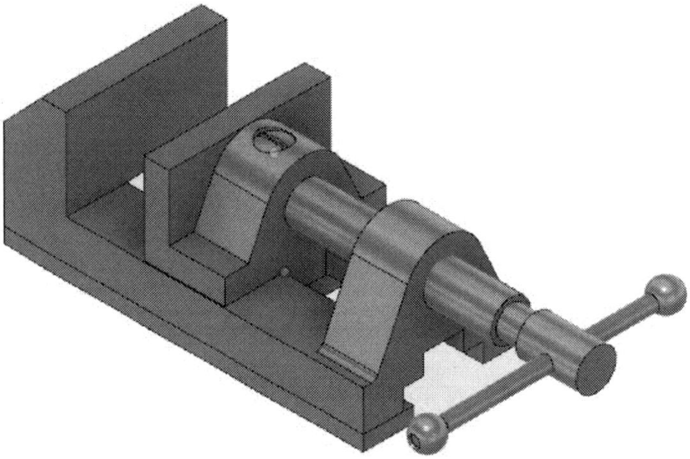

Figure 7-28 *The assembly after assembling Screw Bar and Bar Globes*

Assembling the Clamping Plate

Next, you need to assemble the Clamping Plate with the assembly.

1. Insert the Clamping Plate in the assembly and reposition it in such a manner that there is no overlapping between the components of existing assembly, refer to Figure 7-2.

2. Invoke the **JOINT** dialog box and select the **Rigid** option from the **Type** drop-down list if not selected by default.

3. Next, select the circular edge of the hole of Clamping Plate, as shown in Figure 7-29 and also the inner circular edge of the hole in the Vice Jaw, as shown in Figure 7-30.

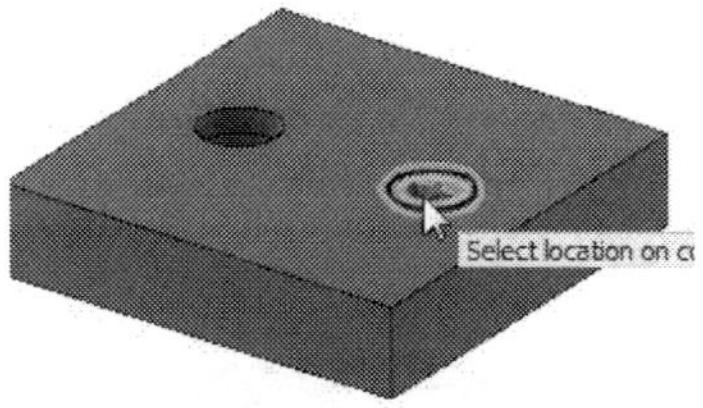

Figure 7-29 *Selecting circular edge of hole*

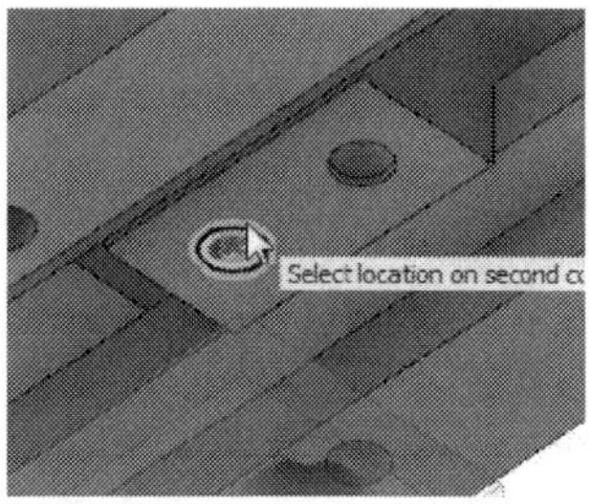

Figure 7-30 *Selecting circular edge of hole*

4. Choose the **OK** button in the **JOINT** dialog box.

5. Similarly, apply Rigid joint between the second hole of Clamping Plate and the Vice Jaw.

Assembling the Screw 1

1. Insert Screw 1 in the assembly and reposition it in such a manner that there is no overlapping between the existing components of the assembly.

2. Choose the **Joint** tool from the **ASSEMBLE** panel; the **JOINT** dialog box is displayed.

3. Select the **Rigid** option from the **Type** drop-down list, if not selected by default.

4. Select the circular edge of Screw1, as shown in Figure 7-31, and circular edge of hole of Clamping Plate by rotating the assembly using ViewCube, as shown in Figure 7-32.

5. Choose the **OK** button from the dialog box.

6. Similarly, assemble screw into the other hole of Clamping Plate.

Figure 7-31 *Selecting circular edge of Screw 1*

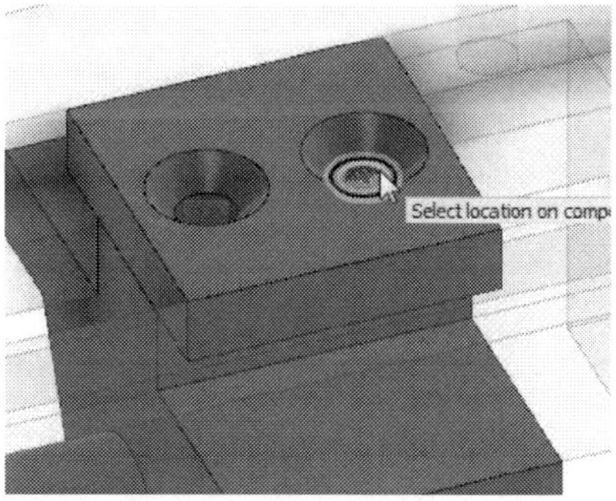

Figure 7-32 *Selecting circular edge of hole of Vice Jaw*

Assembling the Base Plate

Now you need to assemble the Base Plate with the Base.

1. Insert the Base Plate in the assembly and reposition it.

2. Invoke the **JOINT** dialog box and select the **Rigid** option from the **Type** drop-down list.

3. Select circular edge of the hole in the Base Plate, as shown in Figure 7-33, and circular edge of hole of Base by rotating the assembly using ViewCube, as shown in Figure 7-34.

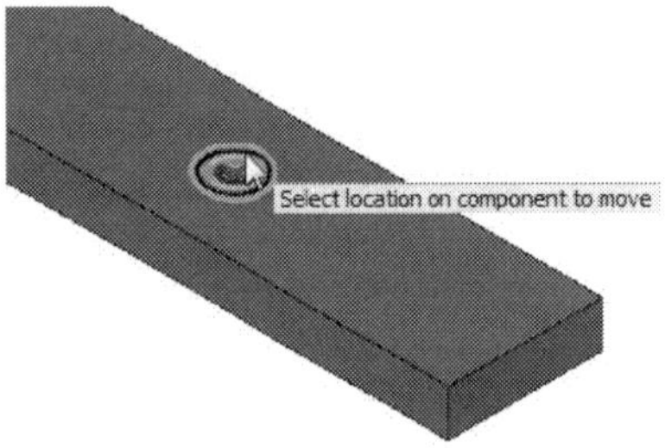

Figure 7-33 Selecting circular edge of Base Plate

Figure 7-34 Selecting circular edge of hole of Base

4. Choose the **OK** button from the dialog box.

5. Similarly, apply Rigid joint between the other circular edge of the hole in the Base Plate and Base of the assembly.

6. Repeat the procedure for applying Rigid joint for the another instance of the Base and Base Plate.

Assembling the Screw 2

1. Insert Screw 2 in the assembly and invoke the **JOINT** dialog box from the **ASSEMBLE** panel.

2. Select the **Rigid** option from the **Type** drop-down list.

3. Select circular edge of Screw 2, as shown in Figure 7-35, and circular edge of hole of Base Plate by rotating the assembly using ViewCube, as shown in Figure 7-36.

4. Choose the **OK** button from the dialog box.

5. Similarly, assemble the other instances of the Screw 2 with the Base Plate.

Figure 7-35 *Selecting circular edge of Screw 2*

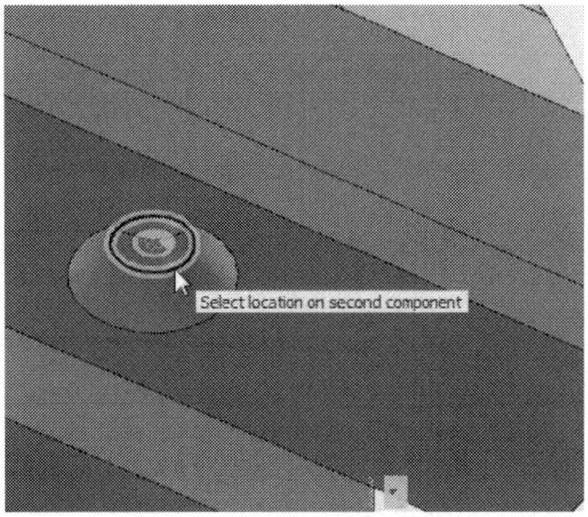

Figure 7-36 *Selecting circular edge of hole*

The assembly after assembling all the components is shown in Figure 7-37.

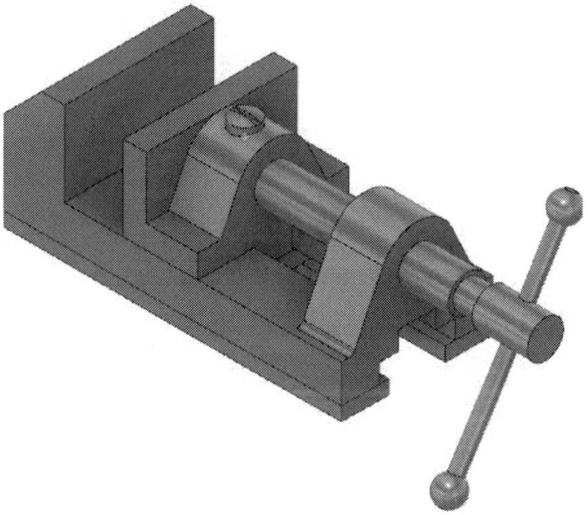

Figure 7-37 *The Final Assembly*

Analyzing the Assembly for Interference

After assembling the components, you will invoke the **INTERFERENCE** dialog box and analyze the assembly for interference. There should be no interference in the assembly.

1. Choose the **Interference** tool from the **INSPECT** panel from the Ribbon; the **INTERFERENCE** dialog box is displayed, as shown in Figure 7-38, and you are prompted to select the bodies or components.

2. Select the **Include Coincident Faces** check box. This check box is selected to show the interference between two coincident faces of components.

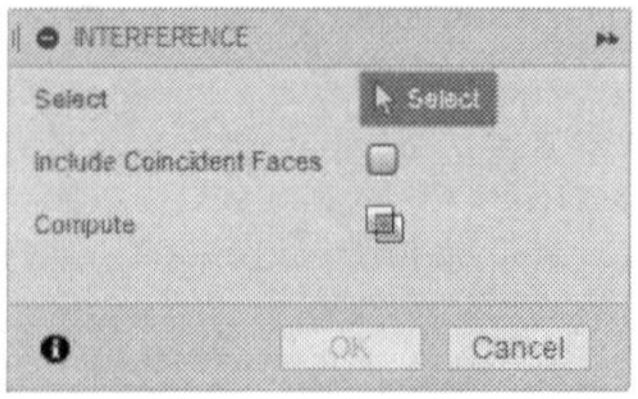

*Figure 7-38 The **INTERFERENCE** dialog box*

3. Select the components of the Bench Vice assembly from the **BROWSER** bar and choose the **Compute** button from the **INTERFERENCE** dialog box; the **Interferences Results** dialog box is displayed, as shown in Figure 7-39, showing the volume of interference between the components.

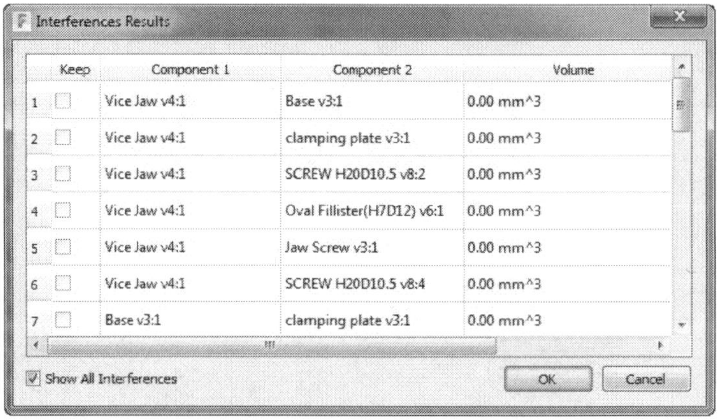

*Figure 7-39 The **Interference Result** dialog box*

> ***Tip***
>
> *If there is an interference in the assembly and the **Include Coincident Faces** check box is not selected in the **INTERFERENCE** dialog box then the interference will be displayed in the graphics window between the faces of the components but the coincident faces will not be displayed.*

Saving the Assembly

Next, you need to save the Assembly.

1. Choose the **Save** option from the **File** menu or the **Application** menu; the **Add Version Description** dialog box is displayed.

2. Choose the **OK** button from the dialog box.

Tutorial 2

In this tutorial, you will create a reciprocating mechanism, as shown in Figure 7-40, by assembling different components using the top-down assembly approach. The dimensions of all the components are shown in Figures 7-41 to 7-44.

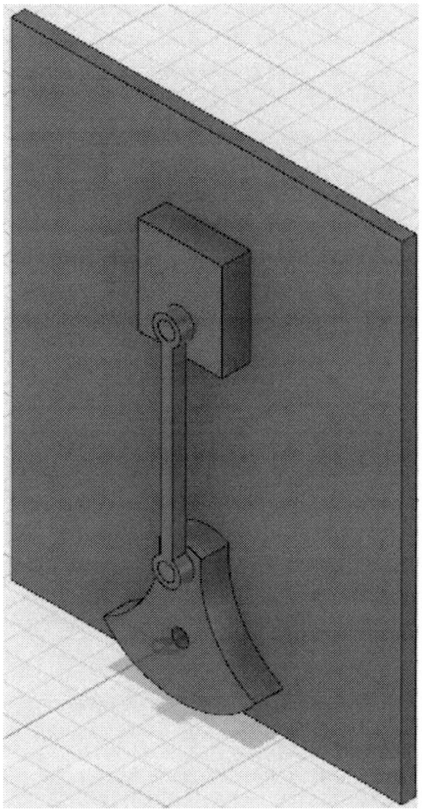

Figure 7-40 *Reciprocating Assembly*

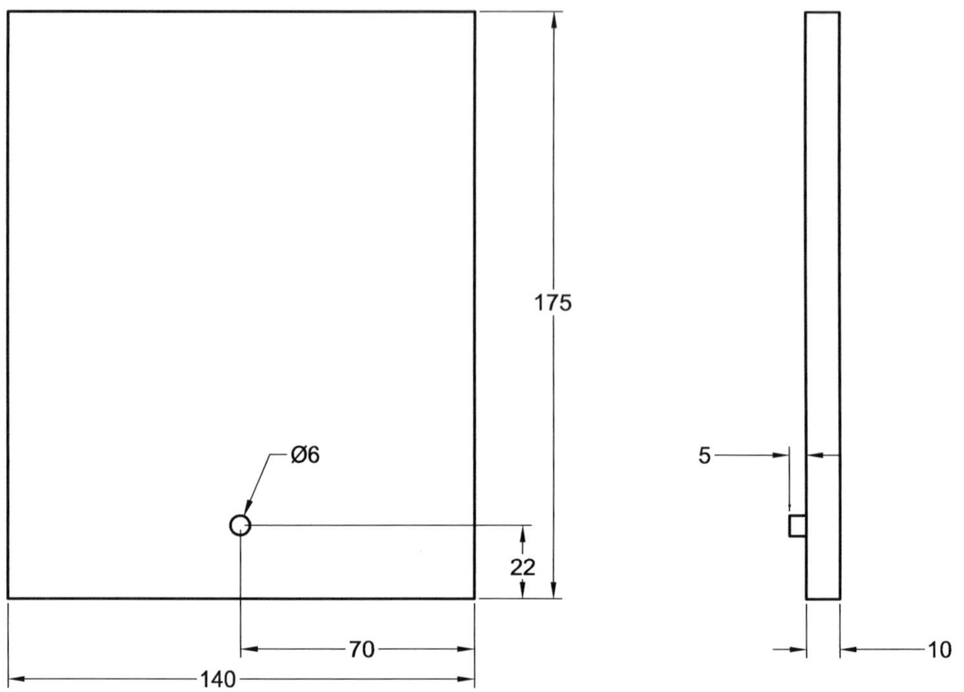

Figure 7-41 *Views and dimensions of the Base Plate*

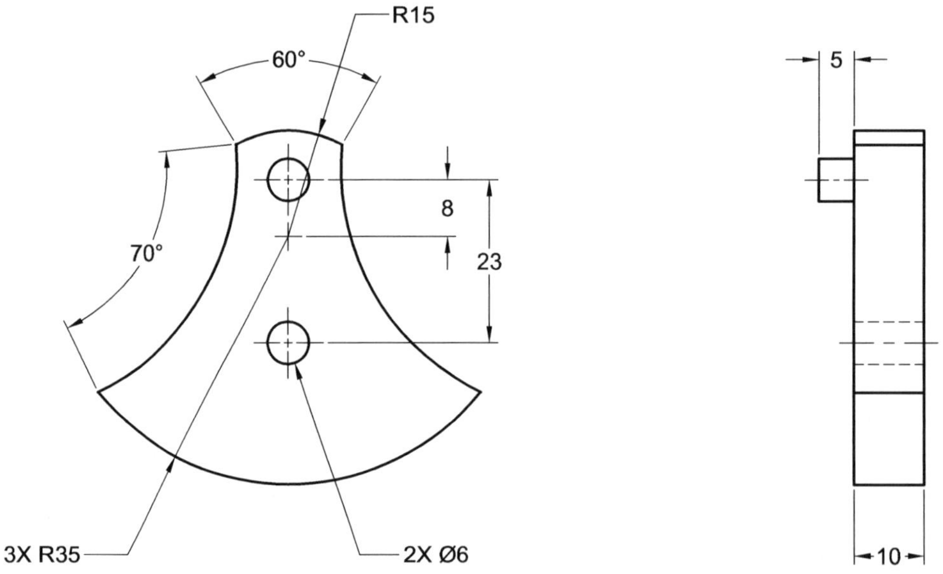

Figure 7-42 *Views and dimensions of the Crank*

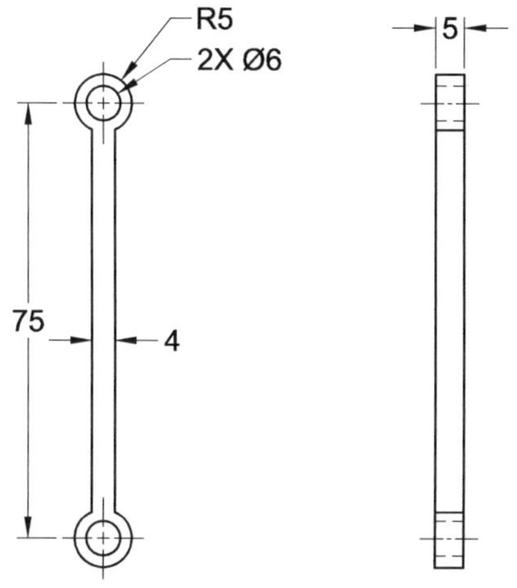

Figure 7-43 Sketch of the Piston rod

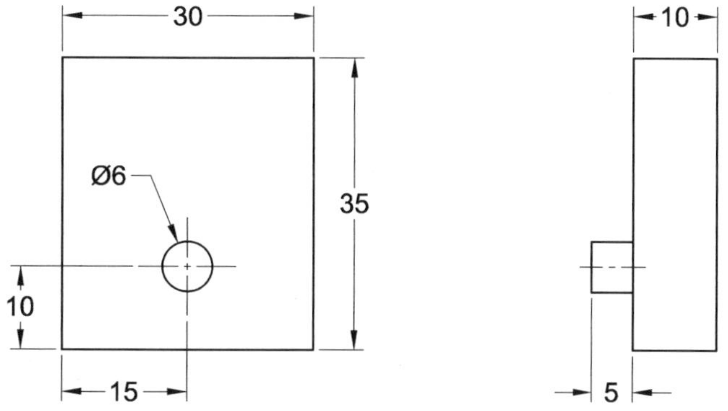

Figure 7-44 Sketch of the Piston Tank

The following steps are required to complete this tutorial:

a. Create the sketches of the mechanism by using the sketch tools.
b. Save the sketches as Reciprocating assembly.
c. Convert sketch into component.
d. Save and close the assembly.

Creating the Sketch of the Components

1. Choose the **Create Sketch** tool from the **SKETCH** panel of the **MODEL** workspace; the default planes are displayed in the graphics window and you are prompted to select a planar face.

2. Create the sketch of the reciprocating mechanism on the XY plane, as shown in Figures 7-45. For dimension refer to Figures 7-41 to 7-44.

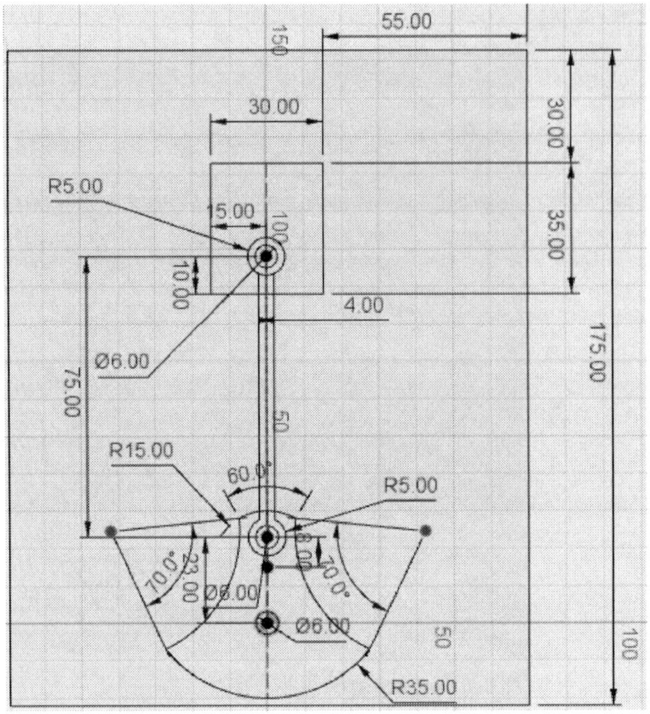

Figure 7-45 Sketch of the Reciprocating mechanism

Converting the Sketches into Components

Now, you need to convert the sketches into parts to create a top-down assembly.

1. Choose the **Extrude** tool from the **CREATE** panel in the **MODEL** Workspace; the **EXTRUDE** dialog box is displayed.

2. Select the profile of the Base Plate, as shown in Figure 7-46.

3. Select the **Offset Plane** option from the **Start** drop-down list and enter **-10** in the **Offset** edit box.

4. Next, enter **-5** in the **Distance** edit box and select the **New Component** option from the **Operation** drop-down list. By doing this, you do not need to convert the body into components. Note that by default, the **New Body** option is selected in the **Operation** drop-down list.

5. Retain rest of the options and choose the **OK** button from this dialog box. The Base Plate is created is shown in Figure 7-47.

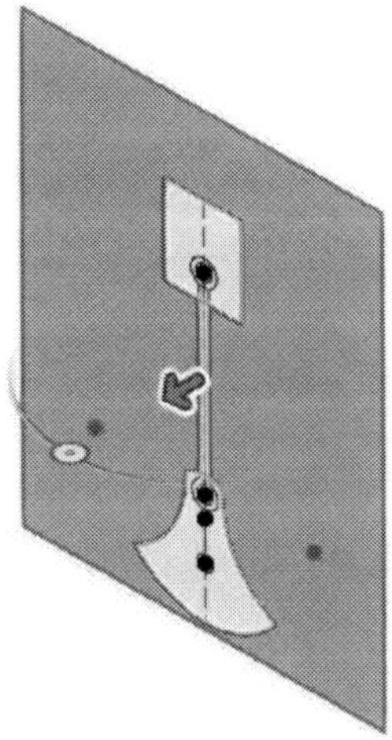

Figure 7-46 Profile of the Base Plate selected

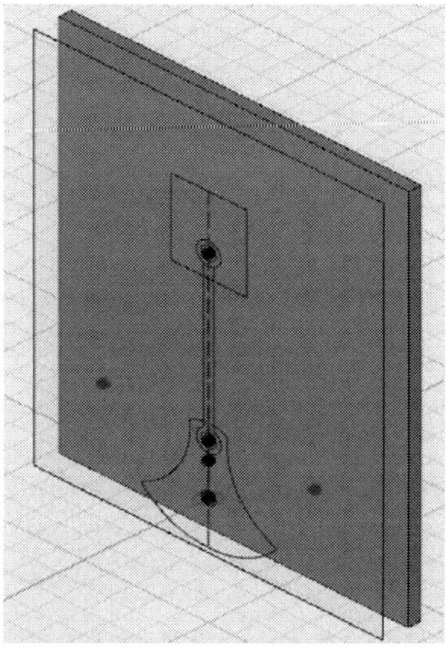

Figure 7-47 Extruded Profile of the Base Plate

Next, you need to extrude the circular profile of the pin of the Base Plate.

6. Select the profile of the pin of the previously created Base Plate from the sketch, as shown in Figure 7-48 and then invoke the **EXTRUDE** dialog box from the **CREATE** panel.

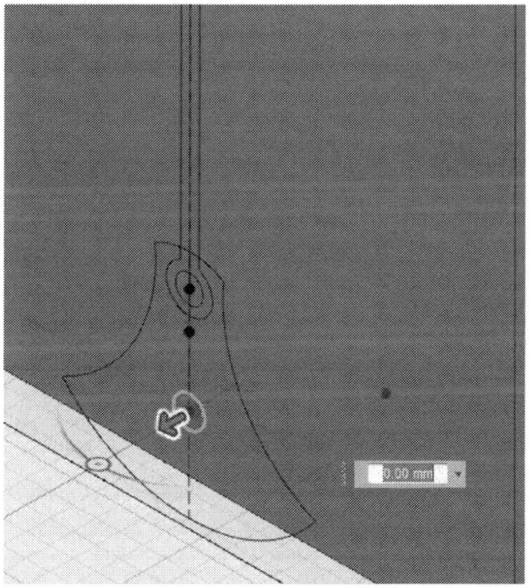

Figure 7-48 Profile of the pin of the Base Plate is selected

7. Enter **-10** in the **Distance** edit box and select the **Join** option from the **Operation** drop-down list.

8. Choose the **OK** button from this dialog box.

   Next, you need to extrude the profile of the Crank.

9. Select the profile of the Crank to extrude, as shown in Figure 7-49, and invoke the **EXTRUDE** dialog box.

10. Enter **-10** in the **Distance** edit box and select the **New Component** option from the **Operation** drop-down list.

11. Retain rest of the settings and choose the **OK** button from the dialog box.

    Next, you need to extrude the profile of the Crank pin.

12. Select the profile of the pin of the Crank from the sketch, as shown in Figure 7-50, and then invoke the **EXTRUDE** dialog box from the **CREATE** panel.

13. Enter **5** in the **Distance** edit box and select the **Join** option from the **Operation** drop-down list.

14. Choose the **OK** button from this dialog box.

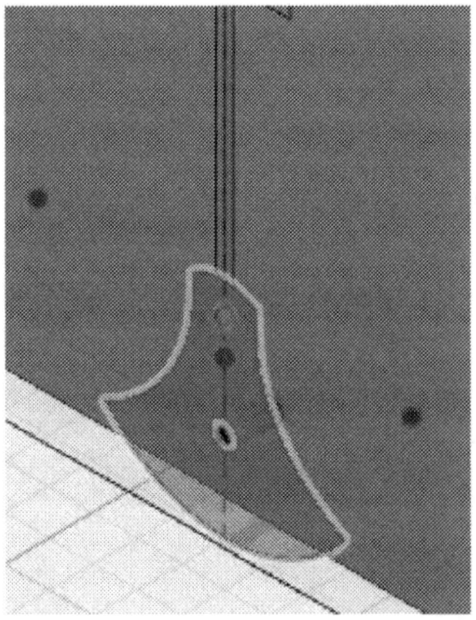

Figure 7-49 *Profile of the Crank selected*

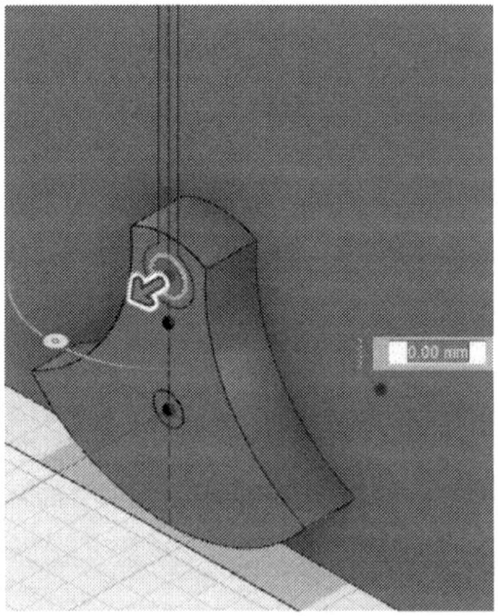

Figure 7-50 *Profile of the Pin of the Crank selected*

Next, you need to extrude the profile of the Piston Tank.

15. Select the profile of the Piston tank from the sketch, as shown in Figure 7-51, and invoke the **EXTRUDE** dialog box.

16. Enter **-10** in the **Distance** edit box; the profile gets extruded in the reverse direction.

17. Select the **New Component** option from the **Operation** drop-down list. Choose the **OK** button from the dialog box.

    Next, you need to extrude the pin of the Piston tank.

18. Select the profile of pin of the Piston tank, as shown in Figure 7-52.

19. Invoke the **EXTRUDE** dialog box and enter **5** mm in the **Distance** edit box. Select **Join** option from the **Operation** drop-down list.

20. Choose the **OK** button from this dialog box.

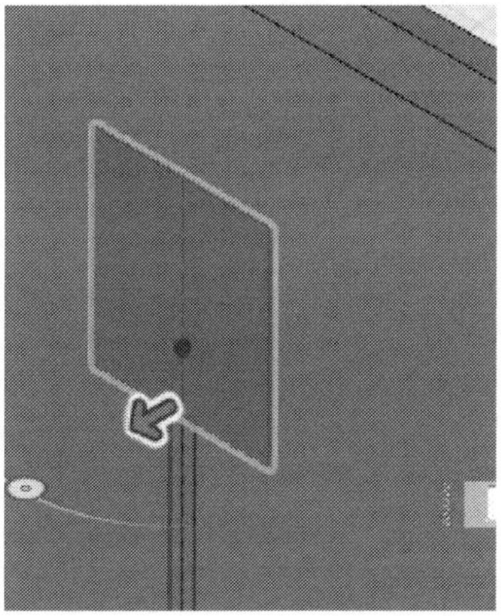

Figure 7-51 *Profile of the Piston Tank selected*

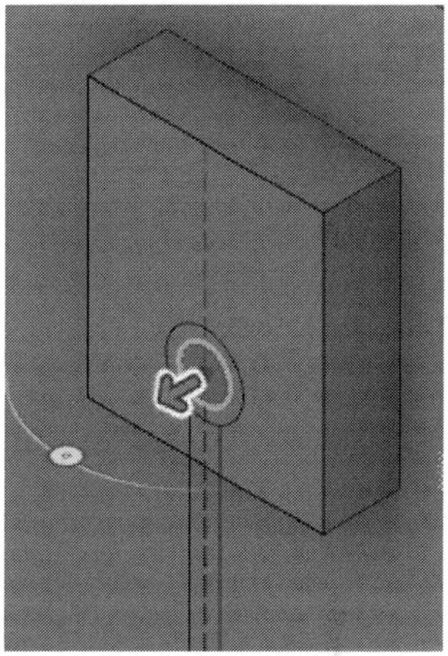

Figure 7-52 *Profile of the pin of the Piston Tank selected*

21. Similarly, select the profile of the Piston rod from sketch, as shown in Figure 7-53, and then choose the **Extrude** tool from the **CREATE** panel.

22. Enter **5** mm in the **Distance** edit box and select the **New Component** option from the **Operation** drop-down list.

23. Choose the **OK** button from the dialog box.

24. Next, you need to rename all the components as mentioned in the table. Figure 7-54 shows all the components for the Reciprocating Mechanism of Top-Down Assembly.

Component 1:1	Base Plate
Component 2:1	Crank
Component 3:1	Piston Tank
Component 4:1	Piston Rod

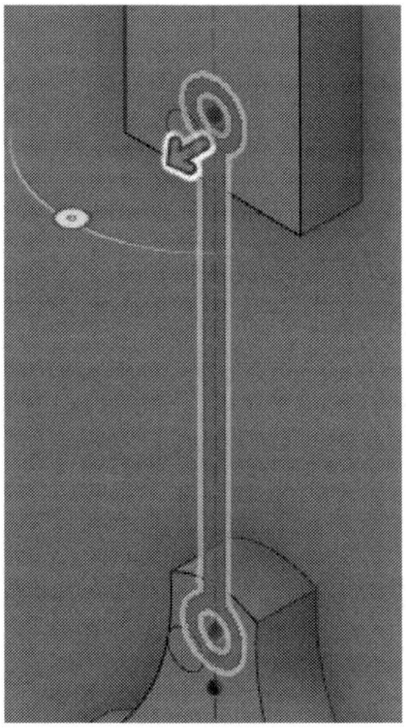

Figure 7-53 *Selecting profile of the Piston Rod*

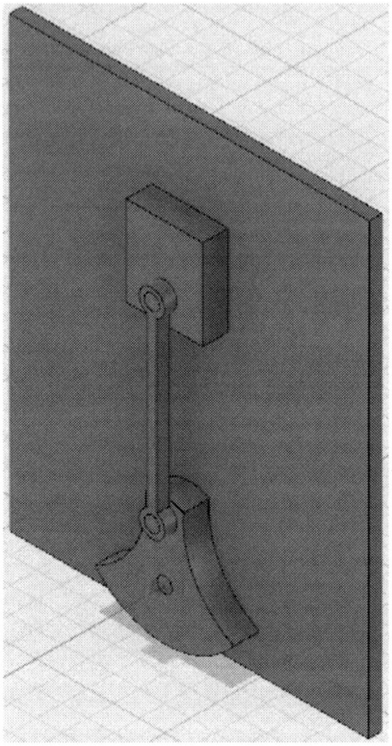

Figure 7-54 *Selecting location on the Crank*

Assembling the Base Plate and Piston Tank

Before assembling the components, you need to move the components to apply the joints easily.

1. Select the component one by one in the design and move them at a certain location, as shown in Figure 7-55.

2. Right-click on the Base Plate in the **BROWSER** bar; a shortcut menu is displayed.

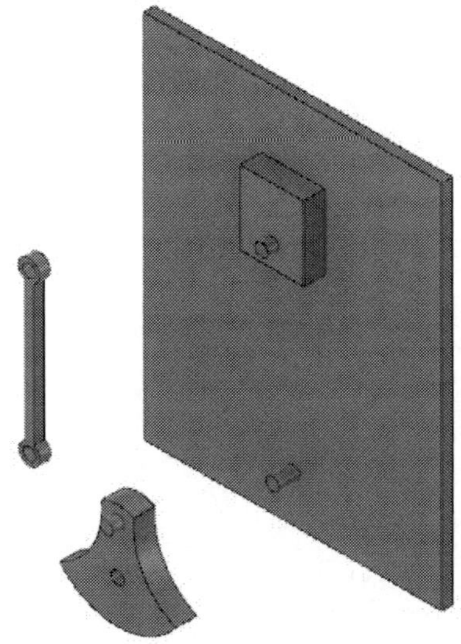

Figure 7-55 *Components of the Reciprocating Assembly being moved.*

3. Choose the **Ground** option from the shortcut menu; the selected component gets grounded in the design.

   By doing this, you have restricted all the DOF of the component.

4. Invoke the **JOINT** dialog box and select the **Slider** option from the **Type** drop-down list.
5. Select edges of the Base Plate and Piston Tank, as shown in Figure 7-56.
6. Also, enter **-55** mm in both the **Offset X** and **Offset Z** edit boxes under the **Alignment** area.
7. Make sure that **Z Axis** is selected in the **Axis** drop-down list.
8. Choose the **OK** button from the dialog box.

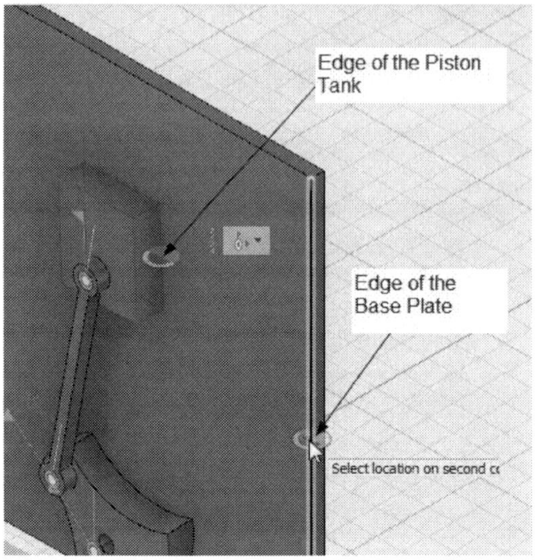

Figure 7-56 Selecting the edges of the Base Plate and Piston Tank

Assembling the Crank with the Piston Rod

1. Change the current view to the isometric view by choosing **Home** button from the graphics window.

2. Choose the **Joint** tool from the **ASSEMBLE** panel of the **MODEL** workspace; a message box is displayed.

3. Choose the **Capture Position** option from the message box; the **JOINT** dialog box is invoked with the **Component 1** option selected by default. Move the cursor in the assembly design; you are prompted to select the location on the component to move.

4. Next, select the location on the Piston rod, as shown in Figure 7-57. On doing so, the **Component 2** option gets selected and you are prompted to select the location on second component.

5. Select the pin of the Crank, as shown in Figure 7-58. Also, select **Revolute** joint from the **Type** drop-down list.

6. Choose the **OK** button from the dialog box.

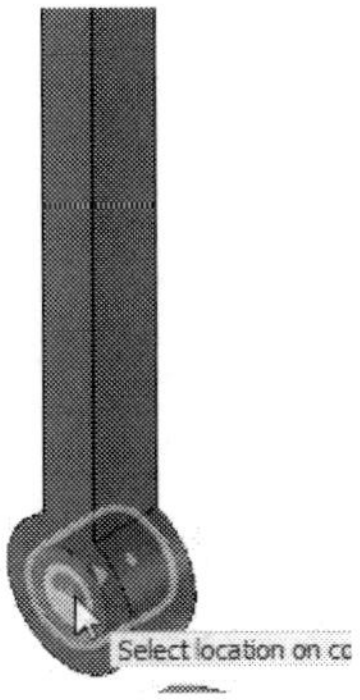

Figure 7-57 *Selecting location on the Piston Rod*

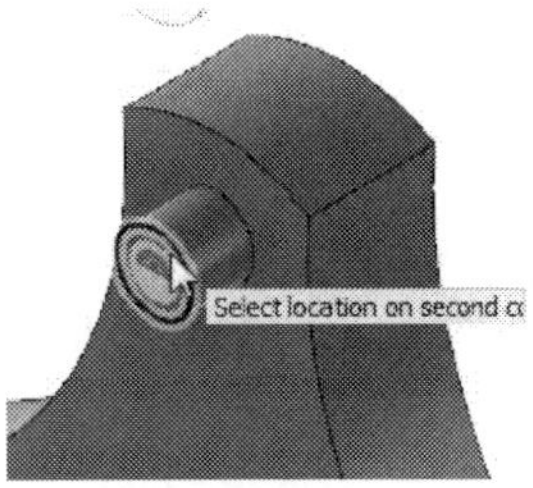

Figure 7-58 *Selecting location on the Crank*

Assembling the Piston Tank and Piston Rod

1. Invoke the **JOINT** dialog box and select the **Revolute** option from the **Type** drop-down list.

2. Select the pin of the Piston tank, as shown in Figure 7-59, and also select the edge of the hole of Piston rod, as shown in Figure 7-60.

3. Choose the **OK** button from the dialog box.

4. Similarly, assemble the Crank and the Base pin.

Figure 7-59 *Selecting pin of the Piston Tank*

Figure 7-60 *Selecting the edge of the Piston Rod*

Self-Evaluation Test

Answer the following questions and then compare them to those given at the end of this chapter:

1. To insert a component into a design, right-click on the component in the __________ and select __________ .

2. The __________ icon is displayed adjacent to the grounded component in the **BROWSER** bar.

3. Invoke the **Interference** tool from the __________ panel.

4. __________ joint has zero degree of freedom preventing relative motion between the components.

5. By default, a component placed in an assembly design is not __________.

6. You can identify any joint by hovering the cursor in the __________.

7. In Autodesk Fusion 360, you can create the bottom-up assembly as well as the top-down assembly. (T/F)

8. An assembly that uses a combination of top-down and bottom-up approaches is called a middle-out assembly. (T/F)

9. You can rotate individual components in the assembly file. (T/F)

10. You can edit the sketch of the component while assembling the components. (T/F)

Review Questions

Answer the following questions:

1. How many types of joints are available in Autodesk Fusion 360?

   (a) 4 (b) 5
   (c) 7 (d) 8

2. The __________ joint allows translational movement along the selected axis.

3. The __________ tool computes interference between selected bodies or components.

4. You can convert bodies into components by right -clicking on the body and choosing the **Create Components from bodies** option from the shortcut menu. (T/F)

5. Which of the following tools is used to rotate as well as translate components in an assembly design?

(a) **Rotate** (b) **Free move**
(c)**Translate** (d) None of these

EXERCISE

Exercise 1

Create the components of the Drill Press Vice assembly and then assemble them, as shown in Figure 7-61. The dimensions of the components are shown in Figures 7-62 through 7-66. Create a folder with the name *Drill Press Vice* in the Data Panel and save all the components in this folder. You can use the bottom-up approach for creating this assembly. **(Expected time: 3 hr 15 min)**

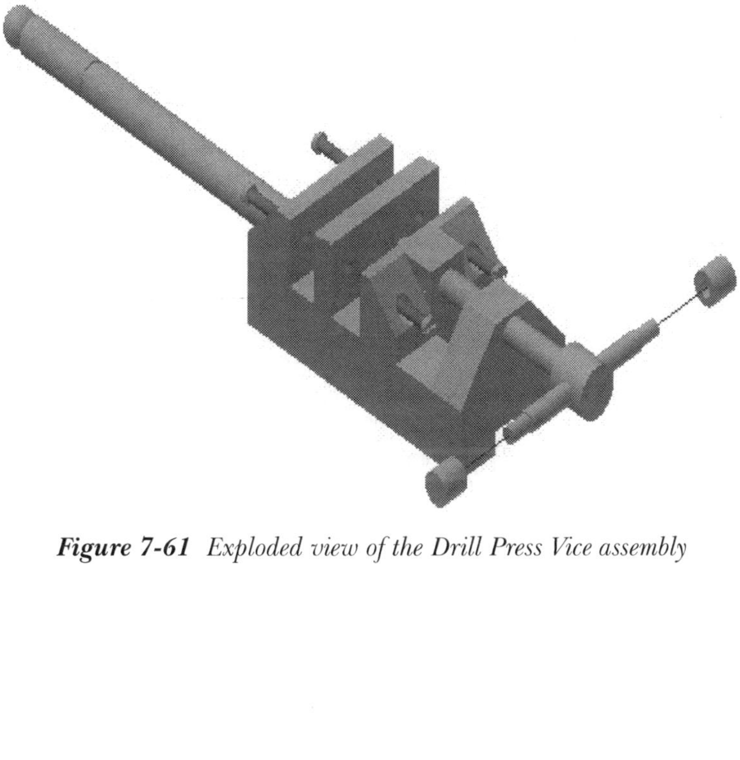

Figure 7-61 *Exploded view of the Drill Press Vice assembly*

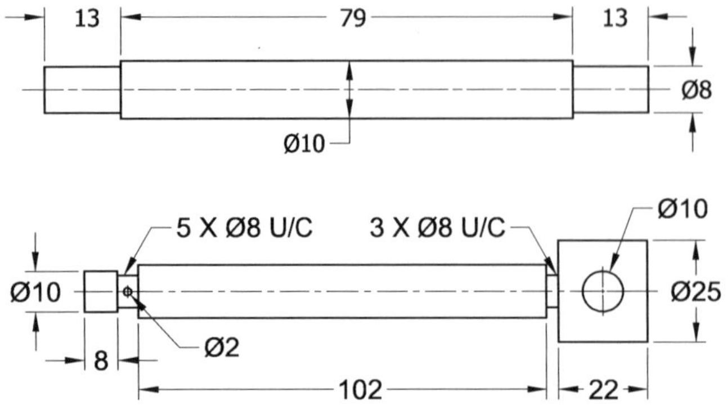

***Figure 7-62** Dimensions of the Clamp Screw Handle and Clamp Screw*

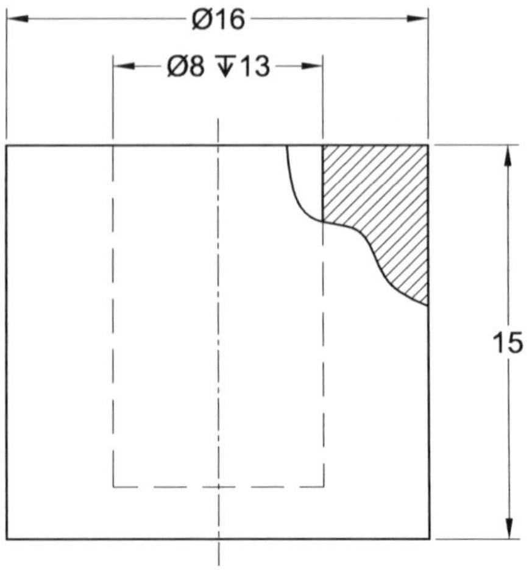

***Figure 7-63** Dimensions of the Handle Stop*

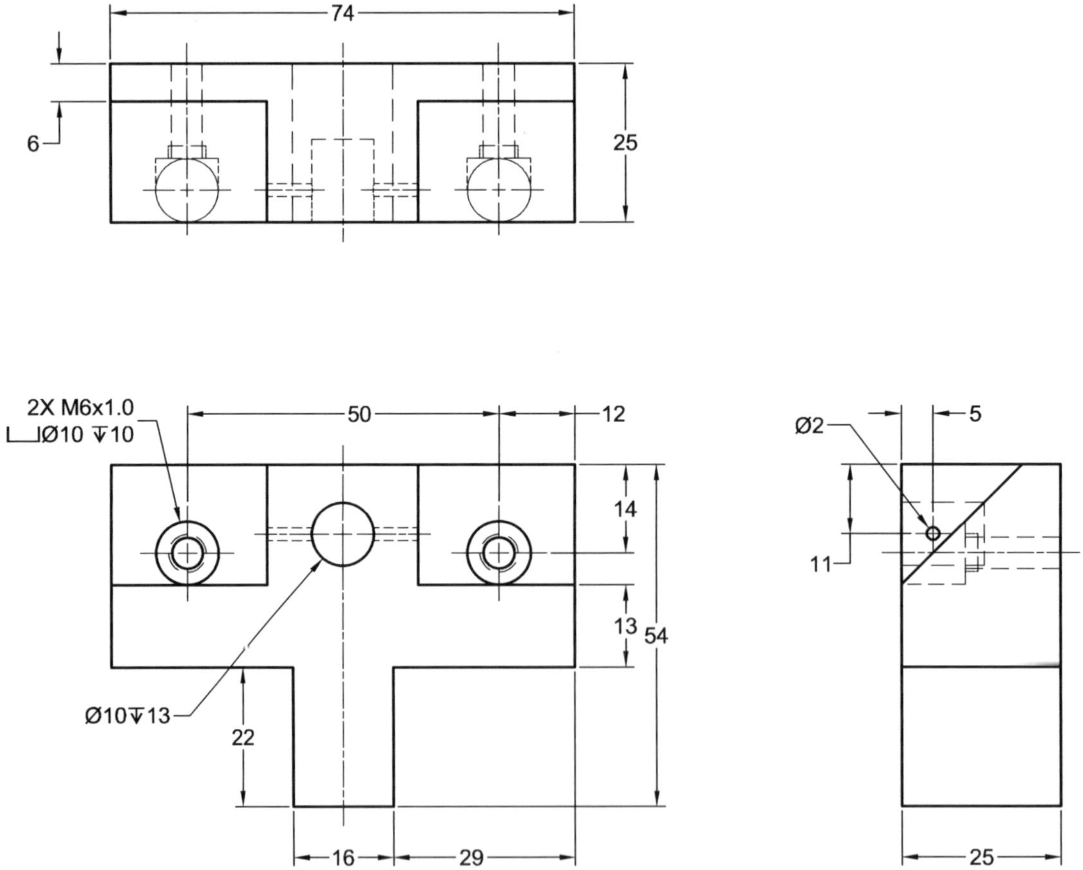

***Figure 7-64** Views and dimensions of the Movable Jaw*

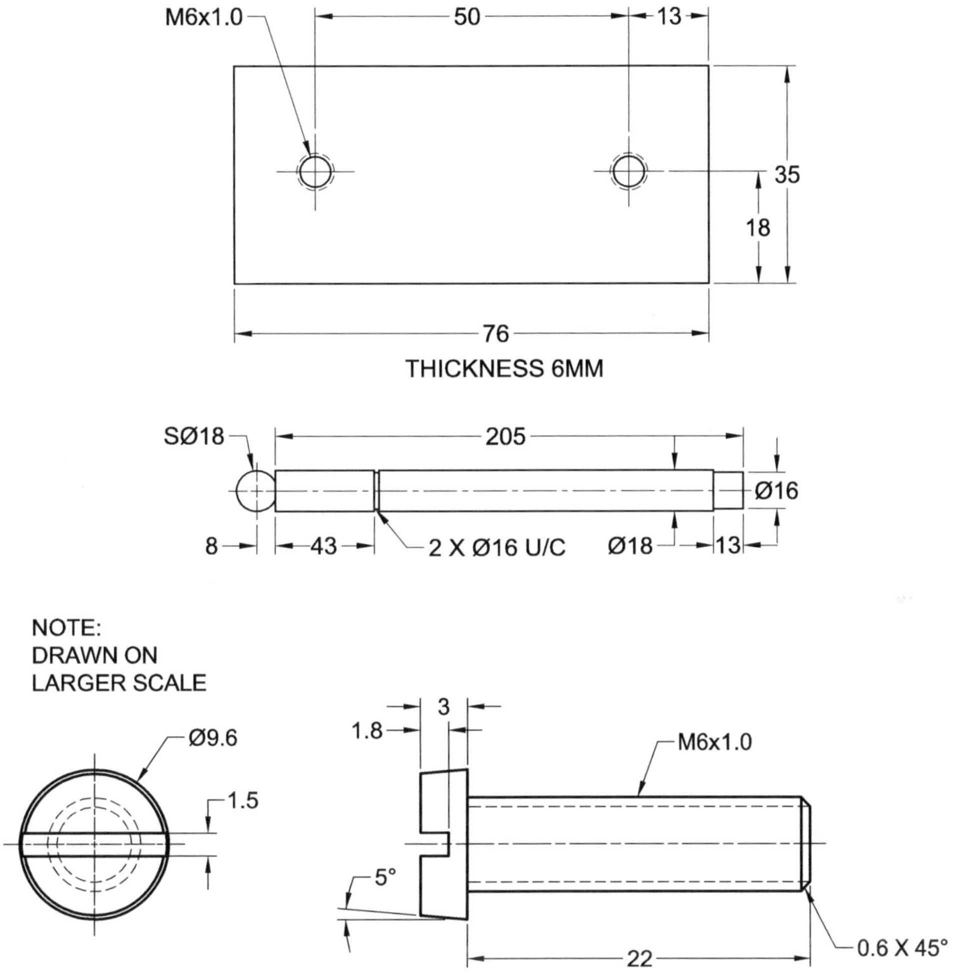

Figure 7-65 *Dimensions of the Jaw Face, Safety Handle, and Cap Screw*

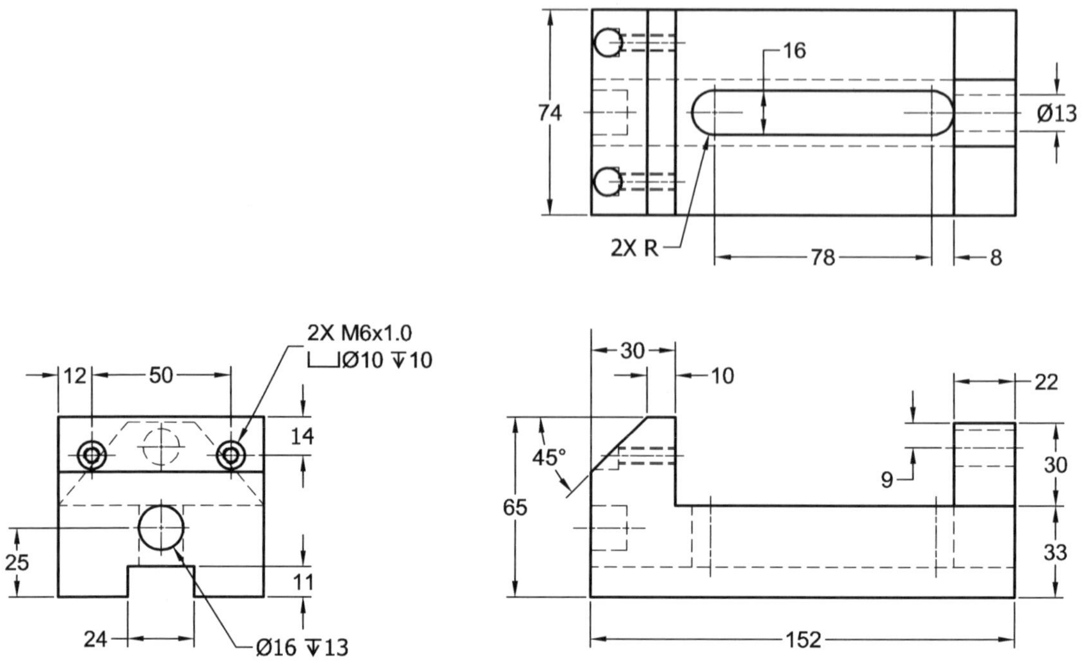

Figure 7-66 Views and dimensions of the Base

Answers to Self-Evaluation Test

1. Data Panel, **Insert into Current Design**, **2.** push pin, **3. INSPECT**, **4. Rigid**, **5.** grounded, **6.** Timeline, **7.** T, **8.** T, **9.** T, **10.** T

Chapter 8

Working with Drawing and Animation Workspace

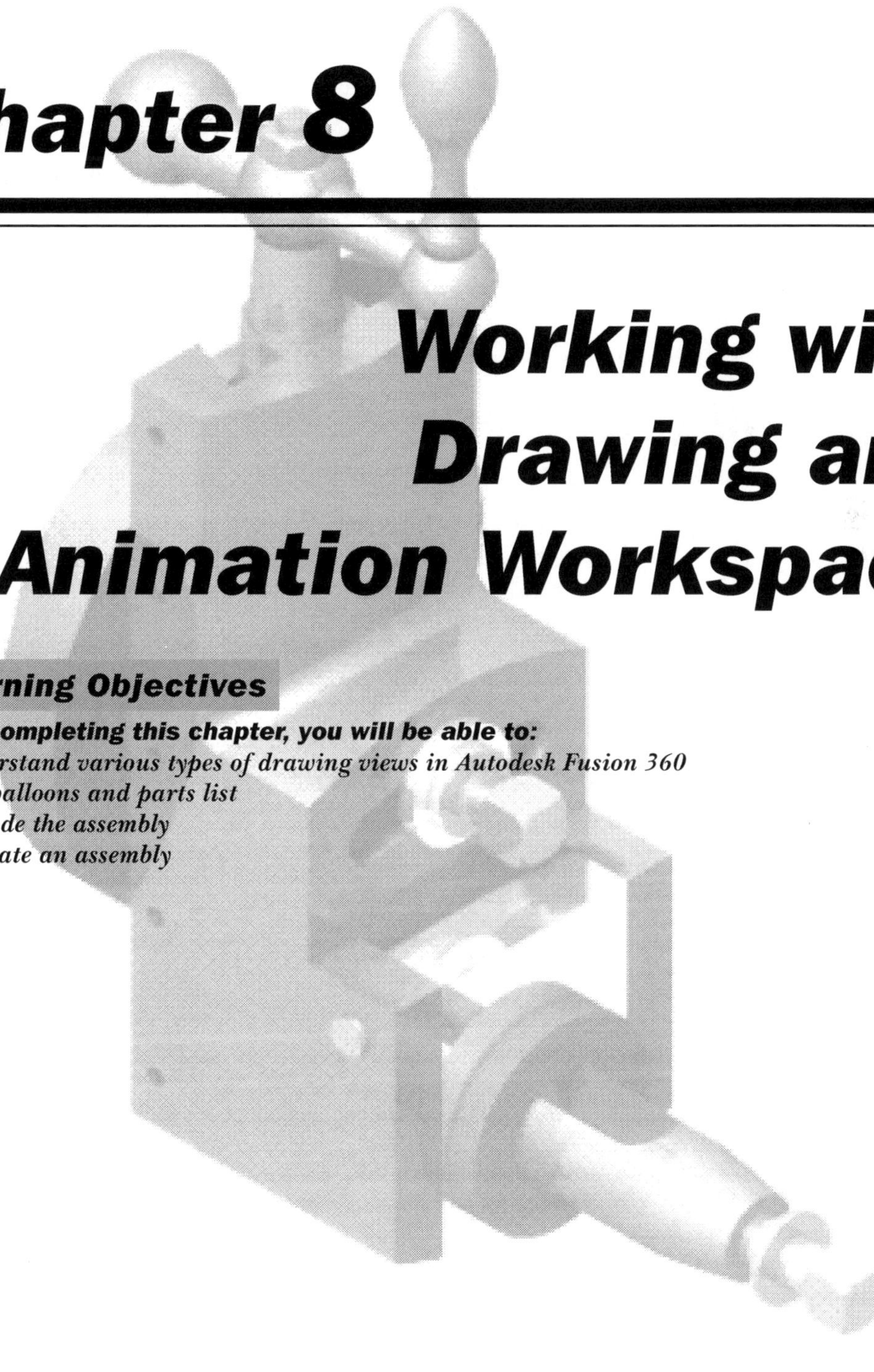

Learning Objectives

After completing this chapter, you will be able to:

- *Understand various types of drawing views in Autodesk Fusion 360*
- *Add balloons and parts list*
- *Explode the assembly*
- *Animate an assembly*

THE DRAWING WORKSPACE

After creating a solid model or an assembly, you need to generate its drawing views. Drawing views are two-dimensional (2D) representations of a solid model or assembly. Autodesk Fusion 360 provides you with a specialized workspace for generating drawing views. This specialized environment is called the Drawing workspace which has all the tools required to generate and modify the drawing views and add dimensions and annotations to them.

The drawing workspace of Autodesk Fusion 360 is associative to the part and assembly workspace. This property ensures that changes made in a part or an assembly are reflected in the drawing views. Autodesk Fusion 360 has predefined drafting standards such as the ISO standard and ASME standard. You can use the required standard file and proceed for generating drawing views.

In Autodesk Fusion 360, 3D modeling tools are available in most of the workspaces, but when you switch to the drawing workspace, the toolbar and marking menu will be updated with new tools.

THE ANIMATION WORKSPACE

You can also animate the assemblies created in the assembly design. The animation of assemblies can be created in the **ANIMATION** workspace. You can use the **ANIMATION** workspace to create exploded views of an assembly. An exploded view is the one in which the assembled components are moved to a defined distance from their original locations.

TUTORIALS

Tutorial 1

In this tutorial, you will generate the front, top, right, aligned section, detail, and isometric views of the model created in Tutorial 4 of Chapter 4. Use Standard A4 sheet format for generating these views. **(Expected time: 30 min)**

The following steps are required to complete this tutorial:

a. Open the model of Tutorial 4 of Chapter 4 and start a new drawing.
b. Select the standard A4 sheet format and generate the parent view.
c. Generate the projected views using the **Projected View** tool.
d. Generate the aligned section view using the **Aligned Section View** tool.
e. Generate the detail view.
f. Save and close the drawing document.

Opening the Model

1. Start Autodesk Fusion 360 and open the model for which you need to generate drawing view from the **Data Panel**.

Starting a New Drawing Document

To generate the drawing views, you need to start a new drawing document. There are two methods to start a drawing document in Autodesk Fusion 360. The first method is to start a drawing document from the **File** menu and the second method is by choosing the **MODEL** Workspace.

1. Choose **Drawing> From Design** from the **MODEL** Workspace, refer to Figure 8-1; the **CREATE DRAWING** dialog box is displayed, as shown in Figure 8-2.

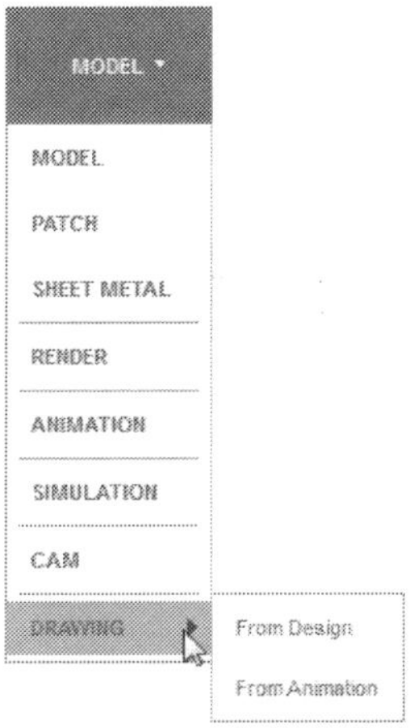

*Figure 8-1 The **From Design** option being chosen from the **MODEL** Workspace*

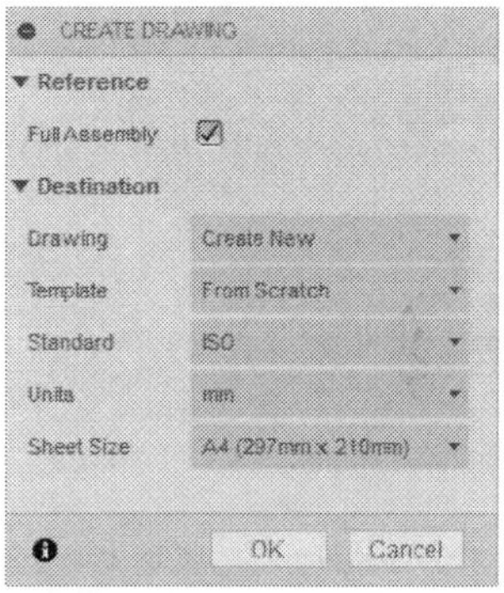

*Figure 8-2 The **CREATE DRAWING** dialog box*

The **CREATE DRAWING** dialog box is used to select the desired format and size of the sheet. You can select the predefined templates from the **Templates** drop-down list. Also, you can select the size of the sheet as per your requirement from the **Sheet Size** drop-down list, drawing standard from the **Standard** drop-down list, and unit from the **Units** drop-down list. In this dialog box, you can select the **Full Assembly** check box to show the views of the entire assembly or you can clear the check box to show the views of the selected components of the assembly.

Tip
*As discussed earlier, you can also start a drawing document by using the **File** menu. To do so, choose the **New Drawing>From Design** from the **File** menu.*

2. Select the **ISO** option from the **Standard** drop-down list in the **CREATE DRAWING** dialog box, if not selected by default.

3. Select **A4(297mm X 210mm)** from the **Sheet Size** drop-down list and choose the **OK** button; the **DRAWING** workspace is invoked. Also, the **DRAWING VIEW** dialog box is displayed, as shown in Figure 8-3.

   The **Orientation** drop-down list is used to select the desired orientation of any component or assembly. You can also select the view style from the buttons available adjacent to is **Style** option and specify the scale of the view in the **Scale** edit box.

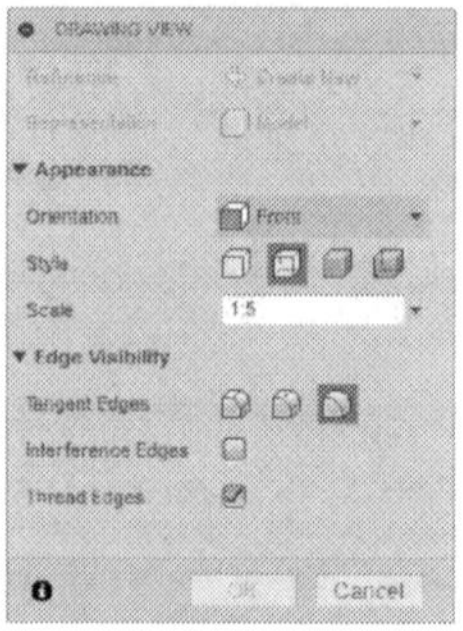

Figure 8-3 *The* ***DRAWING VIEW*** *dialog box*

   The buttons adjacent to the **Tangent Edges** option are used to shorten the tangent display, turn it off, or set it to full length. Select the **Interference Edges** check box to show the interference edges of the selected view. Similarly, select the **Thread Edges** check box to show the edges of the thread of the selected view.

4. Select the **Preference** option from the Account Settings and Preference drop-down list; the **Preference** dialog box is displayed.

5. Select the **Drawing** option from the **General** node; respective options are displayed on the right, as shown in Figure 8-4.

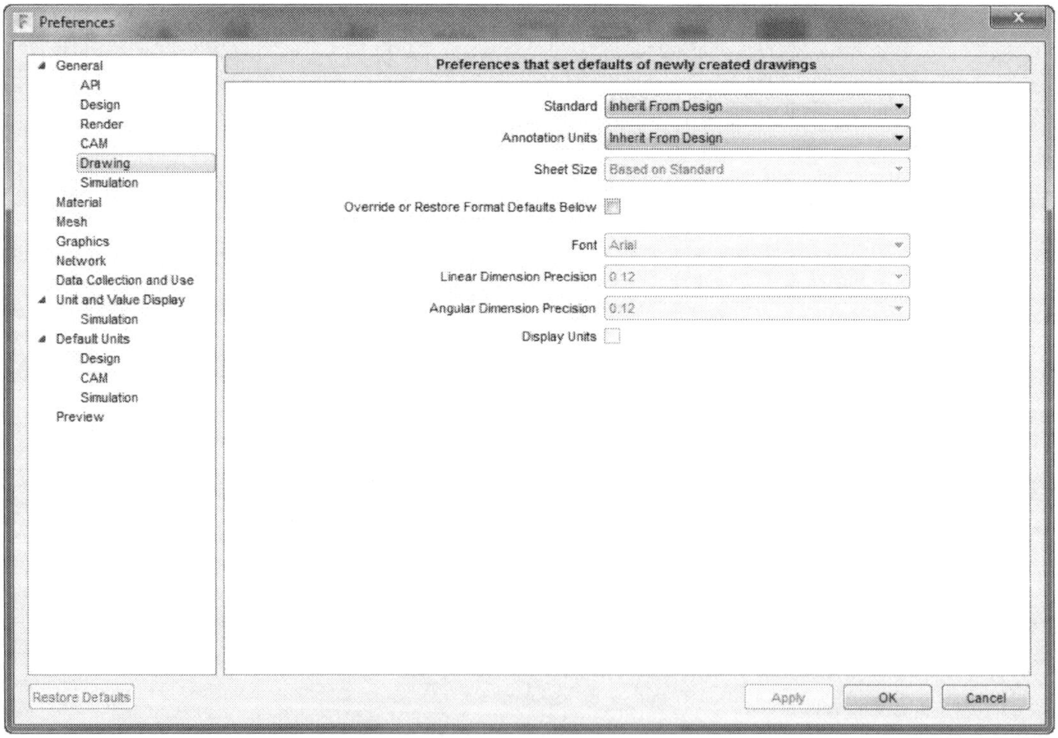

Figure 8-4 *The* ***Preference*** *dialog box*

6. Select **ISO** from the **Standard** drop-down list and **mm** from the **Annotation Units** drop-down list.

7. Select the **Override or Restore Format Defaults Below** check box; the **Project Angle** drop-down list gets activated.

8. Select the **Third Angle** option from the **Project Angle** drop-down list.

9. Choose the **Apply** button from this dialog box.

10. Choose the **OK** button to close the dialog box.

Generating the Parent and Projected Views

1. Select the **Front** option from the **Orientation** drop-down list and the **Visible and Hidden Edges** button adjacent to the **Style** option from the **DRAWING VIEW** dialog box.

2. Set the scale of the drawing to **1:4** ratio in the **Scale** edit box.

3. Move the cursor to the middle-left of the drawing sheet and then click on this location to place it, refer to Figure 8-5.

4. Retain rest of the values in the dialog box and then choose the **OK** button; the front view is generated and placed at this location.

5. Choose the **Projected View** tool from the **DRAWING VIEWS** panel; you are prompted to select the parent view to be projected.

6. Select the parent view from the drawing sheet in the graphics window and move the cursor at the top of the selected view and specify a point to place the top view, refer to Figure 8-5. The top view of the model is generated and a preview of another projected view with front view as the parent view is attached to the cursor.

7. Similarly, move the cursor horizontally toward the right and then move it upward; preview of the isometric view is displayed. Specify a point to place the isometric view.

8. Right-click in the drawing sheet; a marking menu is displayed.

9. Choose the **OK** button from the marking menu.

   The current location of the isometric view of the model is such that it may interfere with the aligned section view that you need to place next. Therefore, you need to move the isometric view close to the top right corner of the drawing sheet.

10. Click on the isometric view; a square symbol is displayed in the middle of the isometric view.

11. Click on the square symbol and drag the view close to the upper right corner of the drawing sheet. The drawing sheet after generating and moving the isometric view is shown in Figure 8-5.

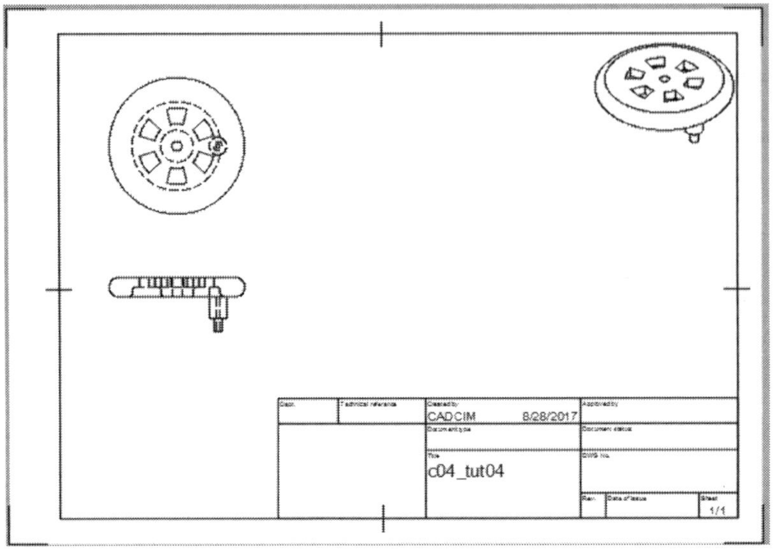

Figure 8-5 *Drawing sheet after generating the front, top, and isometric views*

Note

*In Figure 8-5, the display style of the front and top views is Visible and Hidden Edges whereas the display style of isometric view is Visible Edges. To change the display style of a view, double-click on the desired view; the **Drawing View** dialog box will be displayed. Next, select the required display style adjacent to the **Style** option.*

Generating the Section View

Next, you need to generate the aligned section view using the **Section View** tool.

1. Choose the **Section View** tool from the **DRAWING VIEWS** panel; you are prompted to select the parent view.

2. Select the top projected view as the parent view; you are prompted to specify the start point of the section line and the **DRAWING VIEW** dialog box is displayed.

3. Move the cursor over the top view; the cursor snaps to quadrant, as shown in Figure 8-6. Click to specify the start point of the section line; you are prompted to select the point as second point.

4. Move the cursor vertically downward to specify the second point of the section line, as shown in Figure 8-7. The cursor snaps to quadrant. Now, click to specify the second point of the section line. As soon as you specify the second point, the reference line of the section line will be created.

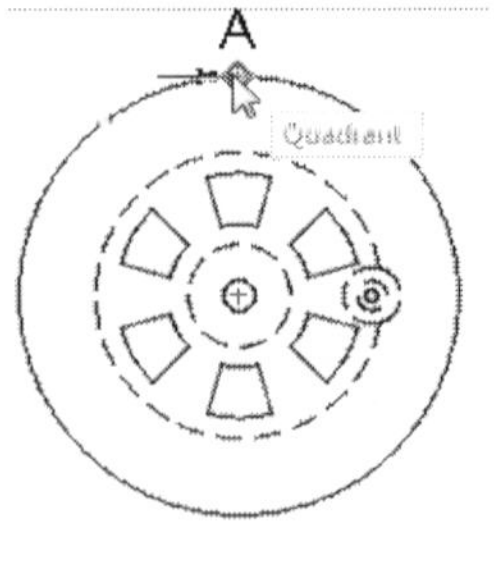

Figure 8-6 *Cursor snapped to a quadrant*

5. Right-click and then choose **Continue** from the Marking menu; a preview of the section view attached to the cursor appears on the sheet. Also, you are prompted to specify the location for the section view.

   Note that hatching lines will not be displayed in the preview of the section view.

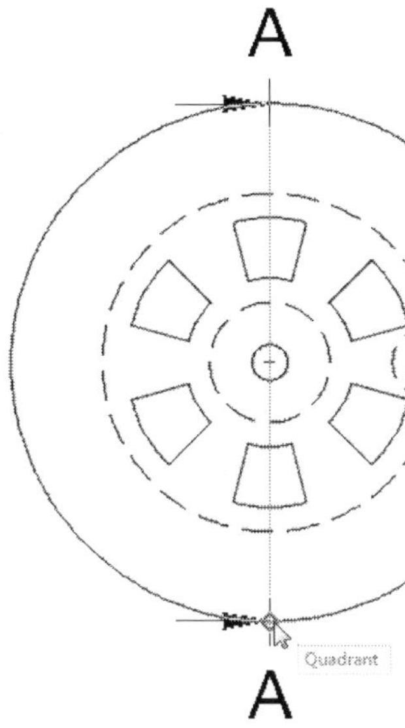

Figure 8-7 *Specifying second point of the section line*

6. Click to specify the location of the section view on the right of the top view, as shown in Figure 8-8. Next, press the ENTER key.

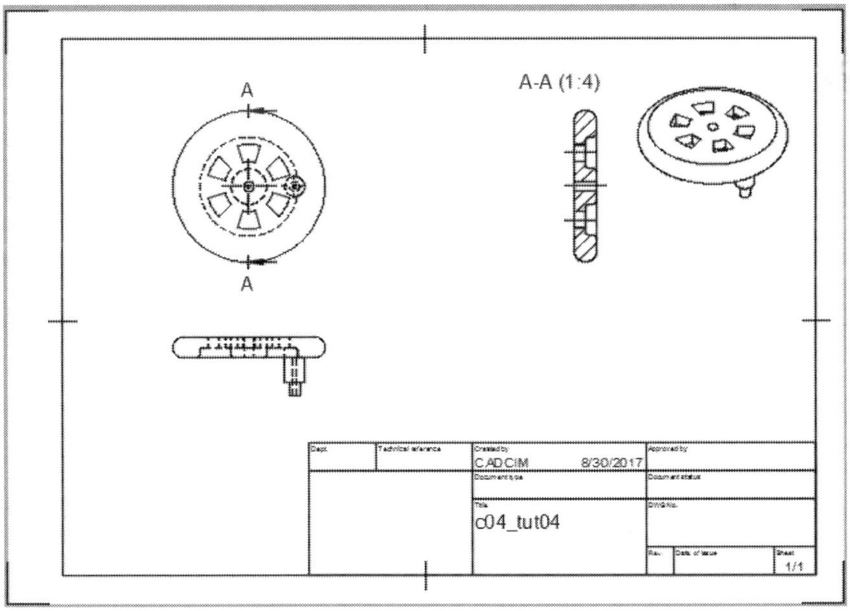

Figure 8-8 *Drawing sheet after generating the section view*

Modifying the Hatch Pattern of the Aligned Section View

1. Double-click on the hatch pattern of the section view; the **HATCH** dialog is displayed, as shown in Figure 8-9.

2. Select the desired hatch pattern from the **Pattern** drop-down list.

   The **Pattern** drop-down list is used to apply the standard hatch patterns to the section view.

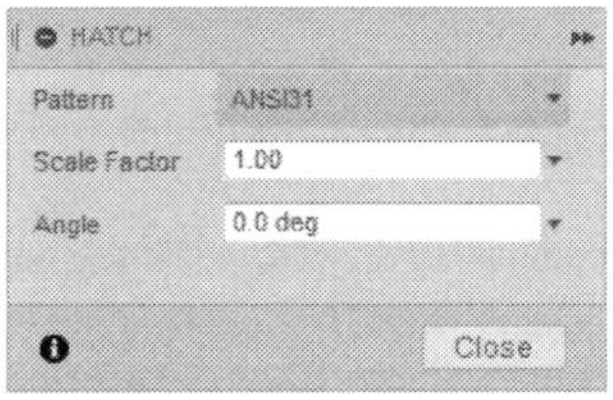

***Figure 8-9** The **HATCH** dialog box*

3. Enter the value **0.5** in the **Scale Factor** edit box; the gap between the hatching lines is changed accordingly.

   The **Scale Factor** edit box is used to specify the scale factor of the standard hatch pattern selected from the **Pattern** drop-down list. When you change the scale factor using this edit box, the preview updates dynamically.

4. Enter **5** in the **Angle** edit box. Choose the **Close** button from the dialog box.

Generating the Detail View

Next, you need to generate the detail view of the circular feature of the model.

1. Choose the **Detail View** tool from the **DRAWING VIEWS** panel; you are prompted to select the parent view.

2. Select the projected top view as the parent view; the **DRAWING VIEWS** dialog box is displayed and you are prompted to specify the center point of the detail boundary

3. Draw a circular boundary in the top view to create a detail view, as shown Figure 8-10. As you draw the circle, the detail view is attached to the cursor and you are prompted to place the detail view.

4. Place the attached detail view on the right side of the drawing sheet above the title block, refer to Figure 8-11.

5. Enter **1:2** in the **Scale** edit box and choose the **OK** button from the **DRAWING VIEW** dialog box.

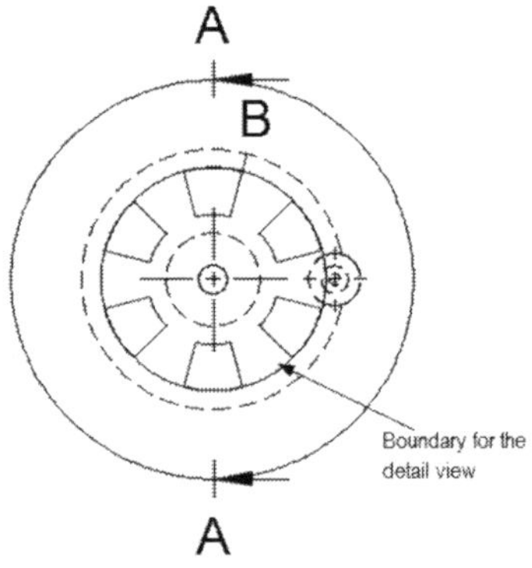

***Figure 8-10** Boundary for the detail view*

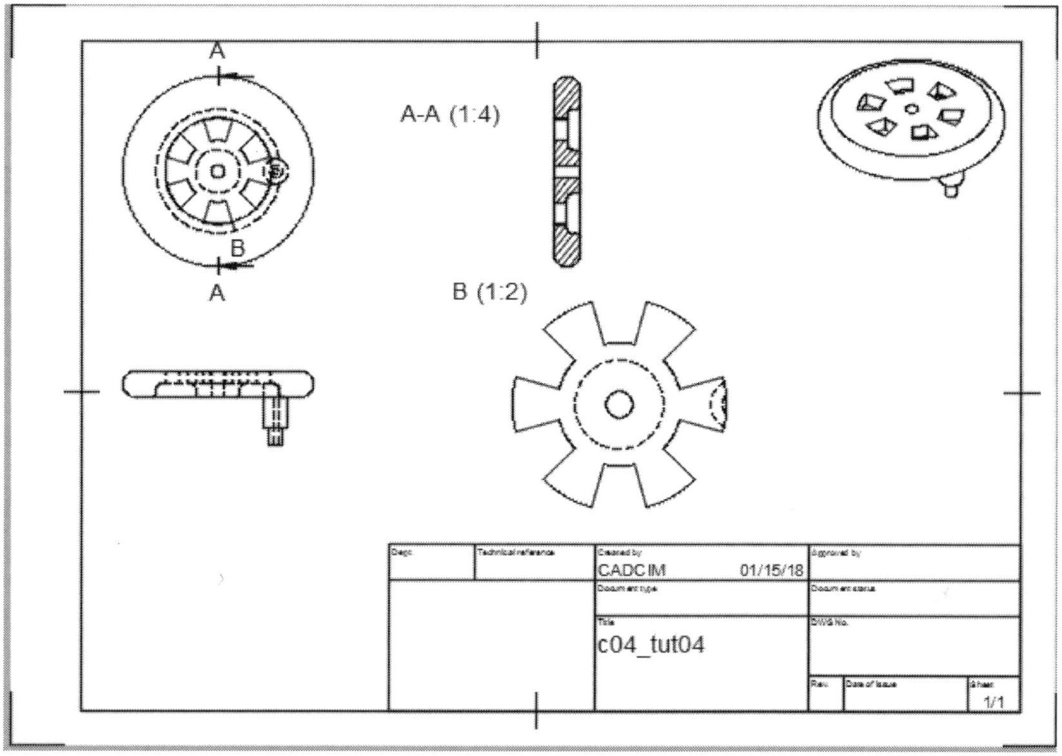

Figure 8-11 *The detail view on the right side of the drawing sheet*

Saving the Drawing

Next, you need to save the drawing file.

1. Choose the **Save** option from the **File** menu; the **Save** dialog box is displayed.
2. Enter **c08_Tut_01** in the **Name** edit box.
3. Click on the down arrow on the extreme right side of the **Location** selection box; the **Save** dialog box gets expanded. 
4. Select **CADCIM** from the **PROJECT** column. Next, choose the **Save** button to save the drawing and exit the **Save** dialog box.

Tutorial 2

In this tutorial, you will generate the drawing view of the Bench Vise assembly created in Tutorial 1 of Chapter 7. You will generate top view, sectioned front view, and isometric view of the assembly. You will also generate the Bill of Materials (BOM) of the assembly and then add balloons to the isometric view. **(Expected time: 45 min)**

The following steps are required to complete this tutorial:

a. Copy the Bench Vice assembly and its components from *c07* folder to *c08* folder in the **Data Panel**.
b. Start a new drawing document from the assembly document using A3 landscape sheet format and generate the top view.
c. Generate the section view using the **Section View** tool.
d. Generate the isometric view.
e. Add balloons to the isometric view.
f. Generate Part List.
g. Save and close the drawing and assembly documents.

Copying the Bench Vice Assembly

As you do not want to modify the assembly created in *c07* folder, you need to copy all components from the folder of the Bench Vice assembly to the *c08* folder. After copying the components, you will open the assembly file and check components for interference. Note that there should be no interference between the components.

1. Start Autodesk Fusion 360 and open the Bench Vise assembly from the **Data Panel**.

2. Select all the components including assembly and then right-click on it.

3. Choose the **Copy** option from the shortcut menu; the **Copy to** dialog box is displayed.

4. Select the **CADCIM** project and create a new folder.

5. Rename this folder as C08 and choose the **Copy** button from the **Copy to** dialog box.

Starting a New Drawing Document

1. Choose **DRAWING>From Design** in the **MODEL** workspace, refer to Figure 8-1; the **CREATE DRAWING** dialog box is displayed, refer to Figure 8-2.

2. Select the **ISO** option from the **Standard** drop-down list in the **CREATE DRAWING** dialog box if not selected by default.

3. Select **A3(420mm X 297mm)** from the **Sheet Size** drop-down list and choose the **OK** button; the **DRAWING** workspace is invoked and a preview of base view gets attached to the cursor. Also, the **DRAWING VIEW** dialog box is displayed.

Generating the Top View

After starting a new drawing document with the standard A3 sheet size, you need to generate the top view of the assembly.

1. Select the **Top** option from the **Orientation** drop-down list under the **Appearance** node in the **DRAWING VIEW** dialog box.

> **Tip**
> *You can change the orientation of the model view even after placing it. To do so, double-click on the required view; the* ***DRAWING VIEW*** *dialog box will be displayed. Select the required view from the* ***Orientation*** *drop-down list; the orientation of the view will be automatically modified.*

2. Set the scale of the drawing to **1:2** ratio in the **Scale** edit box.

3. Choose the **Visible Edges** button adjacent to the **Style** option.

> **Note**
> *If you make changes in any saved file and open a new drawing file without saving the first one, the* ***Add Version description*** *message box is displayed prompting you to save the current version of the updated file.*

4. Move the cursor to the top left of the drawing sheet and then click on this location to place the attached view, refer to Figure 8-12.

5. Retain rest of the default values in the dialog box and then choose the **OK** button; the top view is generated, as shown in Figure 8-12.

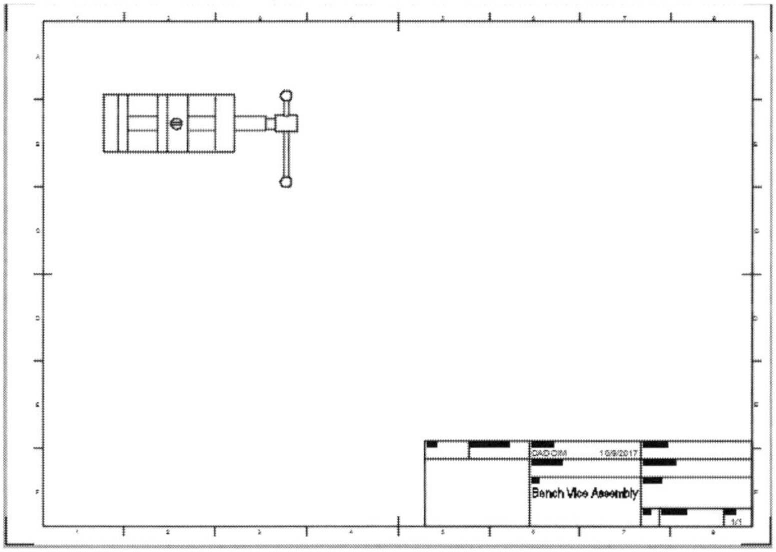

Figure 8-12 *Top view generated*

Creating the Sectioned View

In this section, you need to generate the sectioned front view from the top view.

1. Choose the **Section View** tool from the **DRAWING VIEWS** panel; you are prompted to select the parent view to be sectioned.

2. Select the top view as the parent view from the drawing sheet; you are prompted to specify the start point of the section line.

3. Move the cursor to the extreme left of the top view of the model and create a section line, refer to Figure 8-13.

4. Choose the **Continue** option from the Marking menu; you are prompted to place the sectioned view.

5. Move the sectioned view downward and then place it.

6. Choose the **OK** button from the Marking menu; the section view is generated, as shown in Figure 8-13.

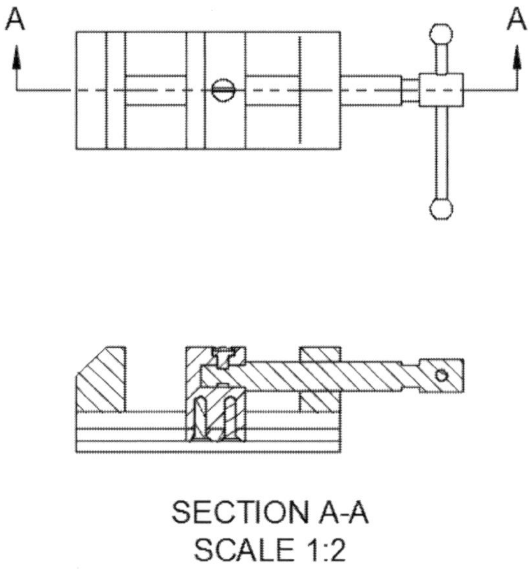

Figure 8-13 *Section view generated using the **Section View** tool*

Tip

While placing the section view, you will observe that the view is aligned to the direction of the arrows on the section line. To remove this alignment, press the SHIFT key and release it. After releasing the key, move the view to the desired location. Now, select a point in the drawing sheet to place the view.

Next, you need to exclude the components from the section cut.

1. Double-click on the sectioned view; the **DRAWING VIEW** dialog box is displayed.

2. Clear the **Screw Bar**, **Jaw Screw**, **Oval Fillister**, **Set Screw1**, and **Set Screw2** check boxes from the expanded **Objects To Cut** node in the **DRAWING VIEW** dialog box. As a result, these components do not get sectioned.

3. Choose the **Close** button from the dialog box to exit the dialog box.

Generating the Right Side View

Next, you need to generate the right side view from the sectioned front view using the **Projected View** tool.

1. Invoke the **Projected View** tool; you are prompted to select the parent view.

2. Select the front sectioned view; you are prompted to place the projected view or press ENTER.

3. Move the cursor to the right of the sectioned front view and place the view.

4. Choose the **OK** button from the Marking menu. The sheet after generating the projected view is shown in Figure 8-14.

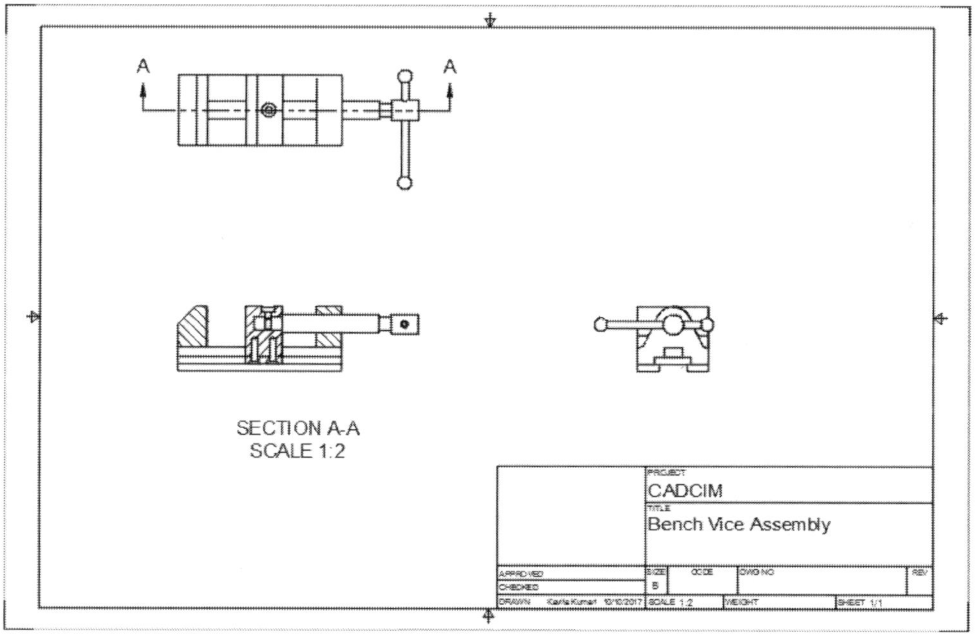

***Figure 8-14** Right side view generated using the **Projected View** tool*

Generating the Isometric View of the Sectioned View

In this section, you will generate the isometric view of the sectioned view.

1. Move the cursor over the sectioned view and right-click when dotted rectangle is displayed; a shortcut menu is displayed.

2. Choose **Drawing Views > Projected View** from the shortcut menu; you are prompted to select the parent view.

3. Select the sectioned view and move the cursor to the right of the sectioned view and then move it upward until the preview of the isometric view appears. Now, specify the location of the view.

4. Right-click on the view; a Marking menu is displayed. Choose **OK** from the Marking menu; the isometric view of the model is generated, as shown in Figure 8-15.

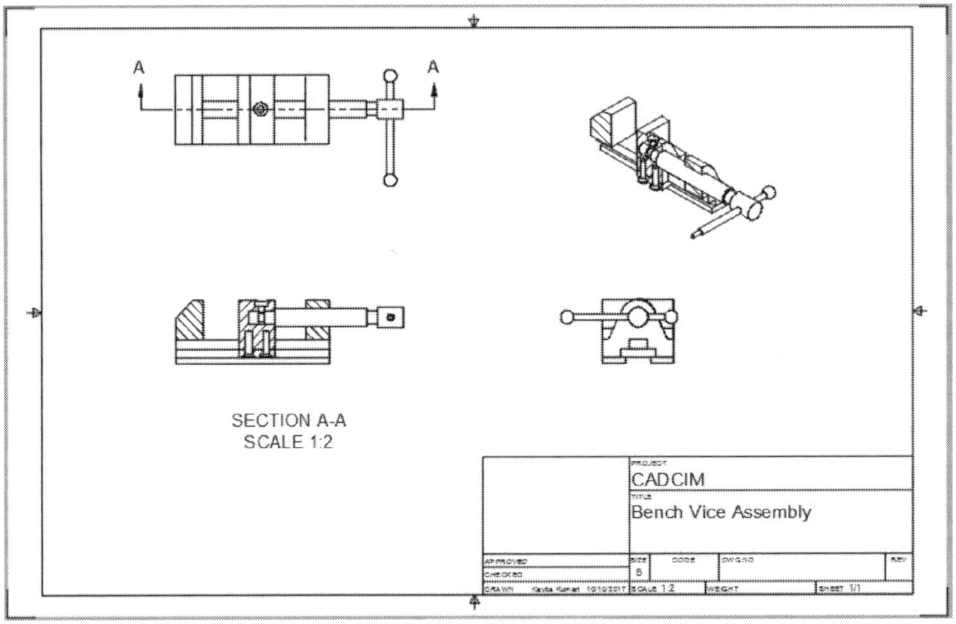

Figure 8-15 *Drawing sheet after generating the isometric view of the model*

Adding Balloons to the Components

The final step in this tutorial is to add balloons to the components in the isometric view. In this assembly, you will add balloons using the **Balloon** tool. Also, you need to drag balloons and place them at proper locations in the drawing sheet. But before generating balloons, you need to perform some modification.

1. Double-click on the sectioned isometric view; the **DRAWING VIEW** dialog box is displayed.

2. Clear the **Inherit Cut** check box and choose the **Close** button; the sectioned isometric view gets converted into full view.

3. Choose the **Balloon** tool from the **TABLES** panel; the **BALLOON** dialog box is displayed and you are prompted to select a component edge.

4. Select the **Standard** option from the **Type** drop-down list, if not selected by default. Next, move the cursor over one of the edges of the Base in the isometric view and click on it to add the balloon. Next, move the cursor away from the Base and place it below the component, refer to Figure 8-16.

5. Choose the **Close** button from the dialog box.

6. Similarly, add balloons to the remaining components, refer to Figure 8-15.

7. Drag balloons to a proper location in the drawing sheet.

   Note that for selecting the components that are not visible in the assembly, you need to change the current visibility by choosing the **Visible and Hidden Edges** button from the **DRAWING** dialog box.

Placing the Parts List

1. Choose the **Table** tool from the **TABLES** panel; the **TABLE** dialog box with parts list attached to the cursor is displayed in the graphics window. Also, you are prompted to place the Part List.

2. Place the **PARTS LIST** above the title block.

Note

If the parts list is placed in the upper half of the drawing space, the parts numbering is done from bottom to top. If the parts list is in the lower half of the drawing space, numbering is done from top to bottom.

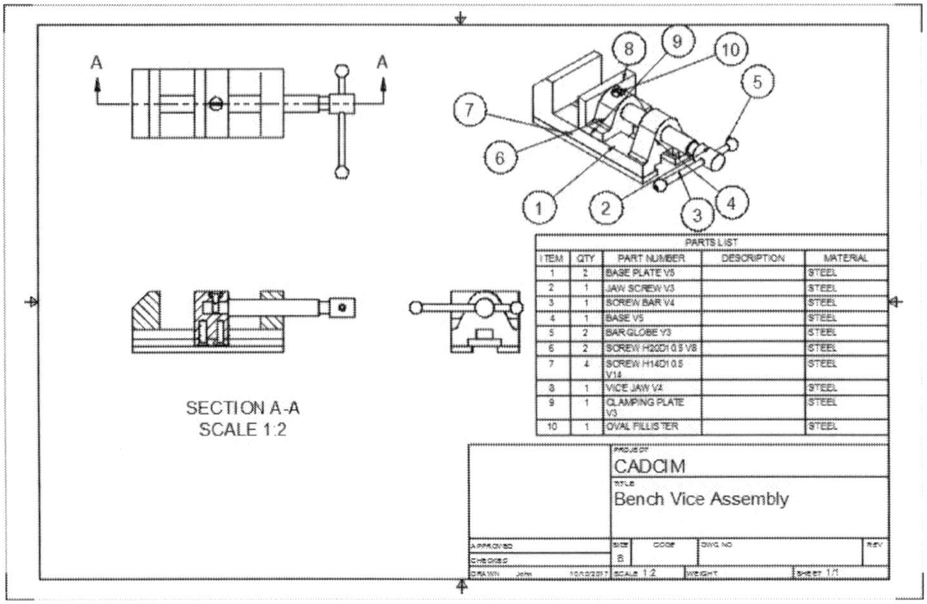

Figure 8-16 Drawing sheet after adding balloons and part list in the isometric view

Saving the Drawing

Next, you need to save the drawing file.

1. Choose the **Save** option from the **File** menu; the **Save** dialog box is displayed.
2. Enter **c08_Tut_02** in the **Name** edit box.
3. Click on the down arrow on the extreme right side of the **Location** selection box; the **Save** dialog box gets expanded. 
4. Select **CADCIM** from the **PROJECT** column and choose the **Save** button to save the drawing and exit the **Save** dialog box.

Tutorial 3

In this tutorial, you will first open the drawing created in Tutorial 1 of this chapter and then generate dimensions and add annotations to it. **(Expected time: 30 min)**

The following steps are required to complete this tutorial:

a. Open the drawing document created in Tutorial 1 of this chapter and save it with a different name.
b. Arrange the dimensions and delete the unwanted ones.
c. Add datum symbols and geometric tolerance to the drawing views.
d. Change the model display state of the drawing views.

Opening and Saving the Drawing Document

You need to open the drawing document created in Tutorial 1 of this chapter in the Autodesk Fusion 360 window and save it with a different name.

1. Open the *c08_tut01* document from the folder of the current chapter in the **Data Panel** of Auodesk Fusion 360.

2. Choose **File > Save As** from the Application bar and save the document with the name *c08_tut03* in the same folder. Figure 8-17 shows the drawing sheet in which you need to add dimensions and annotations.

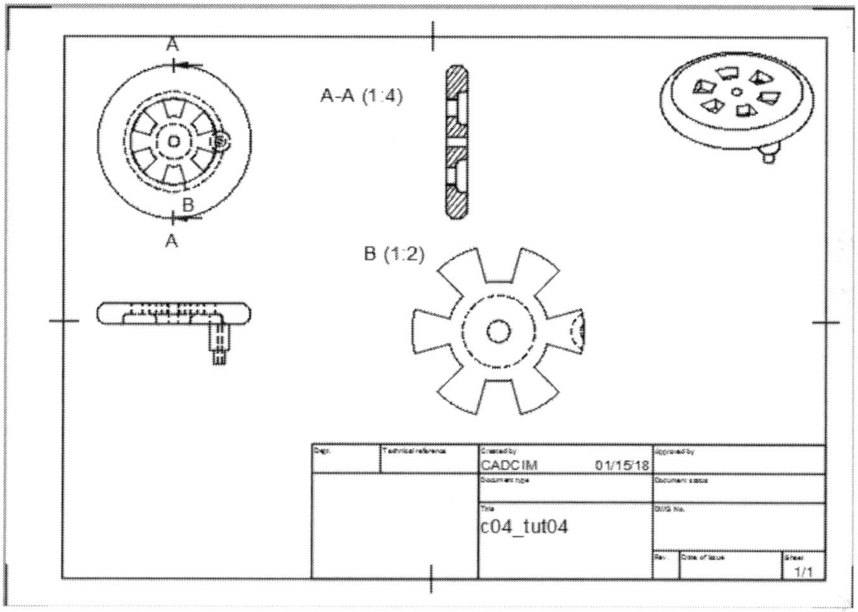

Figure 8-17 *Drawing sheet for adding dimensions and annotations*

Creating Centerlines and Center Marks

Before creating the dimensions of the model, you need to create centerlines and center mark.

1. Choose the **Center Mark** option from the **GEOMETRY** panel in the Ribbon; you are prompted to select the hole or rounded edge.

2. Select the hole, as shown in Figure 8-18. Again, you are prompted to select the hole or rounded edge.

3. Select the other hidden edge of the pin, refer to Figure 8-18 and press the ESC key to exit from the tool. Figure 8-19 shows the center mark added to the drawing view.

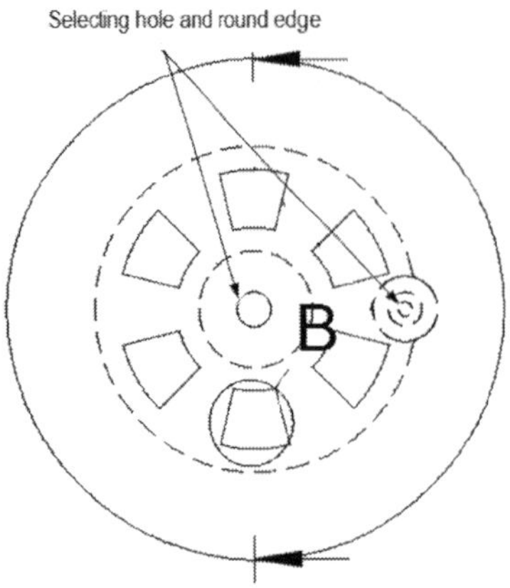

Figure 8-18 *Selecting hole and rounded edge*

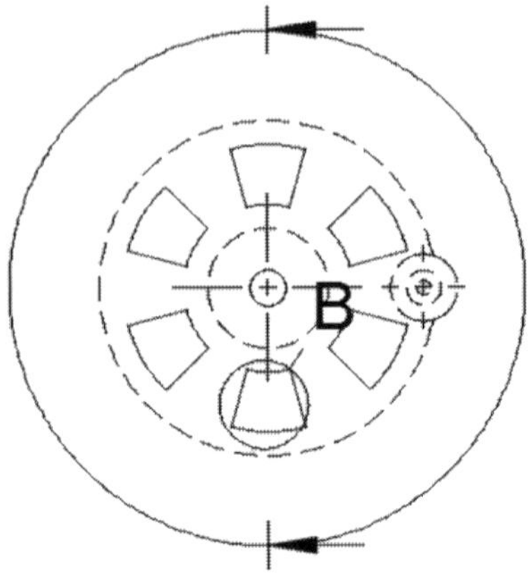

Figure 8-19 *Center mark added to the drawing view*

4. Choose the **Centerline** tool from the **GEOMETRY** panel in the Ribbon; you are prompted to select the edge, as shown in Figure 8-20.

5. Select the edge; you are prompted to select the second edge.

6. Select the second edge, refer to Figure 8-20; a centerline is created between the selected edges of the drawing, as shown in Figure 8-21.

7. Additionally, draw a centerline between the edges of the one of the cutout of the train wheel.

8. Press the ESC key to exit the tool.

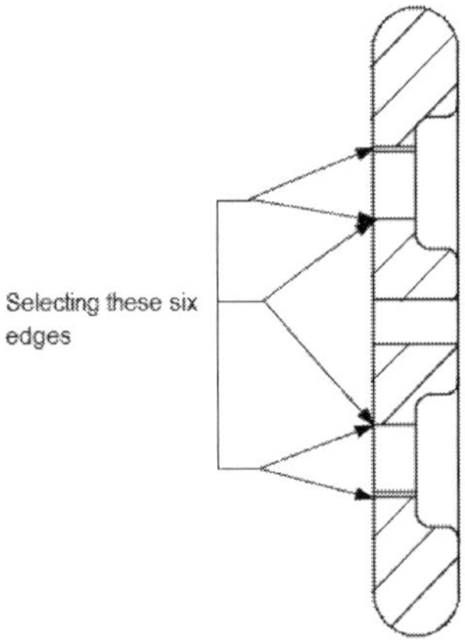

Figure 8-20 *Selecting the edges*

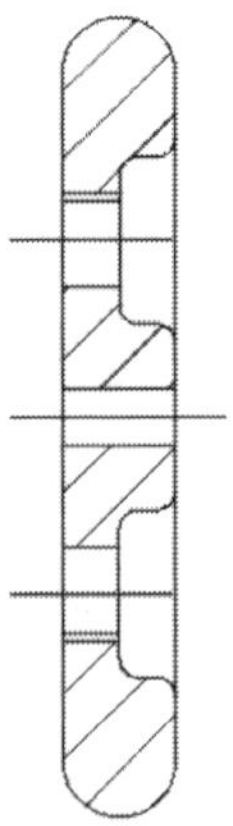

Figure 8-21 *The centerline is created*

Adding Dimensions to the Drawing Views

Next, you need to generate the dimensions using the Dimension tools.

1. Choose the **Diameter Dimension** tool from the **DIMENSION** panel in the Ribbon; you are prompted to select the edge.

2. Select the circular edge of the features and dimension them, as shown in Figure 8-22.

   Next, you need to dimension cut out of the train wheel with respect to the origin.

3. Choose **Angular Dimension** from the **DIMENSIONS** panel in the Ribbon; you are prompted to select the edge.

4. Select center line of the wheel and center line of the cutout of the wheel and place the dimension outside the view, as shown in Figure 8-23.

5. Similarly, apply other angular and radial dimensions, refer to Figure 8-23.

6. Similarly, apply the radial dimension to the cut-out of the wheel by using the **Radial Dimension** tool from the **DIMENSIONS** panel.

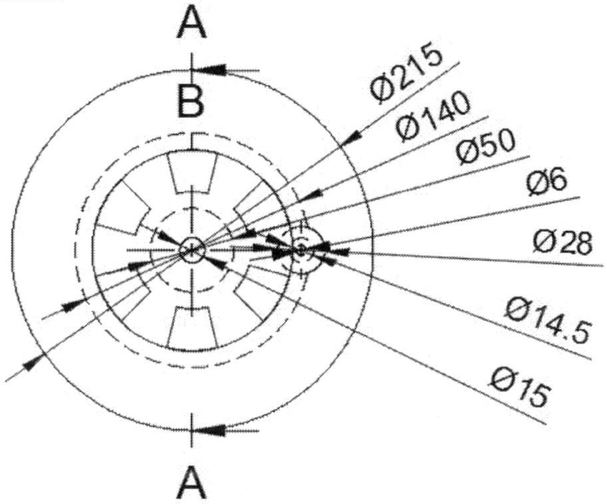

Figure 8-22 Circular edge selected for dimensioning

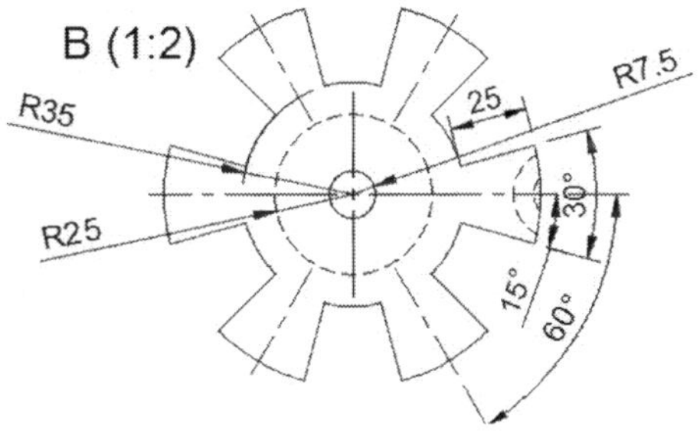

Figure 8-23 Dimension created using center line

Adding the Datum Identifier Symbol to the Drawing View

After generating the dimensions, you need to add the datum feature symbol to the drawing view. The datum feature symbols are used as datum reference for adding the geometric tolerance to the drawing views.

1. Choose **Datum Identifier** from the **SYMBOLS** panel in the Ribbon; you are prompted to select the object. 

2. Select the edge of the outer cylindrical feature from the top view; you are prompted to specify the point.

3. Specify the point at an appropriate location, refer to Figure 8-24.

4. Press ENTER key and choose the **OK** button from the **DATUM IDENTIFIER** dialog box.

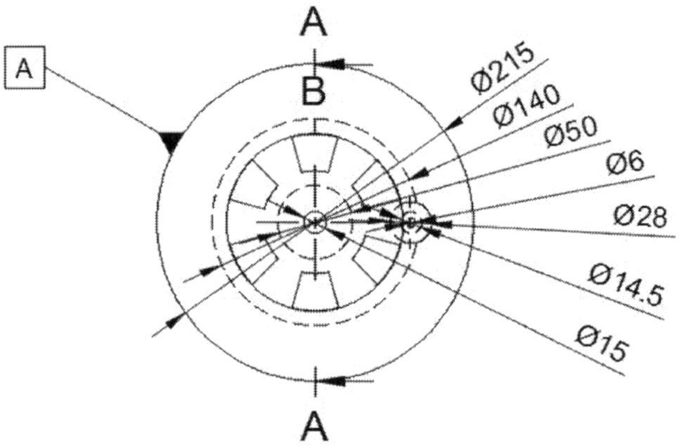

Figure 8-24 *Adding Datum Identifier Symbol*

Adding the Geometric Tolerance to the Drawing View

After defining the datum feature symbol, you need to add the geometric tolerance to the drawing view.

1. Choose the **Feature Control Frame** tool from the **SYMBOLS** panel in the Ribbon; you are prompted to select the object.

2. From the top view, select cylindrical edge having diameter of 140 mm; you are prompted to specify the start point.

3. Specify the point at an appropriate location, refer to Figure 8-23, and press the ENTER key; the **FEATURE CONTROL FRAME** dialog box is displayed.

   The **FEATURE CONTROL FRAME** dialog box is used to specify the parameters of the geometric tolerance.

4. Choose the **Concentricity** symbol under the **First Frame** node adjacent to the **Geometric Symbol** option in the **FEATURE CONTROL FRAME** dialog box.

5. Enter **0.01** in the **First Tolerance** edit box.

6. Enter **A** in the **First Datum** edit box to define the primary datum reference.

   Note that as you specify the parameters in the **FEATURE CONTROL FRAME** dialog box, the preview of the geometric tolerance will be modified dynamically in the drawing sheet.

7. Choose the **OK** button from the **FEATURE CONTROL FRAME** dialog box. You may need to move the geometric tolerance if it overlaps the dimensions. Figure 8-25 shows the drawing view after adding and rearranging the geometric tolerance.

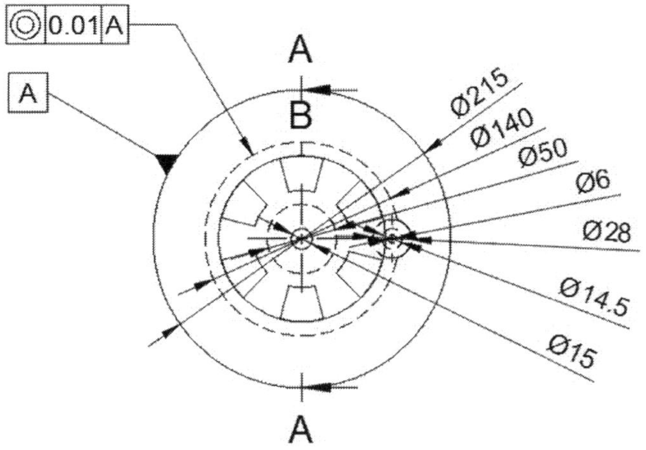

Figure 8-25 *The drawing view after adding and rearranging geometric tolerance*

Saving the Drawing

Next, you need to save the drawing file.

1. Choose the **Save** option from the **File** menu; the **Save** dialog box is displayed.

2. Enter **c08_Tut_03** in the **Name** edit box.

3. Click on the down arrow on the extreme right side of the **Location** selection box; the **Save** dialog box gets expanded. 

4. Select **CADCIM** from the **PROJECT** column and choose the **Save** button to save the drawing and exit the **Save** dialog box.

Tutorial 4

In this tutorial, you will animate the Bench Vice assembly. The animation consists of a linear explode. Next, save the animation in the *.avi* file with the name *Bench Vice assembly*.

(Expected time: 45 min)

The following steps are required to complete this tutorial:

a. Open the **Data Panel**.
b. Invoke the **ANIMATION** Workspace.
c. Create an Animating View.
d. Animate the assembly.

Opening Data Panel and Assembly

1. Click on the **Data Panel** in the Application menu; the **Data Panel** is displayed.

2. Double-click on the **CADCIM** folder and open the Bench Vise Assembly.

Invoking ANIMATION Workspace

1. Choose the **ANIMATION** Workspace from the **MODEL** panel; an **ANIMATION** Workspace is invoked with the Bench Vise Assembly, refer to Figure 8-26.

Creating the Animating View

1. Drag **Playhead** in the **ANIMATION TIMELINE** into the **Scratch Zone**.

   Scratch Zone is used to set the initial state of the model and camera without recording actions.

2. Choose the **Home** button from the ViewCube; the current view changes to isometric view, refer to Figure 8-26.

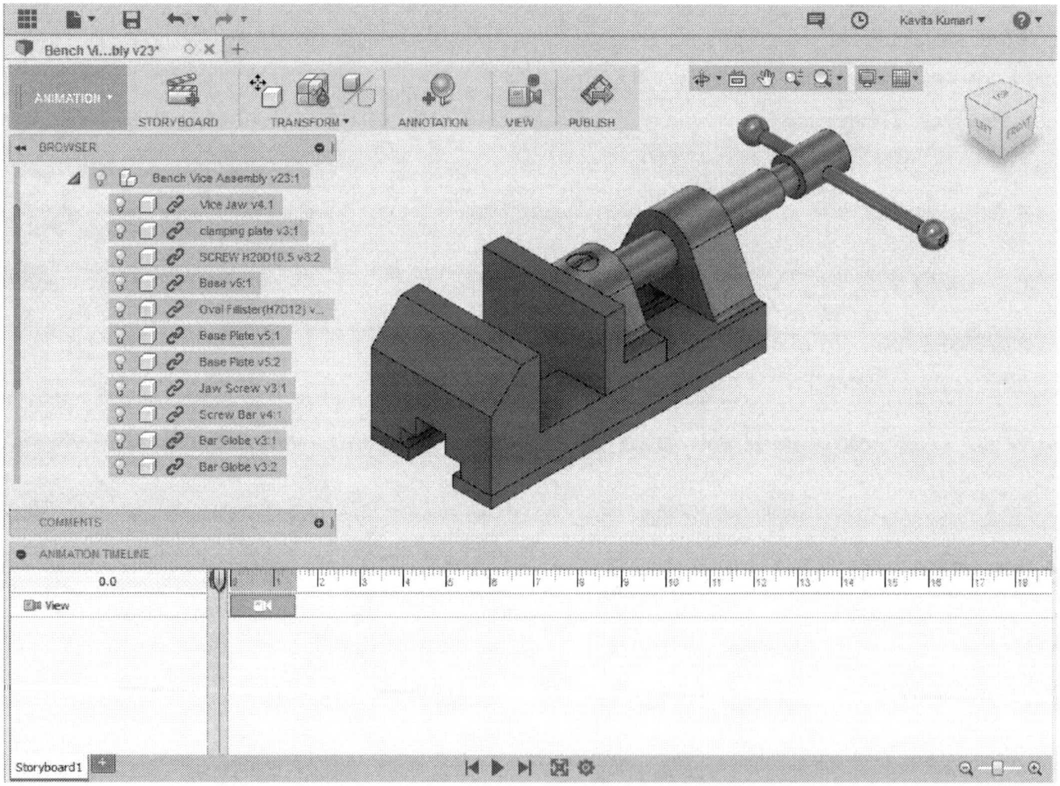

Figure 8-26 Assembly in the ANIMATION Workspace

3. Drag the **Playhead** to around 2 seconds in the **ANIMATION TIMELINE**.

4. Rotate the assembly such that its bottom parts are visible and select all the screws for animation by pressing CTRL key, as shown in Figure 8-27.

5. Right-click on the screw in the **BROWSER** bar; a shortcut menu is displayed.

6. Choose the **Transform Components** option from the shortcut menu; the **TRANSFORM COMPONENTS** dialog box and the triad is displayed on the selected component with the direction vector in the graphics window.

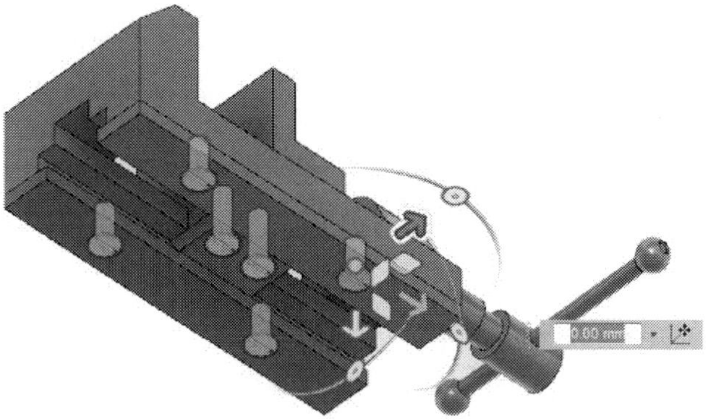

Figure 8-27 *Selecting screw to explode*

7. Select the direction vector of the triad along the Y axis to explode the screws; the **Y** edit box is displayed in the graphics window.

8. Enter **70** in the **Y** edit box; the selected component gets exploded and moved downward, as shown in Figure 8-28. Alternatively, you can also enter the exploded value in the **Y** distance edit box in the **TRANSFORM COMPONENT** dialog box.

9. Choose the **Trail Line Visibility** button from the **TRANSFORM COMPONENT** dialog box; a trail is displayed in the exploded component, refer to Figure 8-28.

   Trails lines are lines that display the path and direction of the exploded components.

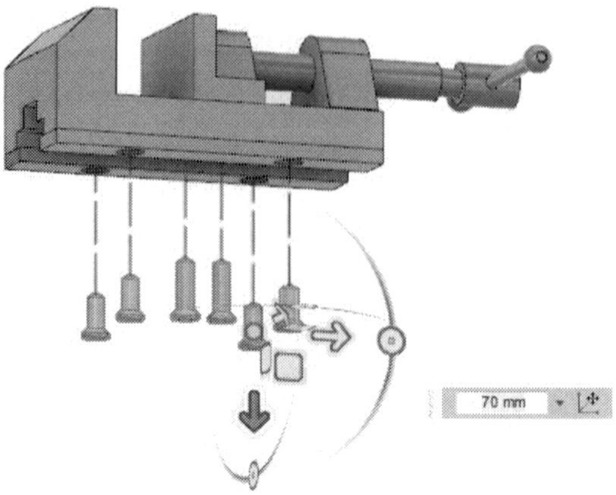

Figure 8-28 Assembly after entering the explosion value

10. Choose the **OK** button from the **TRANSFORM COMPONENTS** dialog box.

    You will notice that move action is created on the Timeline and therefore, it turns green indicating increase in the length of animation. Now, if you perform more action then the length of the animation will further increase.

11. Choose the **Transform Components** tool from the **TRANSFORM** panel in the Ribbon; the **TRANSFORM COMPONENTS** dialog box is displayed.

12. Select Base Plates and Clamping Plate from the graphics window by pressing the CTRL key; all other options become available in the dialog box.

13. Select the direction vector of the triad along which the selected components will explode; the **Z** edit box is displayed in the graphics window.

14. Enter **-50** in the **Z** distance edit box; the selected component gets exploded.

    Note that the direction vector of traid depends on the selection procedure of the components. If you change the sequence of selection then the direction vector of traid is changed.

15. Press and hold the SHIFT key and click on the Base Plate and Clamping Plate one by one. You will notice that the Base Plates and Clamping Plate are no more highlighted in blue color. This indicates that both components have become free from selection set.

16. Select Bar Globes, Screw Bar and Jaw Screw; the selected components are displayed in a blue color along with a triad.

17. Select the direction vector of the triad along which the selected components will explode; the **X** edit box is displayed in the graphics window.

18. Enter **80** in the **X** distance edit box; the selected components get exploded. Again, select these components by pressing the SHIFT key to make them free from the selection set, as shown in Figure 8-29.

19. Choose the **OK** button from the **TRANSFORM COMPONENTS** dialog box.

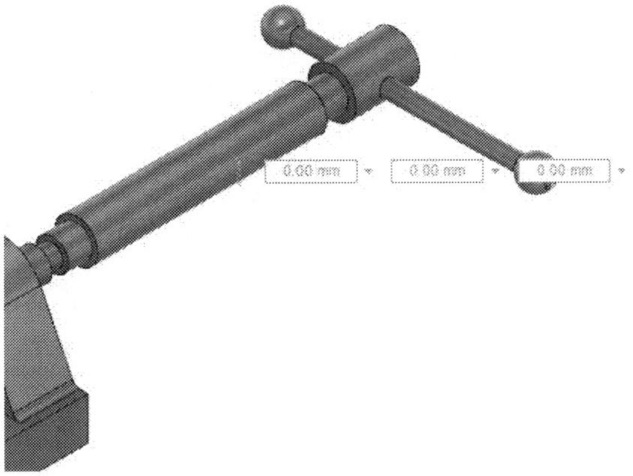

Figure 8-29 *Assembly after exploding three components*

20. Again, invoke the **TRANSFORM COMPONENTS** dialog box and select the Oval Fillister from the graphics window.

21. Select the Y direction vector of the triad along which the selected component will explode; the **Y** edit box is displayed in the graphics window.

22. Enter **60** in the **Y Distance** edit box; the selected component gets exploded. Again, select the Oval Fillister by pressing SHIFT to remove it from the selection set.

23. Select the **Trail Line Visibility** button from the **TRANSFORM COMPONENT** dialog box.

24. Choose the Vice Jaw from the graphics window and enter **40** in the **Y Distance e**dit box; the component gets exploded in the **Y** direction, as shown in Figure 8-30.

25. Select the **Trail Line Visibility** button from the **TRANSFORM COMPONENT** dialog box.

26. Choose the **OK** button from the **TRANSFORM COMPONENTS** dialog box.

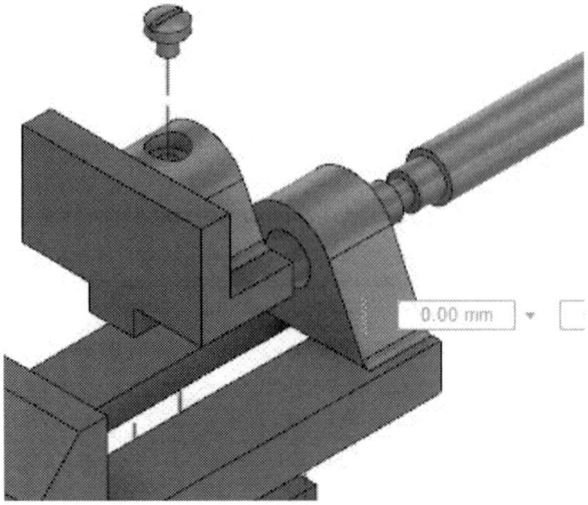

Figure 8-30 *Assembly after exploding Oval Fillister and Vice Jaw*

27. Similarly, explode the Bar Globes one by one in the Z direction. The assembly after exploding all components is shown in Figure 8-31.

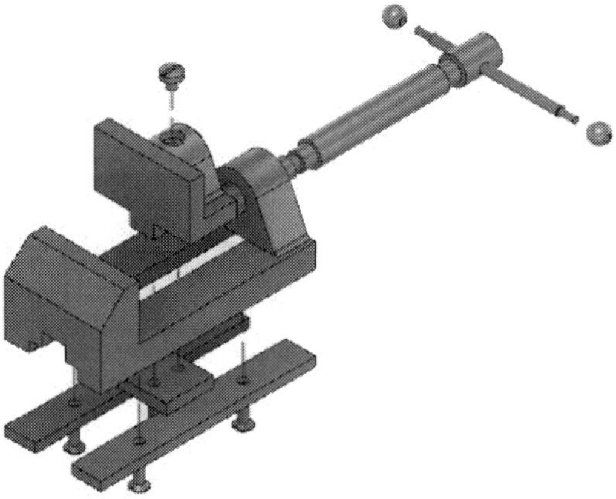

Figure 8-31 *Assembly after exploding all components*

Animating the Assembly

1. Choose the **Play Current Storyboard** button from the bottom of the **ANIMATION TIMELINE**; all the components start moving in the animation in the same order as they were exploded.

2. Choose the **Back to Storyboard Beginning** button from the bottom of the **ANIMATION TIMELINE**; all the exploded components in the animation get assembled in the same order as they were exploded.

3. Choose the **PUBLISH** tool from the **PUBLISH** panel in the Ribbon; the **Video Options** dialog box is displayed, as shown in Figure 8-32.

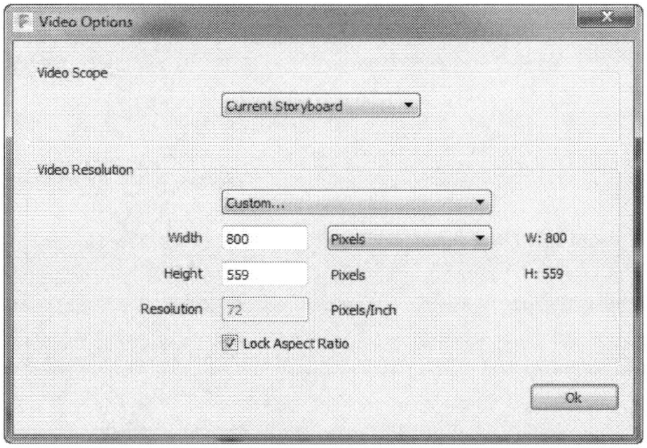

***Figure 8-32** The **Video Options** dialog box*

4. Select the **Current Storyboard** option from the drop-down list in the **Video Scope** area.

5. Choose the **Custom** option from the Window size drop-down list in the **Video Resolution** area.

6. Specify required values in the **Width** and **Height** edit boxes, refer to Figure 8-32.

7. Select the **Lock Aspect Ratio** check box to keep the video aspect ratio fixed when width or height is changed.

8. Choose the **OK** button from the **Video Options** dialog box; the **Save As** dialog box is displayed.

9. Enter the name of the assembly in the **Name** edit box.

10. Select the **Save to a project in the cloud** check box and the **CADCIM** project from the **Project** drop-down list.

11. Choose the **Save** button from the **Save As** dialog box.

Self-Evaluation Test

Answer the following questions and then compare them to those given at the end of this chapter:

1. Which of the following options is used to shorten the tangent display, turn it off, or set it to full length?

   (a) **Interference Edges** (b) **Tangent Edges**
   (c) **Thread Edges** (d) None of these

2. You can add the Parts list to the assembly drawing views using the __________ tool.

3. By default, the display style of the projected views is same as that of the __________.

4. The part of the original view that is sectioned will be displayed with the __________ in the section view.

5 You cannot generate drawing views of an assembly file. (T/F)

6. You can add more sheets for generating drawing views. (T/F)

7. Autodesk Fusion 360 allows you to explode assemblies in a special environment called the **ANIMATION** Workspace. (T/F)

8. By dragging the end point of a traid, the tweak distance or angle can be directly modified in the graphics window. (T/F)

Review Questions

Answer the following questions:

1. Which of the following tools is used to display the details of a portion of an existing view by magnifying that portion and displaying it as a separate view?

   (a) **Detail View** (b) **Base View**
   (c) **Section View** (d) None of these

2. __________ are defined as the parametric lines that display the path and direction of the assembled components.

3. The **Center Mark** tool from the __________ panel is used to add center marks to circles in the drawing views.

4. You cannot prevent dependent views from getting deleted if the parent view is deleted. (T/F)

5. You cannot modify the default parts list. (T/F)

6. You can modify the interval value of an animation. (T/F)

EXERCISE

Exercise 1

Create the components of the V-Block and generate the front, top, right, detail, and isometric views, as shown in Figure 8-33. The views and dimensions of the components of the V-Block are shown in Figures 8-34 and 8-35. **(Expected time: 1 hr)**

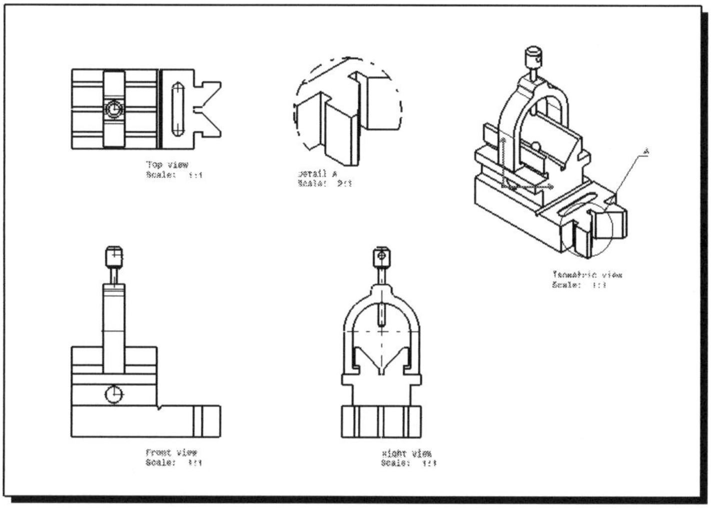

Figure 8-33 Drawing views of the V-Block assembly

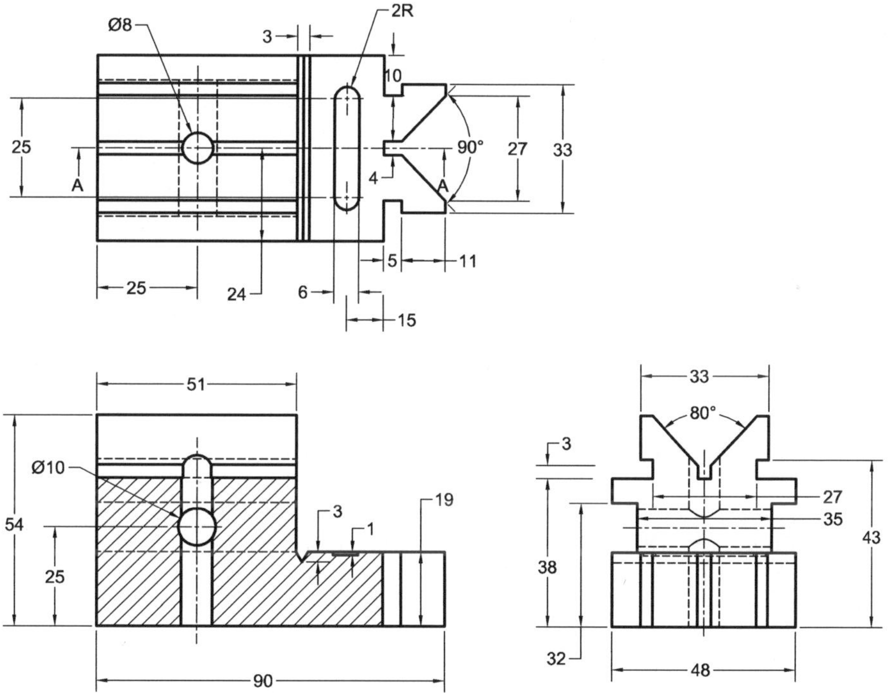

SECTION A-A

Figure 8-34 Views and dimensions of the V-Block body

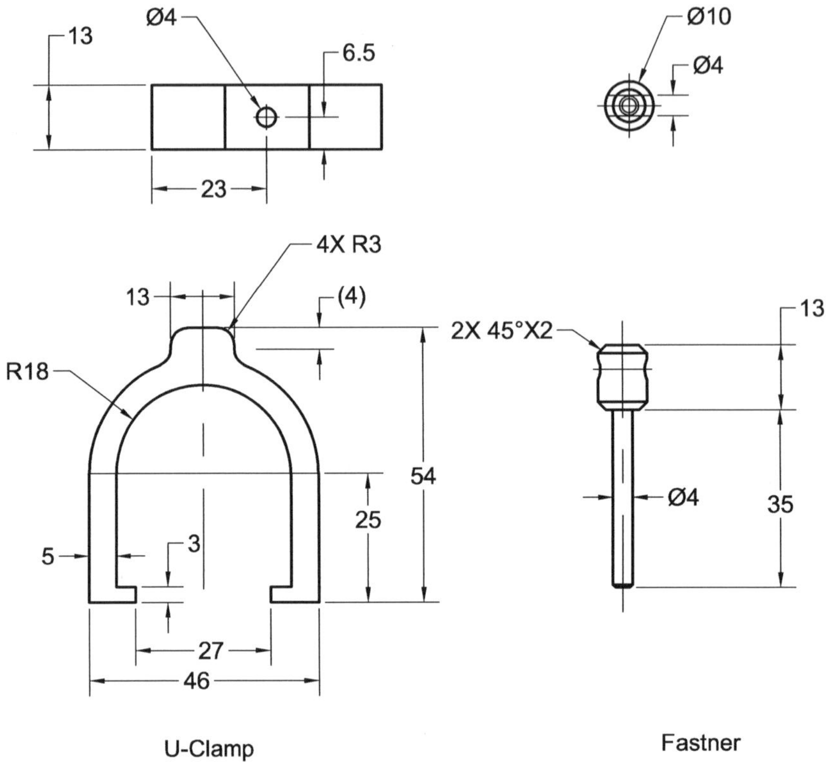

Figure 8-35 Views and dimensions of the U-Clamp and Fastener

Exercise 2

To perform this exercise, you need to download the input file from *www.cadcim.com*. The complete path for downloading the file is:

Textbooks > CAD/CAM > Autodesk Fusion 360: A Tutorial Approach >Input File> c08_Fusion_input.zip

Next, create animation of exploding and unexploding assembly. The exploded view of the assembly is shown in Figure 8-36. **(Expected time: 1 hr)**

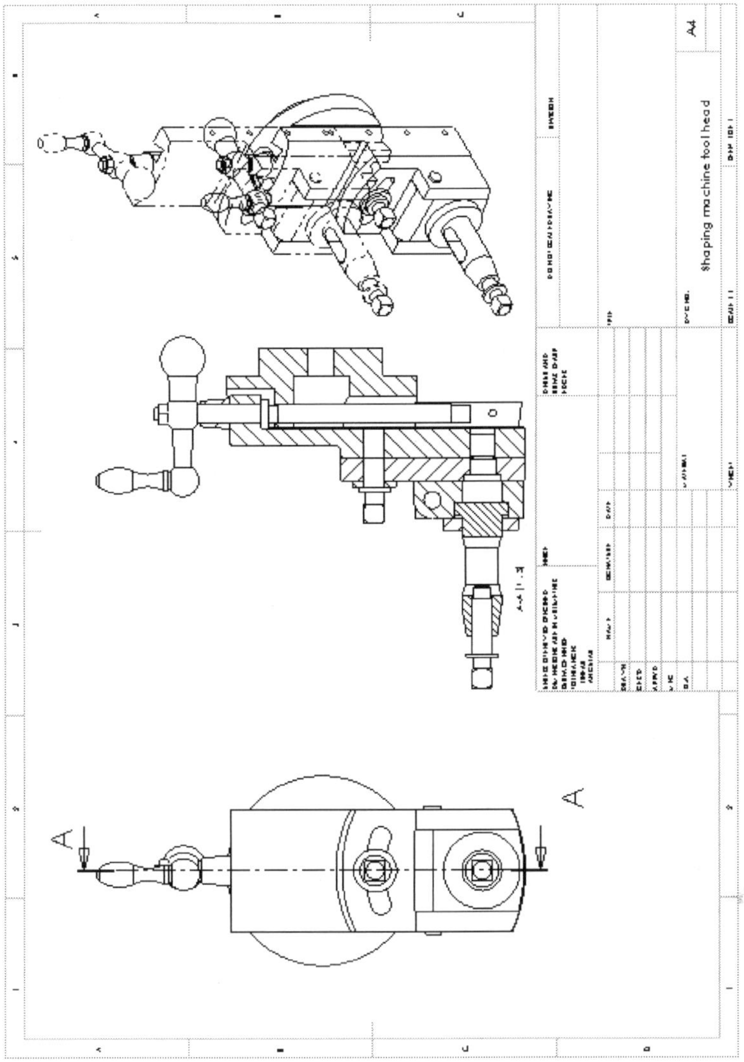

Figure 8-36 *Views to be generated for Exercise 2*

Answers to Self-Evaluation Test

1. Tangent Edges, **2. Table**, **3.** parent view, **4.** hatching lines, **5.** F, **6. T**, **7.** T, **8.** F

Chapter 9

Working with Sheet Metal Components

Learning Objectives

After completing this chapter, you will be able to:

- *Set parameters for creating sheet metal parts*
- *Create base of a sheet metal component*
- *Fold part of a sheet metal part*
- *Add flange to a sheet metal component*
- *Create cut feature in a sheet metal part*
- *Round corners of a sheet metal part*
- *Chamfer corners of a sheet metal part*
- *Create flat pattern of a sheet metal component*

THE SHEET METAL WORKSPACE

Sheet metal is one of the fundamental forms used in metalworking and it can be cut and bent into a variety of shapes. Thicknesses can vary significantly; extremely thin sheets are considered foil or leaf, and pieces thicker than 6 mm (0.25 in) are considered plate. A sheet metal component is created by bending, cutting, or deforming a sheet of metal having uniform thickness, as shown in Figure 9-1.

As it is not possible to machine such a model, therefore after creating a sheet metal component, you need to flatten it for its manufacturing. Figure 9-2 shows the flattened view of the sheet metal component shown in Figure 9-1.

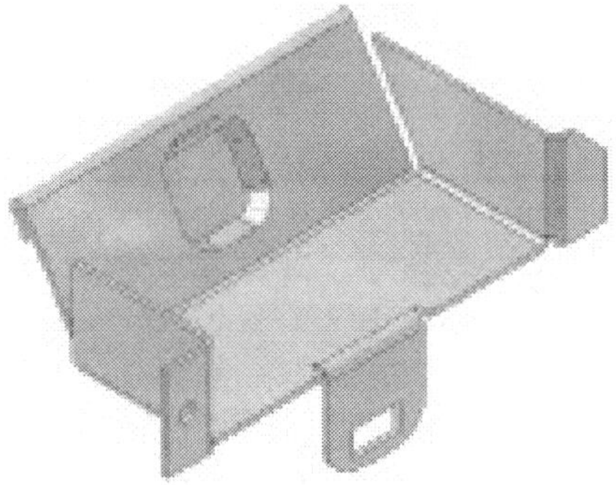

Figure 9-1 *A sheet metal component*

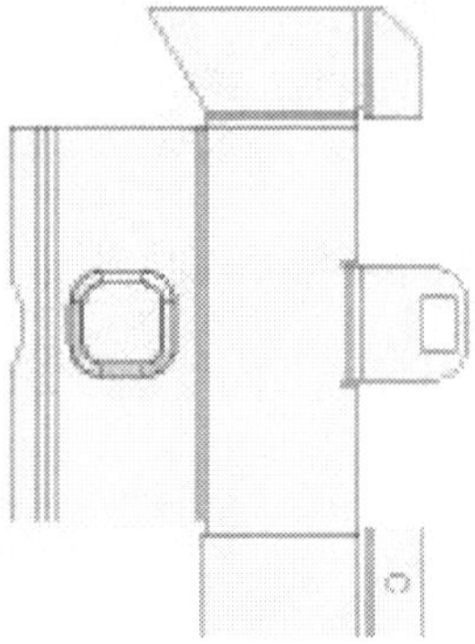

Figure 9-2 *The Flattened view of the sheet metal component*

Autodesk Fusion 360 allows you to create sheet metal components in a workspace called the Sheet Metal workspace. This workspace provides all the tools required for creating sheet metal components. To start a new design in Sheet Metal workspace, choose the **SHEET METAL** option from the Model workspace; the Sheet Metal workspace is invoked.

TUTORIALS

Tutorial 1

In this tutorial, you will create sheet metal component of the Holder Clip shown in Figure 9-3. The thickness of the sheet of the Holder Clip is 1 mm and its views and dimensions are shown in Figures 9-5. Flat pattern of the component is shown in Figure 9-4. **(Expected time: 45 min)**

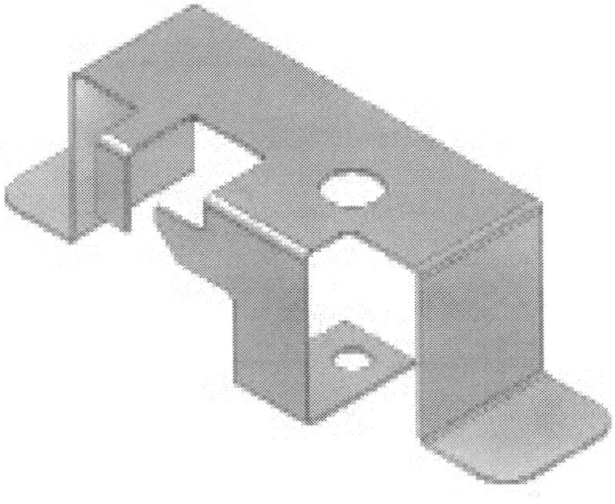

Figure 9-3 *Sheet metal component of the Holder Clip*

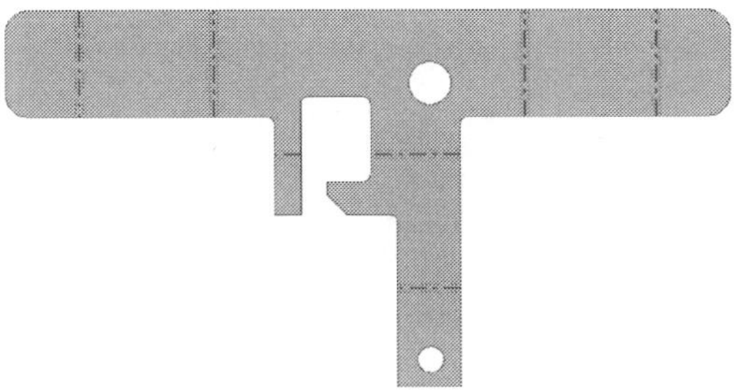

Figure 9-4 *Flat pattern of the component*

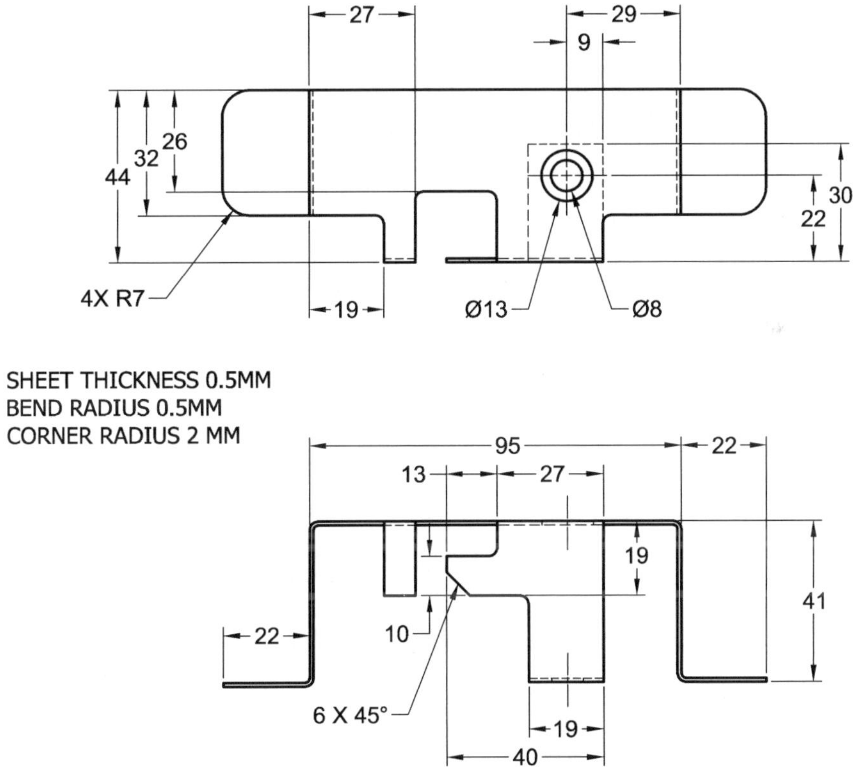

Figure 9-5 *Views and dimensions of the component*

The following steps are required to complete this tutorial:

a. Open a sheet metal workspace.
b. Set parameters for the sheet metal component.
c. Draw the sketch for the base wall.
d. Converting the sketch into a base wall.
e. Creating flanges.
f. Creating rounds and holes

Opening a Sheet Metal Workspace

1. Choose the **SHEET METAL** option from the **MODEL** workspace; the **SHEET METAL** workspace is invoked.

Setting the Parameters for the Sheet Metal Component

1. Choose the **Sheet Metal Rules** tool from the **MODIFY** panel; the **SHEET METAL RULES** dialog box is displayed, as shown in Figure 9-6.

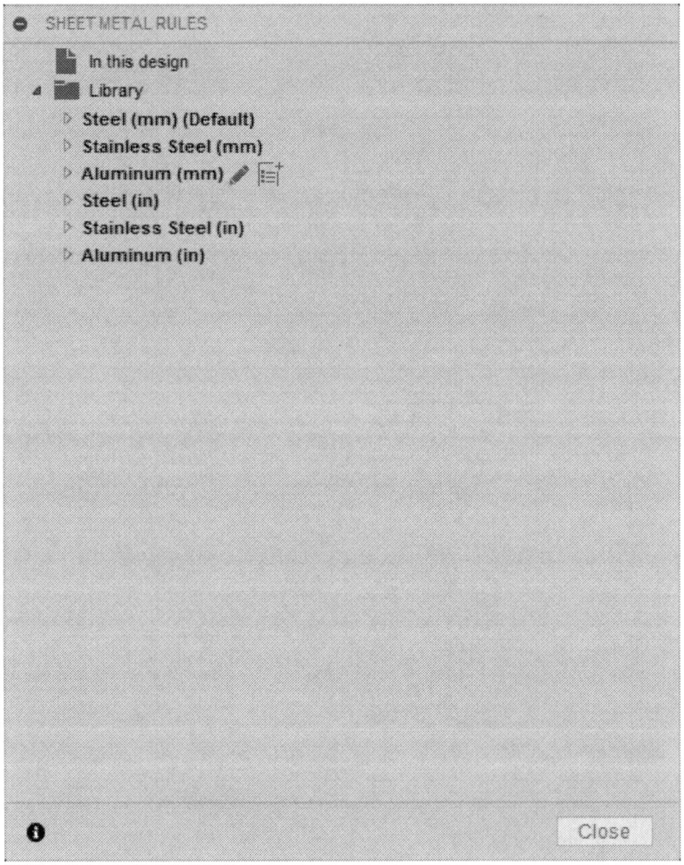

Figure 9-6 *The* ***SHEET METAL RULES*** *dialog box*

2. Hover the cursor on the **Steel (mm) (Default)** node; the **Edit Rule** and **New Rule** options are displayed on the right.

3. Click on the **Edit Rule** option; the **Edit Rule** dialog box is displayed, as shown in Figure 9-7.

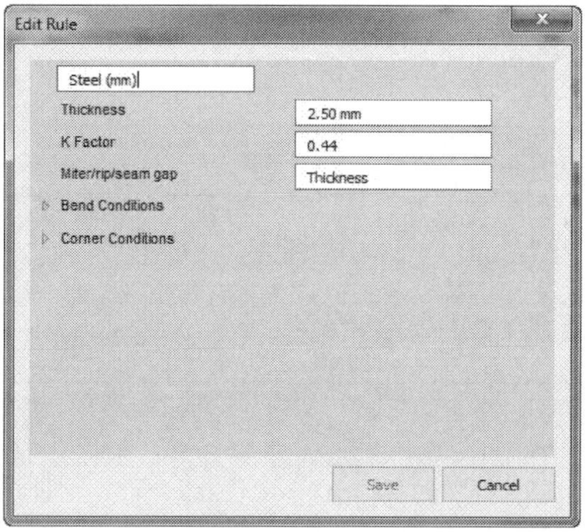

*Figure 9-7 The **SHEET METAL RULES** dialog box*

4. Enter **1mm** in the **Thickness** edit box and expand the **Bend Conditions** node.

5. Enter **2mm** in the **Bend radius** edit box.

6. Keep rest of the values as default and then choose the **Save** button from the **Edit Rule** dialog box and the **Close** button from the **SHEET METAL RULES** dialog box.

Drawing the Sketch for the Base Wall

1. Choose the **Line** tool from the **SKETCH** panel and select the **XZ** plane from the **BROWSER** bar.

2. Create the sketch for the base wall, as shown in Figure 9-8.

3. Choose the **STOP SKETCH** tool from the Ribbon.

Converting the Sketch into a Base Wall

1. Choose the **Flange** tool from the **CREATE** panel; the **FLANGE** dialog box is displayed. 

2. Select the sketch from the graphics window; the preview of the base feature is displayed.

3. Choose the **OK** button from the dialog box; the base wall is created. The final feature of the base wall is shown in Figure 9-9.

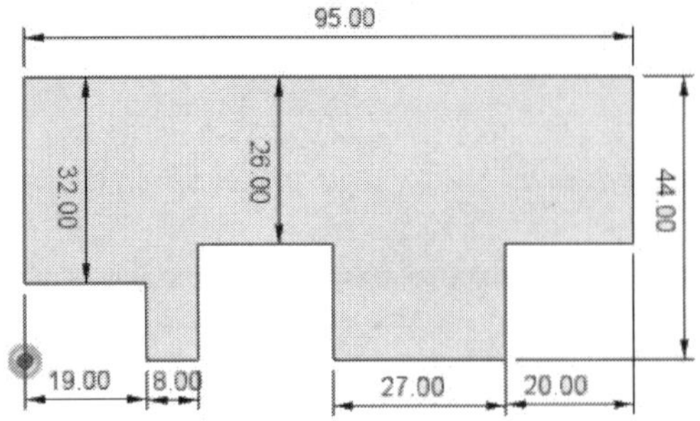

Figure 9-8 *Sketch of the top face of the Holder Clip*

Figure 9-9 *Isometric view of the Holder Clip*

Creating the First Contour Flange

A contour flange is created with the help of a sketched contour. Therefore, first you need to sketch the contour so that it can be used to create a flange.

1. Choose the **Line** tool from the **SKETCH** panel of the Ribbon and then select the face, as shown in Figure 9-10 as the sketching plane.

2. Create sketch of the profile of the contour flange, as shown in Figure 9-11. Refer to Figure 9-5 for dimension.

Figure 9-10 *Face to be selected as the sketching plane*

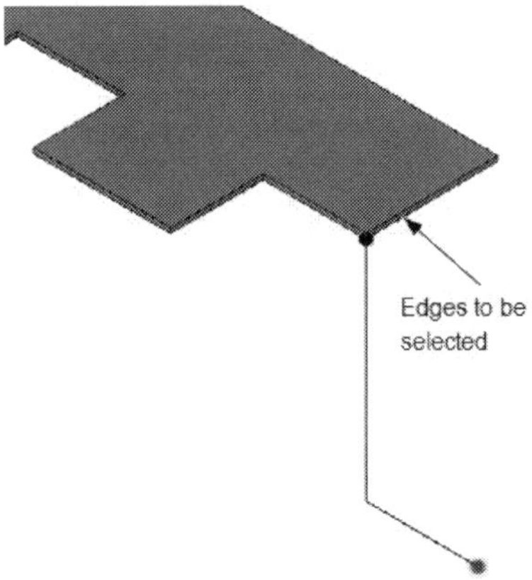

Figure 9-11 *Sketch of the contour flange*

3. Choose the **STOP SKETCH** tool from the Ribbon. Next, choose the **Flange** tool from the **CREATE** panel; the **FLANGE** dialog box is displayed.

4. Select one of the two sketched lines as the profile for creating the flange. As the other line is a part of the same sketch, it gets automatically selected and turns blue and also the **FLANGE** dialog box gets modified. In this dialog box, the name of the selected profile is displayed in the display box.

5. Select the edge on the right of the top face to create the flange, refer to Figure 9-11.

6. Accept the remaining default options and choose **OK** to create the flange; a bend is automatically created between the base sheet and the flange. The parameters of this bend are taken from the parameters defined in the **SHEET METAL RULES** dialog box.

7. Similarly, create the profile of the second contour flange on the other side of the top face.

8. Select the edge on the left of the top face to create the flange. The sheet metal model of the Holder Clip after creating the two contour flanges is shown in Figure 9-12.

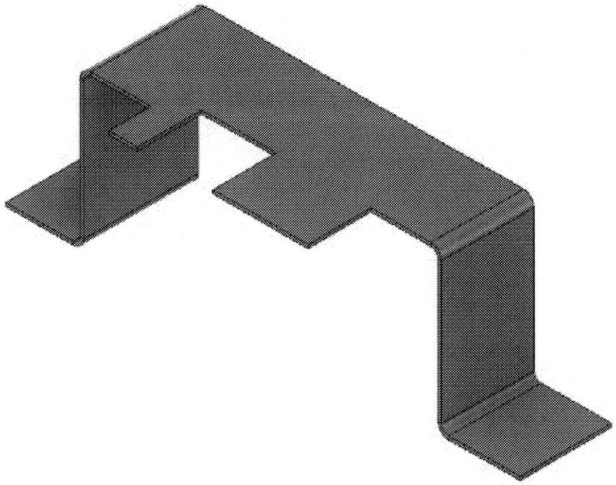

Figure 9-12 *Sheet metal component after creating the two contour flanges*

Creating the Third Contour Flange

1. Define a new sketch plane on the planar face of the base feature and then create the sketch for the contour flange, as shown in Figure 9-13.

2. Choose the **STOP SKETCH** tool from the Ribbon. Next, choose the **Flange** tool from the **CREATE** panel; the **FLANGE** dialog box is displayed.

3. Select the sketch created in step1 as the contour for creating the flange; the selected profile gets added in the Display box.

4. Enter **-27** in the **Distance** edit box of the **FLANGE** dialog box.

5. Select the horizontal edge on the top face to create the flange, refer to Figure 9-13.

6. Accept the remaining default options and choose **OK** to create the flange. The sheet metal component after creating the third contour flange is shown in Figure 9-14.

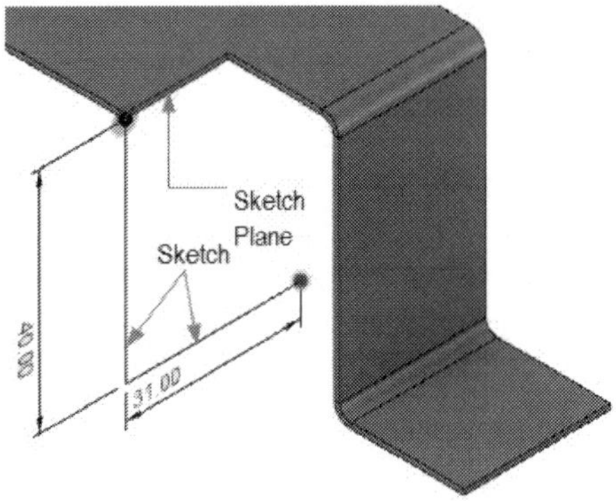

Figure 9-13 The sketch plane and the sketch for the contour flange

Figure 9-14 Sheet metal component after creating the contour flange

Creating a Cut and a New Face on the Front Face of the Third Contour Flange

1. Define a new sketch plane on the front face of the third contour flange and create the sketch of the cut feature, as shown in Figure 9-15. After creating the sketch, choose the **STOP SKETCH** tool from the Ribbon.

2. Invoke the **EXTRUDE** tool from the **MODEL** Workspace and then create the cut feature by selecting the **All** option from the **Extent** drop-down list. Next, choose the **Flip** button to reverse the direction. The sheet metal component after creating the cut is shown in Figure 9-16.

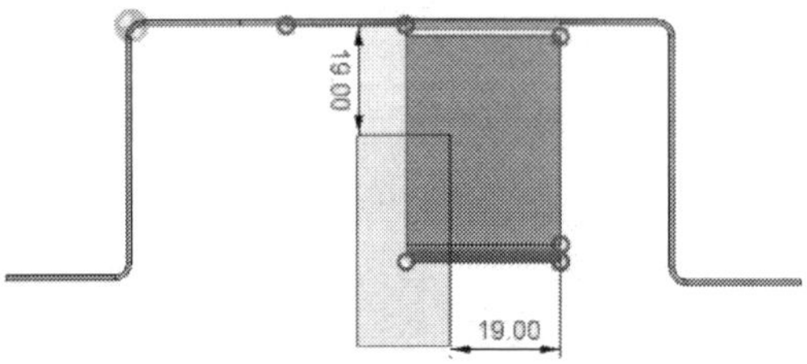

Figure 9-15 The sketch plane and the sketch for the cut feature

Figure 9-16 Sheet metal component after creating cut feature

3. Similarly define a sketch plane on the front face of the third contour flange and create a new rectangular face by using the **Extrude** tool. Refer to Figure 9-5 for dimensions.

4. Make sure that the **Join** option should be selected in the **Operation** drop-down list.

5. Add the chamfer by using the **Chamfer** tool in the **MODIFY** panel, as shown in Figure 9-17, refer to Figure 9-5 for dimensions.

Figure 9-17 *Sheet metal component after creating the new face and the chamfer*

Creating the Last Flange

1. Choose the **Flange** tool from the **CREATE** panel in the Ribbon; the **FLANGE** dialog box is displayed.

2. Select the edge on the top face of the base feature, as shown in Figure 9-18.

3. Enter **19** in the **Distance** edit box; the size of the flange in the preview is modified.

4. Choose the **Flip** button from the **FLANGE** dialog box to flip the direction of the flange.

5. Accept the remaining default options and choose the **OK** button to create the flange and exit the dialog box. The sheet metal component after creating the flange is shown in Figure 9-19.

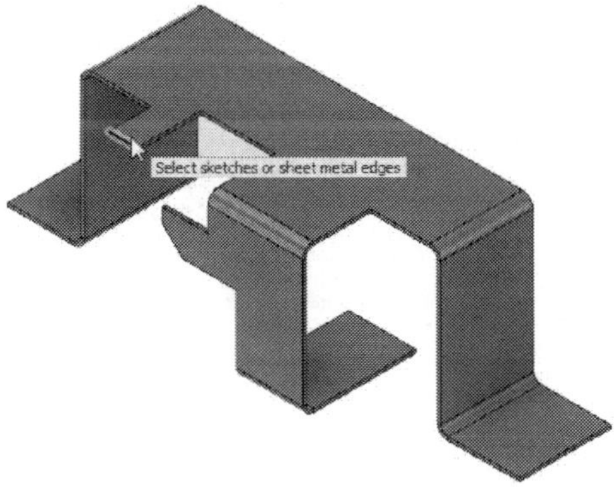

Figure 9-18 *Selecting the edge to create the flange*

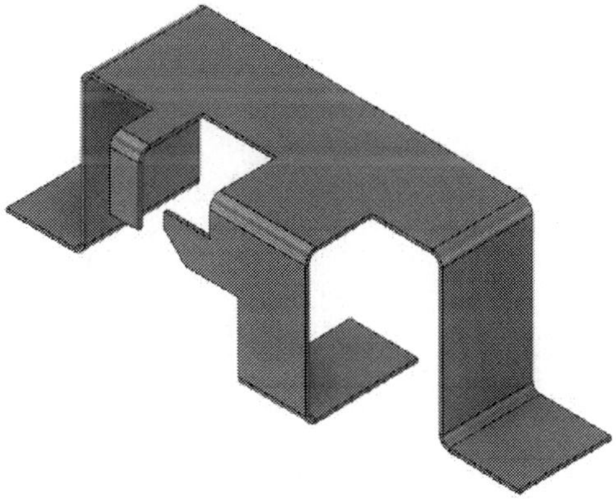

Figure 9-19 *Model after creating the flange*

Creating Rounds and Holes and Saving the Components

1. Create all rounds by using the **Fillet** tool. Refer to Figure 9-5 for dimensions.

2. Create two holes by using the **Hole** tool from the **CREATE** panel of the Ribbon. Refer to Figures 9-5 for dimensions.

   The final sheet metal component of the Holder Clip is shown in Figure 9-20.

3. Save the sheet metal component with the name **c09_Tut_01** under the **CADCIM** project.

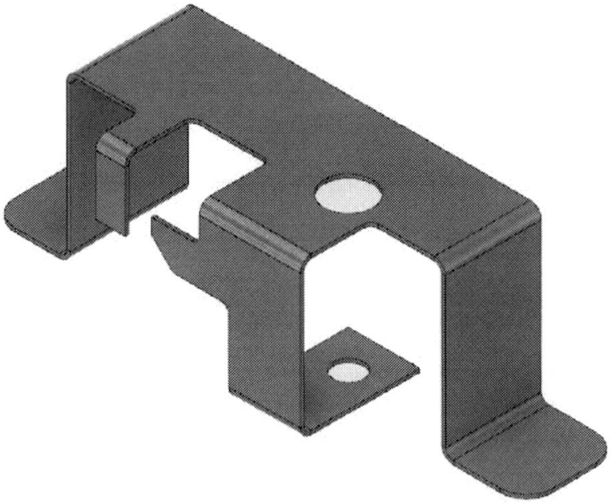

Figure 9-20 *Final model of the Holder Clip*

Tutorial 2

In this tutorial, you will create the sheet metal component shown in Figure 9-21. The flat pattern of the sheet metal component is shown in Figure 9-22. After creating the model, you need to generate its drawing views, as shown in Figure 9-23. **(Expected time: 1 hr)**

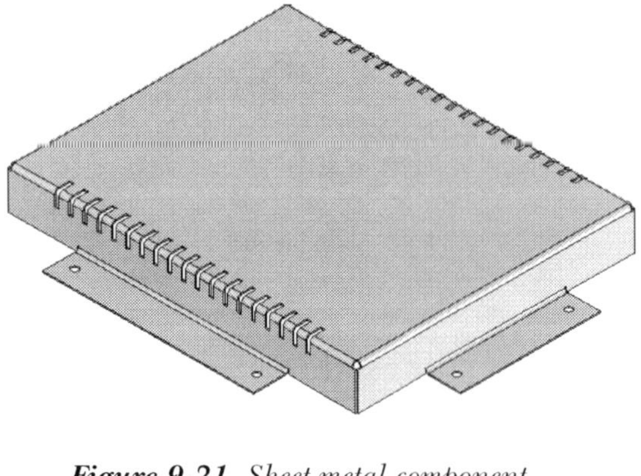

Figure 9-21 *Sheet metal component*

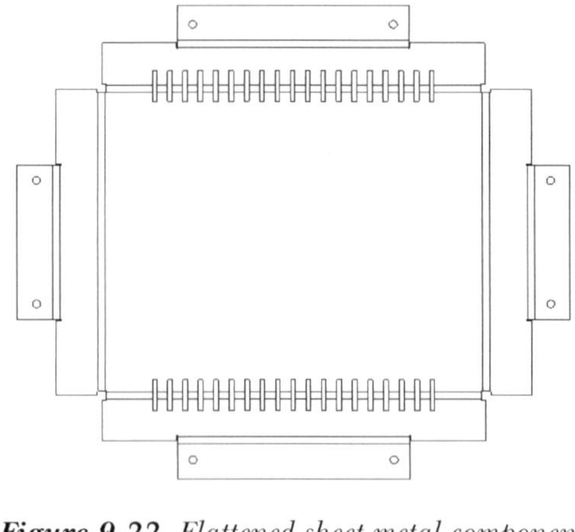

Figure 9-22 *Flattened sheet metal component*

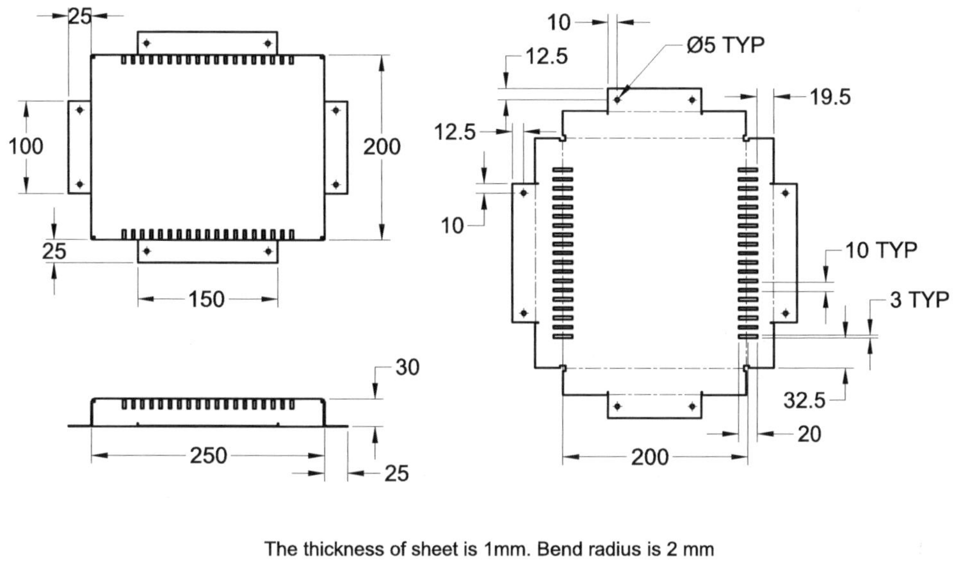

Figure 9-23 *Views and dimensions for Tutorial 2*

The following steps are required to complete this tutorial:

a. Create the base flange of the sheet metal component.
b. Add other required flanges to the sheet metal component.
c. Create the flat pattern of the sheet metal component.
d. Create slots using the extrude feature and pattern them on the sides of the flattened sheet metal component.
e. Refold the sheet metal component.
f. Generate the drawing views of the sheet metal component.
g. Save the model.

Creating the Base Flange

For creating this sheet metal component, you first need to create the base flange. The base flange will be created by using a rectangular sketch drawn on the Top Plane.

1. Invoke the **SHEET METAL** workspace and choose **Create Sketch** from the **SKETCH** panel in the Ribbon.

2. Select the **XZ** plane and create a rectangle of 250x200 mm as the sketch of the base flange

3. Choose **STOP SKETCH** from the Ribbon. Next, choose the **Flange** tool from the **CREATE** panel; the **FLANGE** dialog box is displayed.

4. Select the sketch from the graphics window and choose the **OK** button to create the flange.

Creating Four Flanges

1. Choose the **Flange** tool from the **CREATE** panel and select the edges of the base flange, as shown in Figure 9-24.

2. Enter **30** in the **Height** edit box.

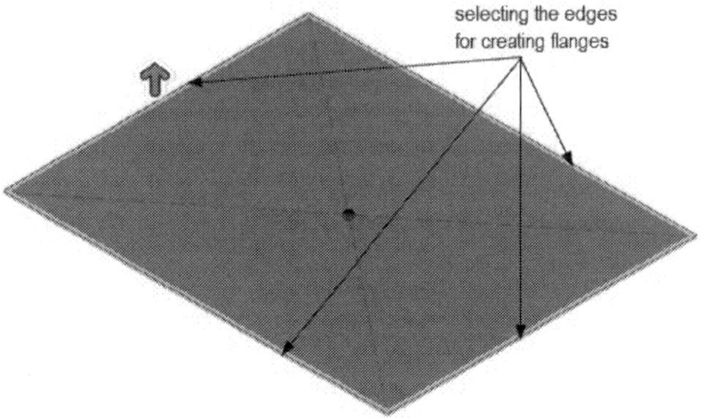

Figure 9-24 *Selecting the edges for creating a flange*

3. Choose the **Flip** button to reverse the direction of flange. Next, choose the **OK** button from the **FLANGE** dialog box; the flange is created, as shown in Figure 9-25.

Figure 9-25 *Model after creating flanges*

Creating Wall on Edges

1. Choose the **Flange** tool from the **CREATE** panel and select the bottom edge of the front flange, as shown in Figure 9-26.

2. Select the **Symmetric** option from the **Full Edge** drop-down list in the selection box of the **FLANGE** dialog box.

3. Enter **75** in the **Distance** edit box and **25** in the **Height** edit box.

4. Choose the **OK** button from the dialog box; the side wall feature is created, as shown in Figure 9-27.

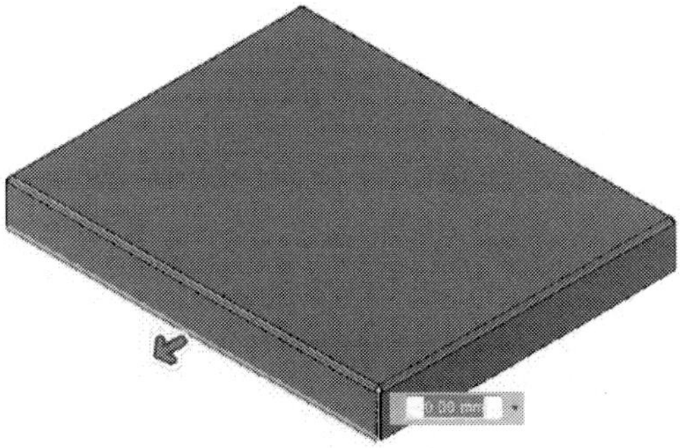

Figure 9-26 Selecting the edge for creating the wall on the flange

Figure 9-27 The side wall feature created

5. Similarly, create the other three side walls on the edges of rest of the flanges. For dimension, refer to Figure 9-23. The model after creating the four walls is shown in Figure 9-28.

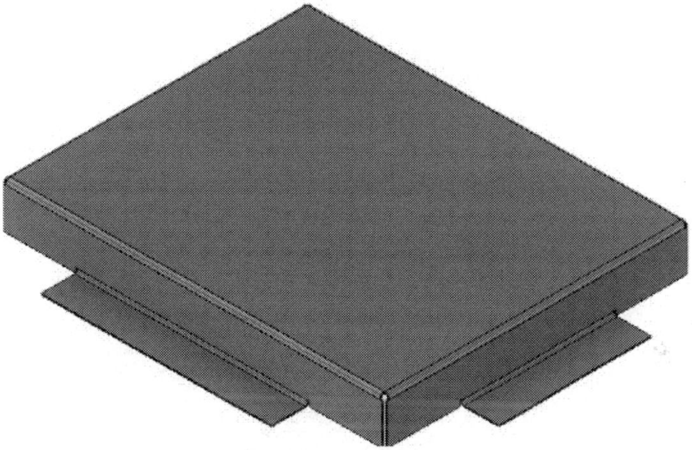

Figure 9-28 Model after creating four walls

Creating Cuts Across the Bends

The next feature that you need to create is the cut feature across the bends. For creating this feature, you need to first unfold the sheet metal component.

1. Choose the **Unfold** tool from the **MODIFY** panel in the Ribbon; the **UNFOLD** dialog box is displayed and you are prompted to select the stationary entity.

2. Select the top face as the stationary entity; you are prompted to select the bends to unfold.

3. Select the **Unfold all bends** check box and then choose the **OK** button from the **UNFOLD** dialog box. Alternatively, you can select one by one all the bends in the model in the graphics window. The unfolded view of the model is shown in Figure 9-29.

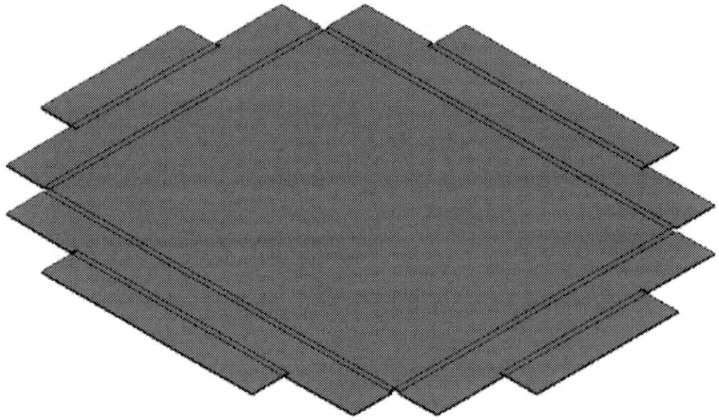

Figure 9-29 *Unfolded view of the model*

4. Select the top face of the model and choose the **2-Point Rectangle** tool from the **SKETCH** panel.

   Next, draw the sketch of the cut feature, as shown in Figure 9-30.

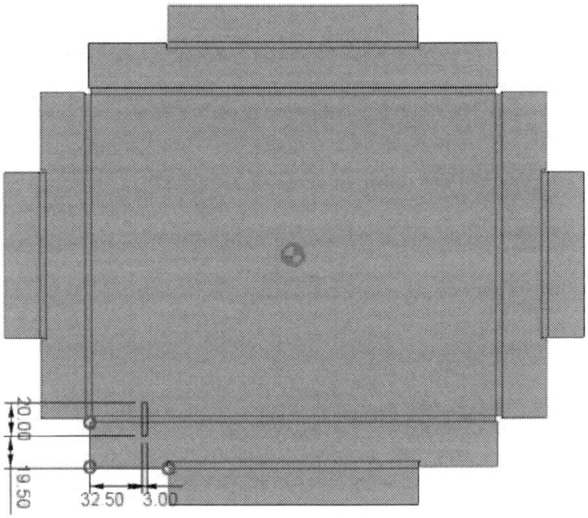

Figure 9-30 *Sketch of the cut feature*

5. Invoke the **EXTRUDE** dialog box and create the cut feature. The model after adding the cut feature is shown in Figure 9-31.

6. Pattern the cut feature. The model after patterning the cut feature is shown in Figure 9-32.

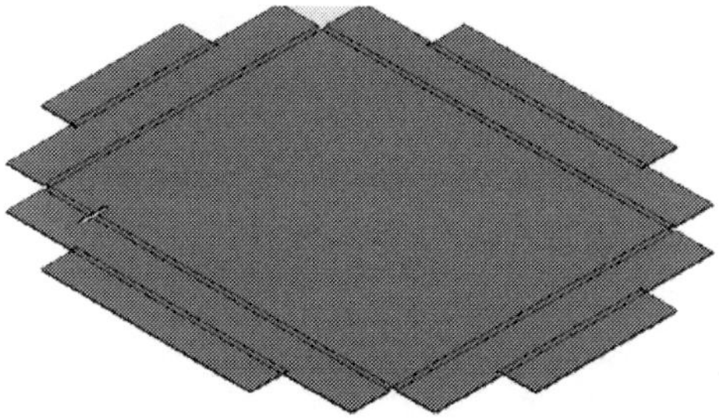

Figure 9-31 Model after adding the cut feature

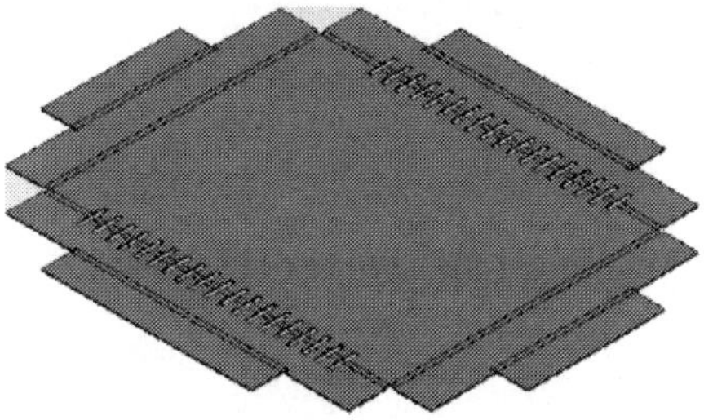

Figure 9-32 Model after patterning the cut feature

Refolding the Sheet Metal Part and Creating holes

After creating the sheet metal component, you need to refold the sheet and create holes.

1. Choose the **REFOLD FACES** tool from the Ribbon; the sheet metal component gets refolded.

2. Create holes on all four walls. For dimensions, refer to Figure 9-23. The sheet metal component after refolding and creating holes is shown in Figure 9-33.

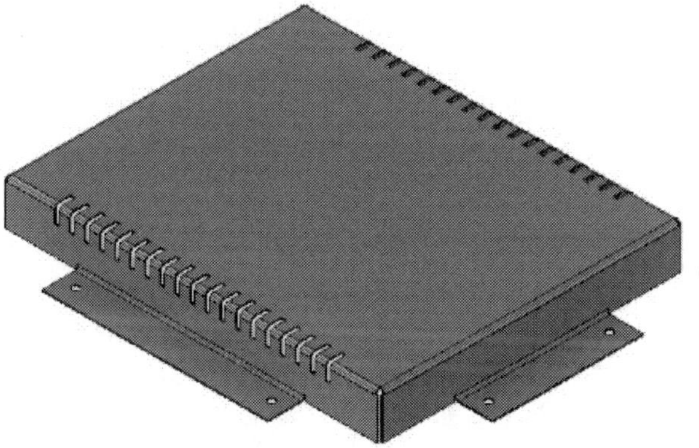

Figure 9-33 Sheet metal component after refolding and creating holes

3. Save the sheet metal component with the name **c09_Tut_02** under the **CADCIM** project.

Self-Evaluation Test

Answer the following questions and then compare them to those given at the end of this chapter:

1. In Autodesk Fusion 360, the __________ tool in the **CREATE** panel is used to create a base flange.

2. You can edit the bend radius under the __________ node.

3. You can set the sheet metal parameters in the __________ dialog box.

4. To create flat pattern of a sheet metal component, you need to choose the **Create Flat Pattern** tool. (T/F)

5. The bend radii and relief sizes are same throughout the part for manufacturing purposes. (T/F)

Review Questions

Answer the following questions:

1. The __________ tool in the **MODIFY** panel is used to unfold a bend.

2. The __________ button is used to add flanges to the selected edges in the **FLANGE** dialog box.

3. The __________ check box in the **UNFOLD** dialog box is used to unfold all the bends of a component.

4. Choose the __________ tool from the Ribbon to refold sheet.

5. If you modify a value in the **EDIT RULE** dialog after creating the sheet metal component, the changes will reflect in the sheet metal component when you exit the dialog box after saving the changes. (T/F)

EXERCISE

Exercise 1

Create the sheet metal component shown in Figure 9-34. The dimensions of the model are shown in Figure 9-35. **(Expected time: 30 min)**

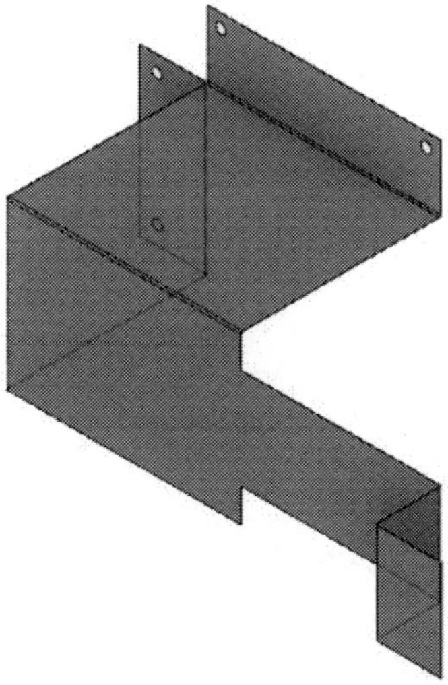

Figure 9-34 *Sheet metal part for Exercise 1*

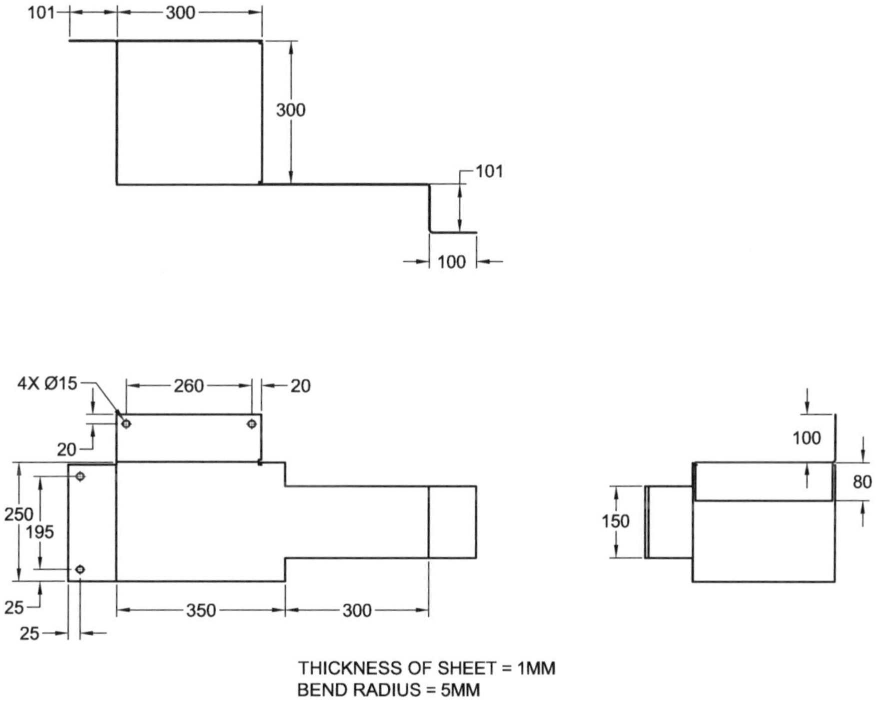

Figure 9-35 *Orthographic views and dimensions of the component*

Exercise 2

Create the sheet metal component shown in Figure 9-34. The dimensions of the model are shown in Figure 9-35. **(Expected time: 30 min)**

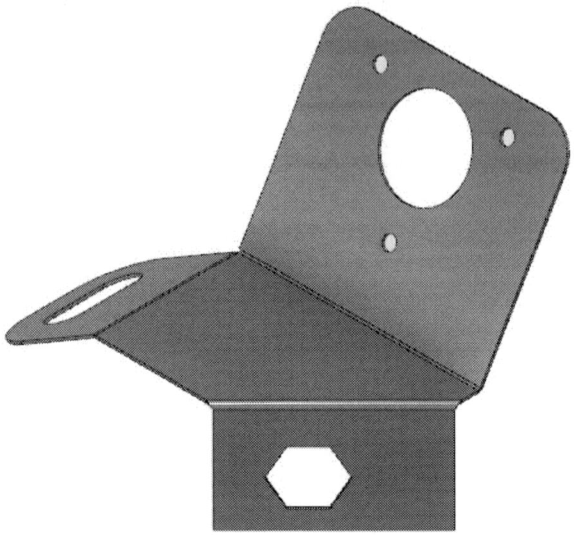

Figure 9-34 *Sheet metal part for Exercise 2*

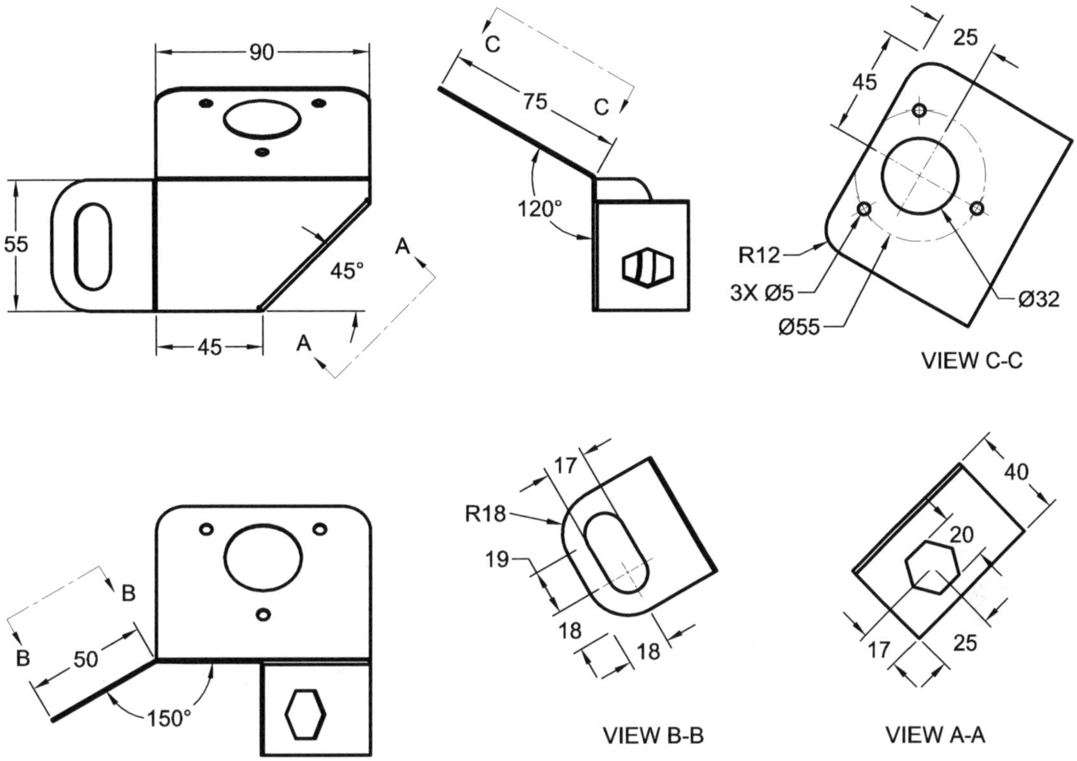
90
55
45°
45
A
A
C
C
75
120°
25
45
R12
3X Ø5
Ø55
Ø32
VIEW C-C
B
B
50
150°
17
R18
19
18
18
VIEW B-B
40
20
17
25
VIEW A-A

Answers to Self-Evaluation Test

1. Flange, 2. Bend Conditions, 3. SHEET METAL RULE, 4. T, 5. T

Chapter 10

Managing and Collaborating on the Cloud

Learning Objectives

After completing this chapter, you will be able to:

- *Understand the cloud integration process of Autodesk Fusion 360*
- *Manage and share projects through A360*
- *Organize live review Sessions*
- *Manage different design versions*

MANAGING AND COLLABORATING IN FUSION 360

Autodesk Fusion 360 is different from other CAD/CAM/CAE software packages available in the market because of its A360 cloud sharing and collaborating capabilities. It enables the users to access their design from any browser even without having Autodesk Fusion 360 installed on their system. The design files are accessed through a secure link that users send via email or chat. In Autodesk Fusion 360, users can collaborate their projects with the other group members and can also do preliminary investigations on any previously created design. Figure 10-1 shows a diagram of distributed design and Cloud-Sharing process of Fusion 360.

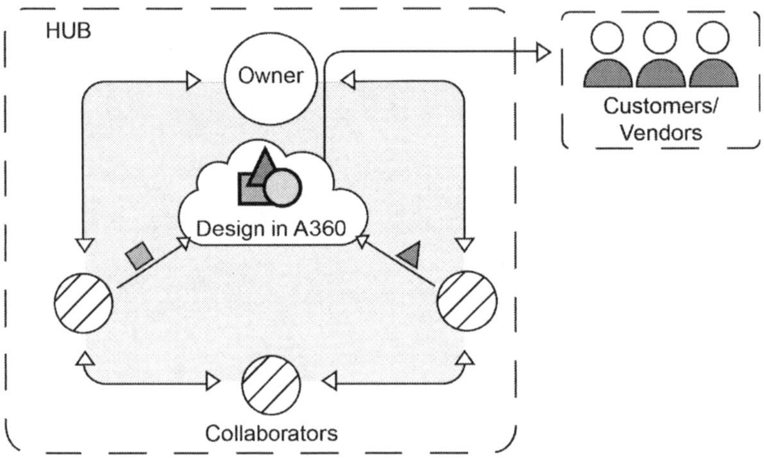

Figure 10-1 *Distributed design and Cloud-Sharing process of Fusion 360*

A360: Cloud-Based Workspace

A360 is a cloud-based central workspace developed by Autodesk for all its software and, as of now, has been fully integrated with Fusion 360 and Revit. This workspace perfectly stores all the designs, drawings, and rendered images that you will create in Fusion 360. Also you can upload other supported files such as .dwg, .png, and .pptx directly to A360.

While using Fusion 360, the design files generated by the users get stored on the cloud. Therefore, users need not carry their systems to access their design files. To access the project library for general viewing and discussion purposes, the user can simply log into their A360 account using their Autodesk ID from any device, from anywhere in the world. Several features of A360 are listed next:

- Project creation
- Member addition to projects
- Project content and information management and sharing
- 2D and 3D designs viewer (A360 Viewer)
- A360 Viewer, Design Library, Comments, and Design History
- Direct design review and mark-up

Note

To access this cloud directory, sign in to your Autodesk account at https://a360.autodesk.com/. The A360 home page for Fusion 360 shows all the project of which you are owner as well as of which you are member.

The A360 workspace, also known as a Hub, is the main control station from where you can create projects and folders and view your projects, ongoing project activity, as well as notifications, and other settings.

TUTORIALS

Tutorial 1

In this tutorial, you will use the collaboration tools offered by Autodesk Fusion 360 to share your designs with groups and/or individuals. Note that, several figures in the tutorial have been provided for demonstration purposes for the clarity of the tutorial. The reader is encouraged to invite friends/colleagues to collectively explore collaborative capabilities of the software.

(Expected time: 1 hour)

The following steps are required to complete this tutorial:

a. Open and explore your design in A360
b. Add collaborators to your project
c. Share your designs with collaborators and outsiders
d. Import and reference external components to your design
e. Start a live review session of your design

Opening the Design in A360

1. Visit https://a360.autodesk.com/; **A360** window is displayed, as shown in Figure 10-2.

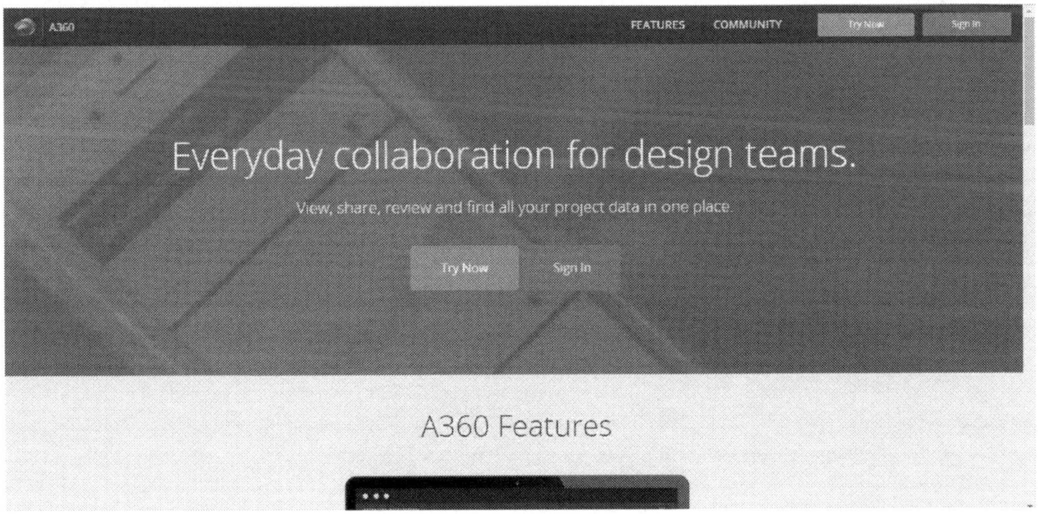

Figure 10-2 The A360 window

You will be able to access all projects that you are a part of whether as an owner or as a group member from the Autodesk Fusion 360 cloud window. On the right side of the Autodesk Fusion 360 cloud window in the Activity panel, you can also view all the updates on the latest activity within your center such as designs, members added, administrator nominations, upcoming events.

2. Click on the **Sign In** button; the **Sign In** page is displayed.

3. Sign in to Autodesk account using Autodesk ID and password; the Autodesk Fusion 360 cloud window is displayed, as shown in Figure 10-3.

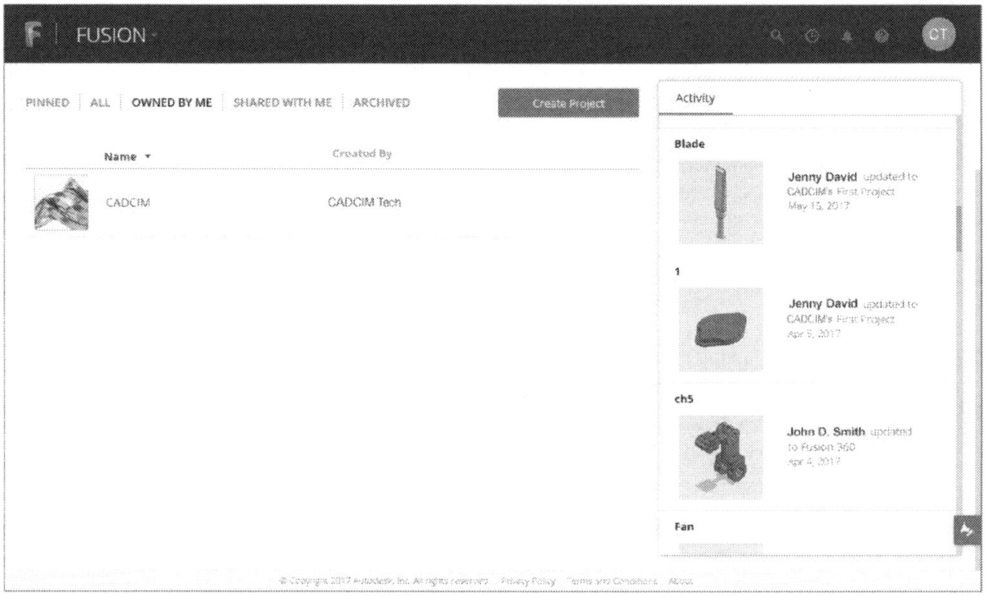

Figure 10-3 *The Autodesk Fusion 360 cloud window*

4. Choose the **Create Project** button from the Autodesk Fusion 360; the **Create Your New Project** dialog box is displayed, as shown in Figure 10-4.

5. Type **Fusion 360** in the **Project Name (required)** field.

6. Choose the **Upload Your Own** button from the **Choose a Project Avatar(required)** area in the dialog box; the **Open** dialog box is displayed.

7. Select the image file from your system and choose the **Open** button from the **Open** dialog box; the selected image gets added to the dialog box along with the existing image.

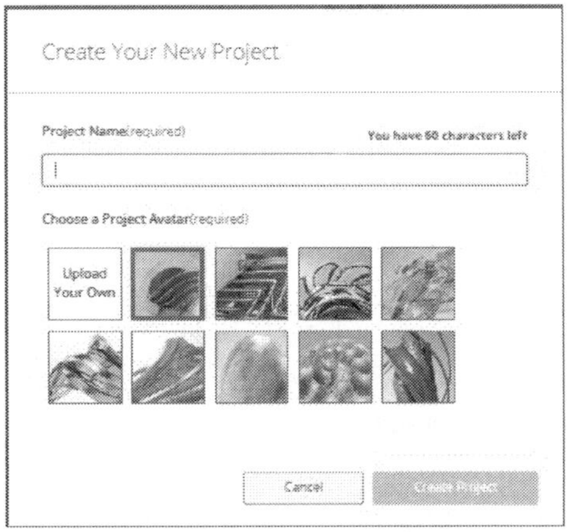

Figure 10-4 *The* ***Create Your New Project*** *dialog box*

8. Choose the **Create Project** button from the **Create Your New Project** dialog box; the Fusion 360 project environment is displayed, as shown in Figure 10-5.

   In this environment, you can upload your project as folders, files, assembly, and drop box file. Also, you can create new folder and invoke the Fusion 360 software by choosing the respective option from the **New** flyout.

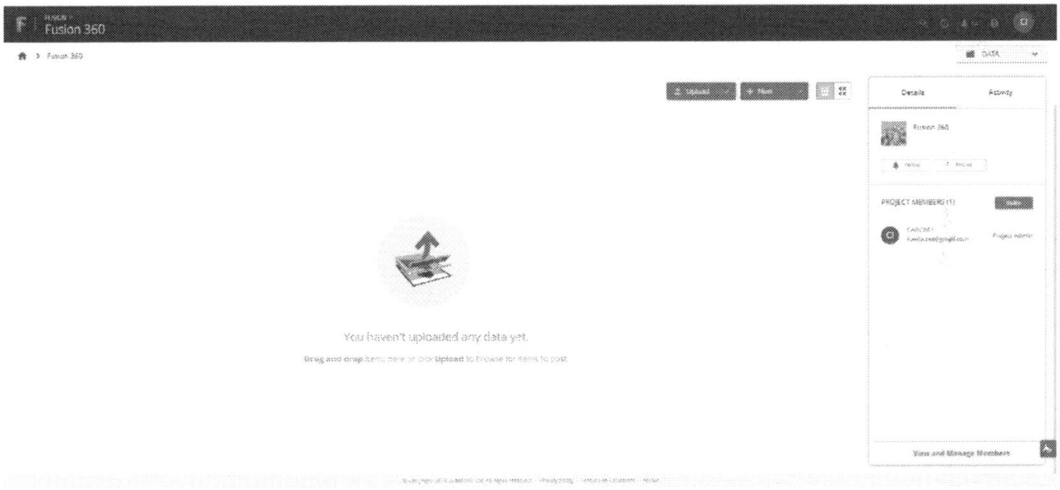

Figure 10-5 *The* ***Fusion 360*** *project window*

9. Select the Home button from the top-left corner of the Project area; the Autodesk Fusion 360 cloud window is displayed. This window lists all the files saved in the project, which can be designs, drawings, rendered images, and so on.

10. Open the **CADCIM** project from the Autodesk Fusion 360 cloud window by clicking on the project.

11. Right-click on **c06_tut02**; a shortcut menu is displayed.

12. Choose the **Move** option from this shortcut menu; the **Move** dialog box is displayed.

13. Select the **Fusion 360** project folder from the **Move** dialog box as destination folder to move this file.

14. Choose the **Move** button from the **Move** dialog box; the **c06_tut_02** file from the **CADCIM** project moved to the Fusion 360 project folder.

15. Open the **Fusion 360** project folder and select the design file named **c06_tut02**; the **Design References** window is invoked, refer to Figure 10-5.

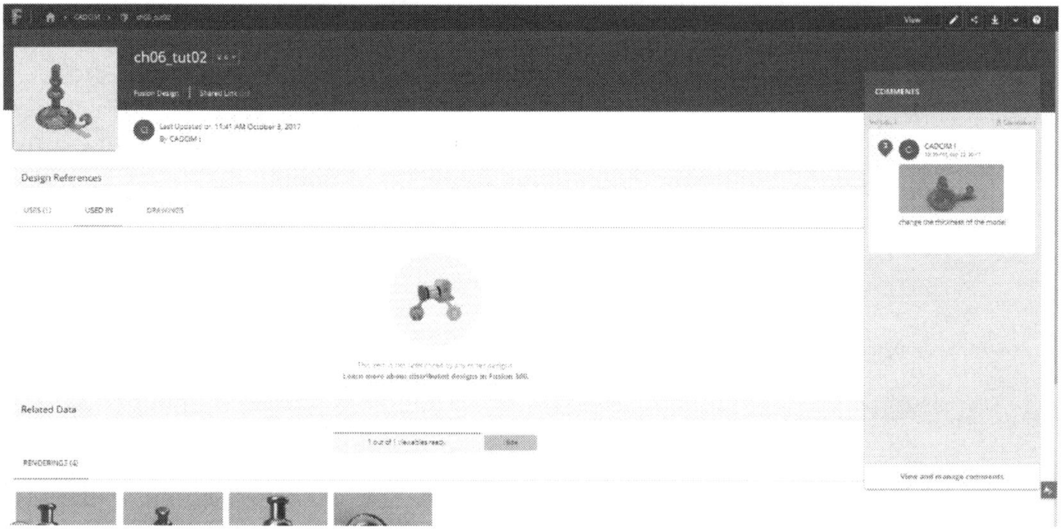

Figure 10-5 *The partial view of the* ***Design References*** *window*

In the **Design References** window, you can edit your design by using the **Edit** button at the right side of the top most of this window and add comments to your designs. You can also share the link of your design via e-mail and also download the design file in many formats such as *Inventor 2016, IGES, SAT, SMT, DWG*. By using **Design References**, window you can also view the design reference as well as generated rendering image of the design.

Viewing and Expecting Designs

The A360 Viewer is a platform where you can view your designs directly from your web browser and make some preliminary inspections.

1. Choose the **View** button from the top right corner of the **Design References** window; the design opens in online viewer called the A360 Viewer window, refer to Figure 10-6.

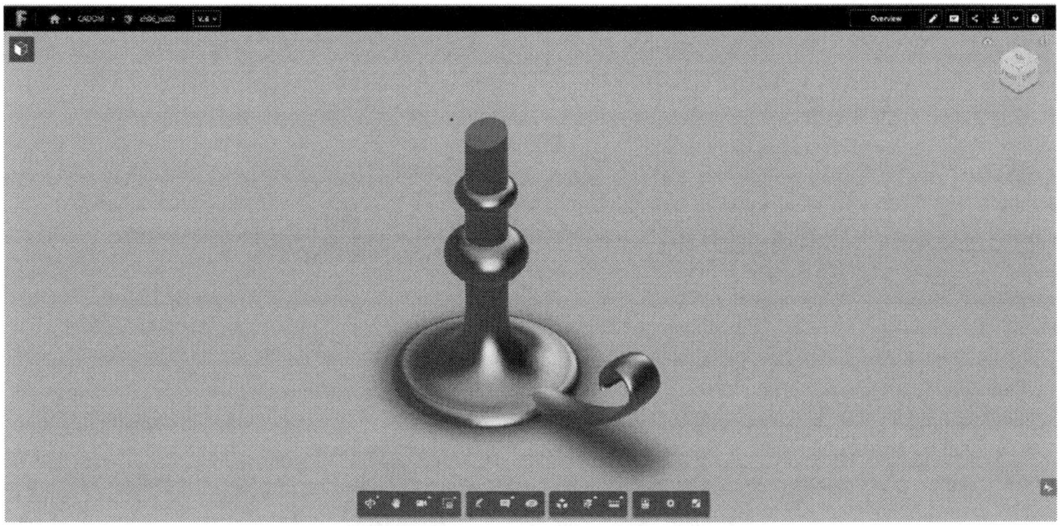

Figure 10-6 *A360 Viewer window*

The A360 Viewer is a platform where you can view your designs directly by using the **Design** button from your web browser. You can navigate the design by using the tools available in the View toolbar, mark and comments on the design with the help of Markup and Comments toolbar. The preliminary inspections of the design are made by using the Examine toolbar. You can manage the visualization of the designs such as Display lines, Display points, Environment and Lighting Selection by using the options available in the Settings flyout. Also, you can view the properties of the designs such as material, mass, volume.

2. Select the **Orbit** tool from the **View** toolbar of the A360 Viewer. Next, move the cursor towards the design.

3. Drag the cursor vertically or horizontally in the graphics area; the camera view of design gets reoriented.

> **Tip**
> *The view can also be manipulated with the help of the ViewCube on the right corner of the graphics window.*

4. Select the **Comments** tool from the Markup and Comment toolbar in the A360 Viewer; the Comments flyout is displayed.

5. Choose the **Comment on Points** option from the flyout; a red horizontal/vertical infinite line is attached to the cursor.

6. Hover the cursor on the model; the red infinite line will turn blue.

7. Click anywhere on the design; a small Comment on Point icon is attached to the chosen point and the Comment dialog box is displayed next to it, as shown in Figure 10-7.

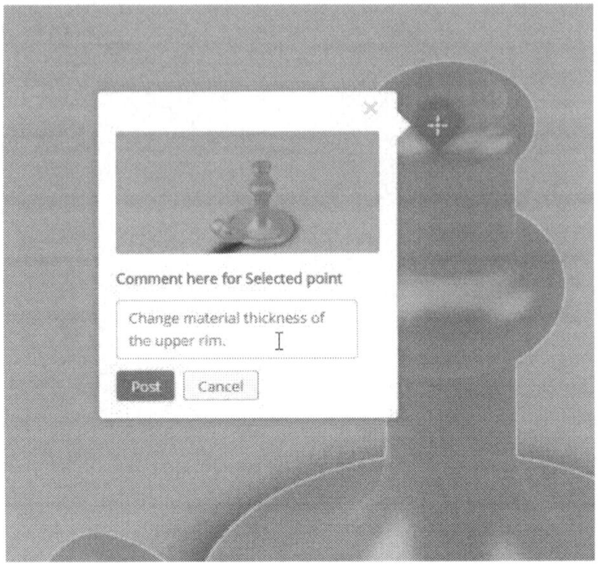

Figure 10-7 The Comments dialog box

The infinite line helps the user in locating the exact point where you wish to make a comment on the design.

8. Type **Change material thickness of the upper rim** in the comment box.

9. Choose the **Post** button from the **Comments** dialog box; the comment is displayed in the **COMMENTS** side bar of the **Design Reference** window, as shown in Figure 10-8. To switch back to the **Design Reference** window, you need to choose the **Overview** button from the A360 View window.

   Note that when a Project Manager accesses any comment left by some other member collaborator, the design view will be automatically changed to the view in which the comment was made. Also, all the comments are arranged numerically in ascending order from oldest to most recent.

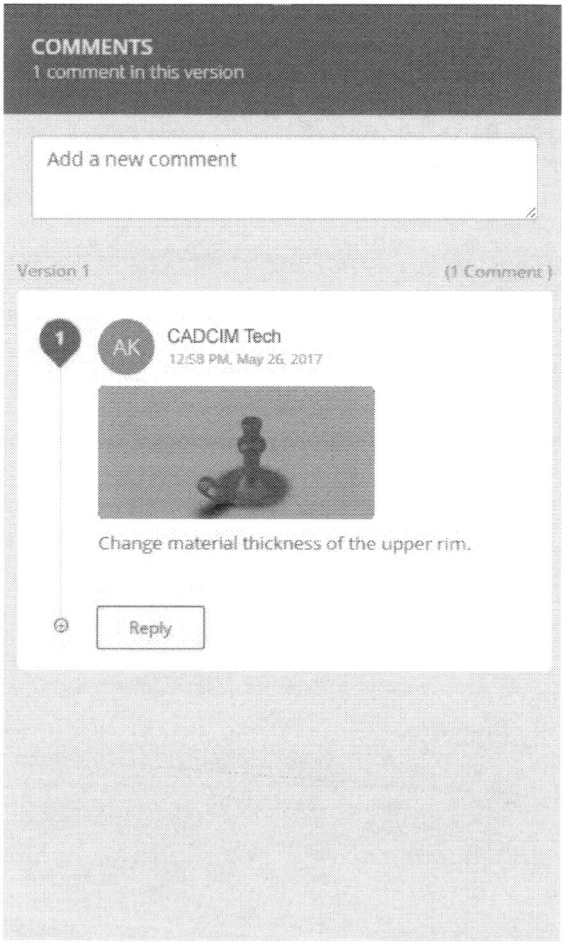

Figure 10-8 The COMMENTS side bar

10. Choose the **Markup** tool from the Markup and Comment toolbar; the **Markup** contextual toolbar is displayed, as shown in Figure 10-9.

*Figure 10-9 The **Markup** contextual toolbar*

11. Choose the **Arrow** tool from the **Line Tool** flyout.

12. Choose the **Line Style** tool; the Line Style toolbox is displayed, as shown in Figure 10-10.

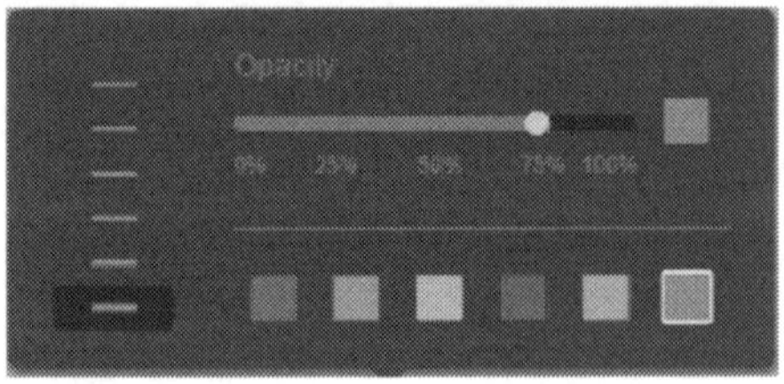

Figure 10-10 *The Line Style toolbox*

The tools available in the pop up window are used to change the line weight, color and opacity of the design. If you want to set the background color of any of the markups then reduce the opacity to zero.

13. Set the opacity to approximately 75%, the color to green and the line weight to the heaviest stroke, refer to Figure 10-10.

14. Draw an arrow pointing away from the design and choose the **Text** tool from the Markup contextual toolbar.

15. Add a text box near the arrowhead and type a comment in it. Next, click anywhere in the graphics window to exit the text box.

    You can edit the formatting of the text using the **Text Style** tool in the Markup contextual toolbar. To manually adjust the position of these markups, just click and drag the individual items to the desired location.

16. Choose the **Save** button from the Markup flyout. Now, the comment can be seen on the **COMMENTS** side bar of the **Design Reference** window.

> ***Tip***
> *While working in a group, you can access the comments of your Project member from the* ***COMMENTS*** *side bar of the* ***Design References*** *window and reply to them directly from there.*

You will now measure the total height of the candlestick design.

17. Select the **Measure** tool from the **Examine** toolbar; the **MEASURE** dialog box is displayed.

18. Select edges 1 and 2, refer to Figure 10-11; the two edges are highlighted in blue and the distance between the two edges and the radii of the circular edges are displayed, as shown in Figure 10-12.

    To make another measurement, choose the **Restart** button from the **MEASURE** dialog box; all previous selections are cleared out.

Figure 10-11 The edges to be selected

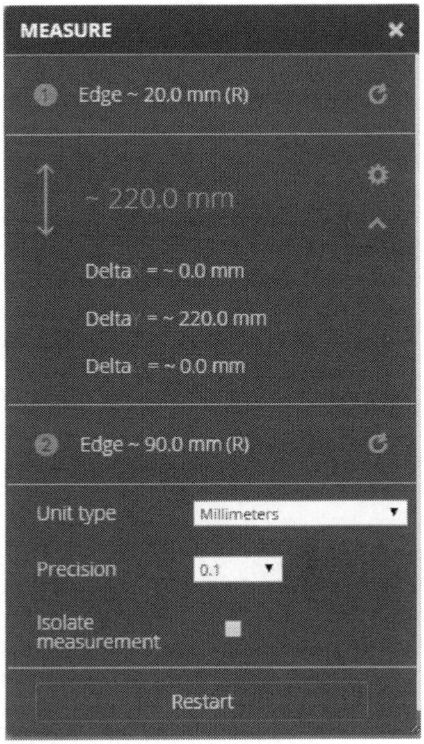

***Figure 10-12** The **MEASURE** dialog box*

Note

Apart from the model design, on the basis of the design file attributes, the Design browser menu on the left hand side of the viewer will display additional options like CAM details, Simulation results, and Animation, refer to Figure 10-13. Select the required option to view the desired details in the A360 viewer.

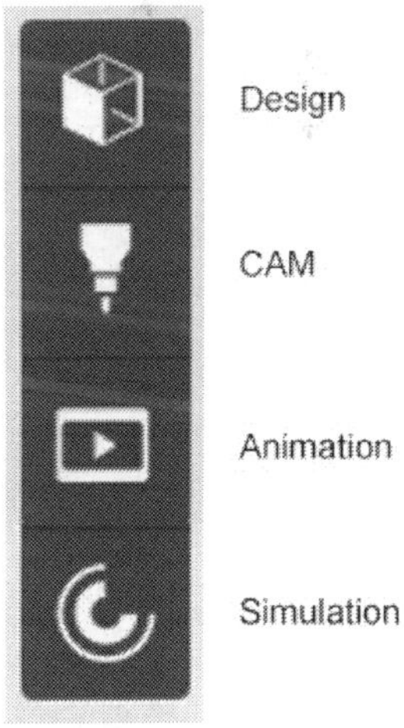

***Figure 10-13** The Design Browser options*

Sharing Designs and Adding Collaborators

The Fusion 360 offers you powerful and flexible sharing capabilities to send your files to the collaborators, members, or vendors.

1. Choose the **Share** button from the top right corner of the window; a flyout is displayed.

2. Choose the **Share** option from the **Share** flyout; the **Share** dialog is displayed, as shown in Figure 10-14.

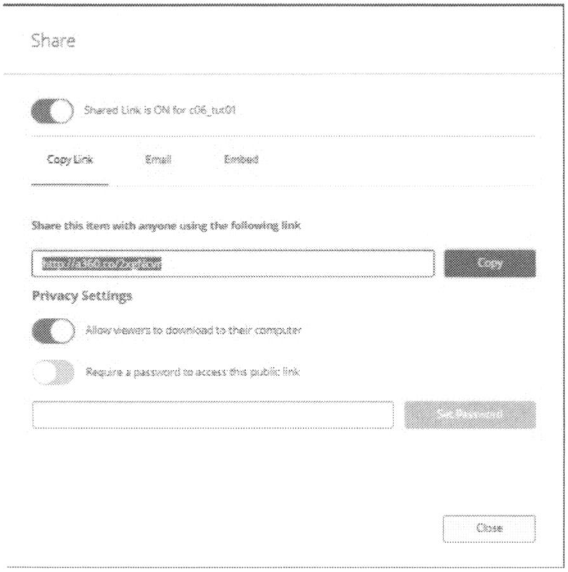

Figure 10-14 *The **Share** dialog box*

3. Choose the **Copy** button from the **Copy Link** tab to copy the design link to your clipboard. Now, you can share the file by sending this link via any medium, such as e-mail, chat conversation, or a website post to anyone.

   You need to restrict access to your design so that only people having viewing permission can have its access.

4. Choose the **Require a password to access this public link** toggle button from the **Privacy Settings** area of the **Share** dialog box; the **Set Password** text edit box gets activated.

5. Enter a password into the **Set Password** edit box and then choose the **Set Password** button. Now, only the people who know the password can access the shared design.

   Now, you need to add collaborators to your project. For this, you have two options - Add members directly from Fusion 360 using the **Data Panel** or from **A360** by using the Activity Panel found in the Fusion 360 Project environment.

6. Invoke the **Fusion 360** project folder page and choose the **View and Manage Members** button from the **Details** panel; the **MANAGE PROJECT MEMBERS** window is displayed, as shown in Figure 10-15.

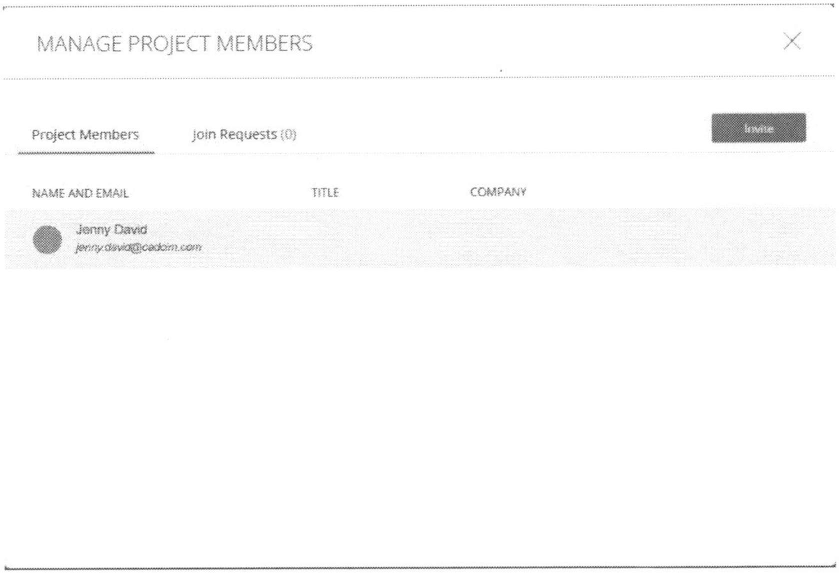

***Figure 10-15** The **MANAGE PROJECT MEMBERS** dialog box*

7. Choose the **Invite** button from the top right corner of the **MANAGE PROJECT MEMBER** dialog box; the **Who do you want to invite to this project?** dialog box is displayed, refer to Figure 10-16.

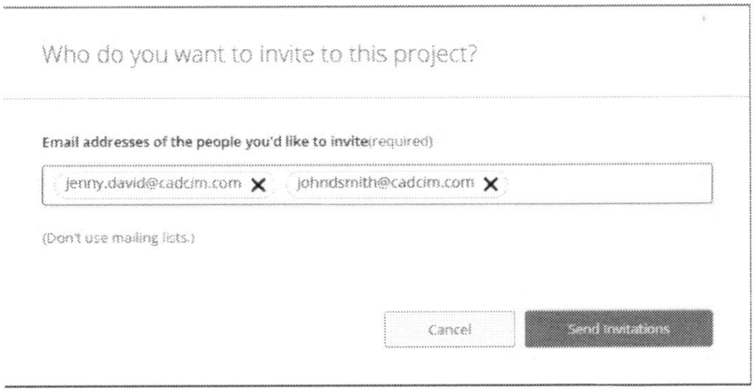

***Figure 10-16** The **Who do you want to invite to this project?** dialog box*

8. Enter the e-mail of the invitees separated by commas in the dialog box, refer to Figure 10-16.

9. Choose the **Send invitations** button from the dialog box; an invitation e-mail with an **Access Project** link is sent out to invitees and they get added to your project.

Alternatively, open the **Data Panel**. Next, open the desired project folder from the

PROJECTS list to which you want to add members. Then, select the **People** tab and enter the e-mail address of the member in the **Enter email addresses** text box and choose the **Invite** button, as shown in Figure 10-17.

Note that to invite multiple people at the same time, enter the email addresses separated by commas.

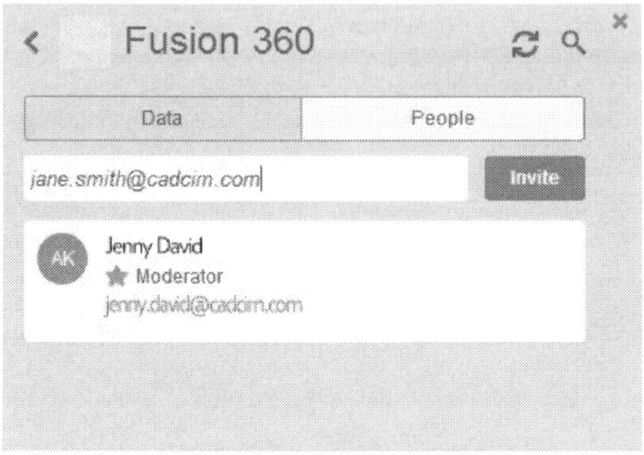

Figure 10-17 *Inviting a collaborator*

Note

You can also delete a collaborator using the ***MANAGE PROJECT MEMBERS*** *dialog box. To delete a collaborator, choose the* ***Delete*** *icon in the row of the collaborator; a message is displayed asking you to remove the members you would like to remove from the project. Also two buttons are added to the dialog box. Choose the* ***Remove*** *button; collaborator will be removed from the project.*

Creating Distributed Design

The designs created using Fusion 360 have another powerful feature that is Distributed Design. This feature enables a collaborator to use the design components created by others in their group and integrate them seamlessly with own design. For instance, if a group is working on creating a bike design in Fusion 360, the bike frame could be made by collaborator A, the bike seat by collaborator B, the clamps made by collaborator C, and so on.

1. Open the **c06_tut01** design in A360.

2. Choose the **Edit design** button from the **Design References** window, refer to Figure 10-5; a flyout is displayed.

3. Choose the **Edit on Desktop** option from the flyout; the design is displayed in the Autodesk Fusion 360 software.

4. Choose **File>New Design** and create the design as shown in Figure 10-18 and save the design with the name Candle in the **Data Panel**.

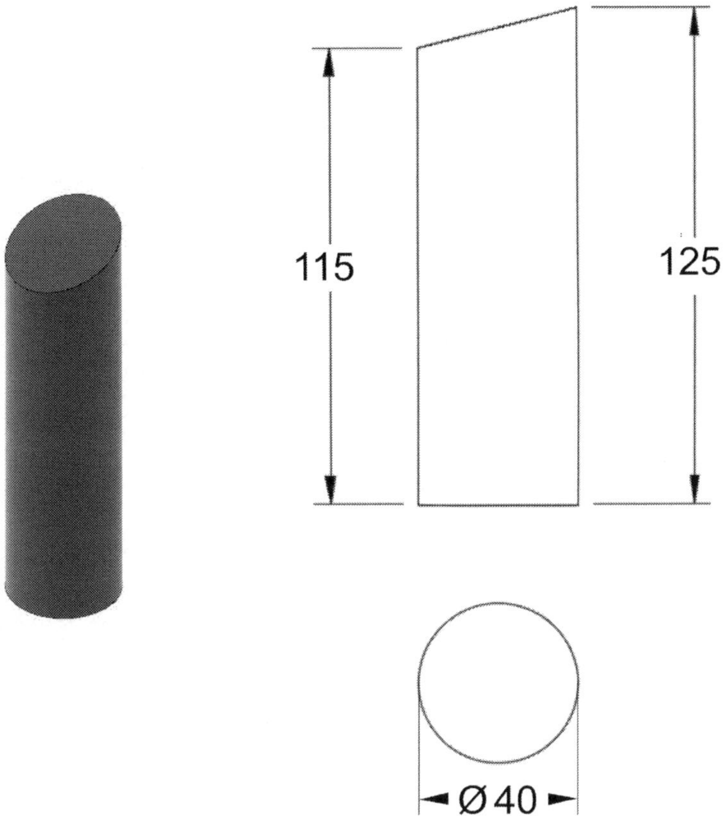

Figure 10-18 Model and model dimensions

5. Switch to the **c06_tut02** design file and open the Data Panel.

6. Drag the newly created Candle design to the graphics window and invoke the **JOINTS** dialog box from the **ASSEMBLE** panel.

7. Apply the requisite joints between the two bodies to obtain the design, as shown in Figure 10-19 and save the design.

8. Select the Fusion 360 folder from the very top of the Data Panel; the Fusion 360 is opened in the web browser.

9. Open the **c06_tut02** design; the Fusion 360 **Design References** window is displayed, refer to Figure 10-20.

Figure 10-19 Design of Candle Stick

Here, you will notice that Candle design is displayed in the **Design Reference** area under the **USES** tab.

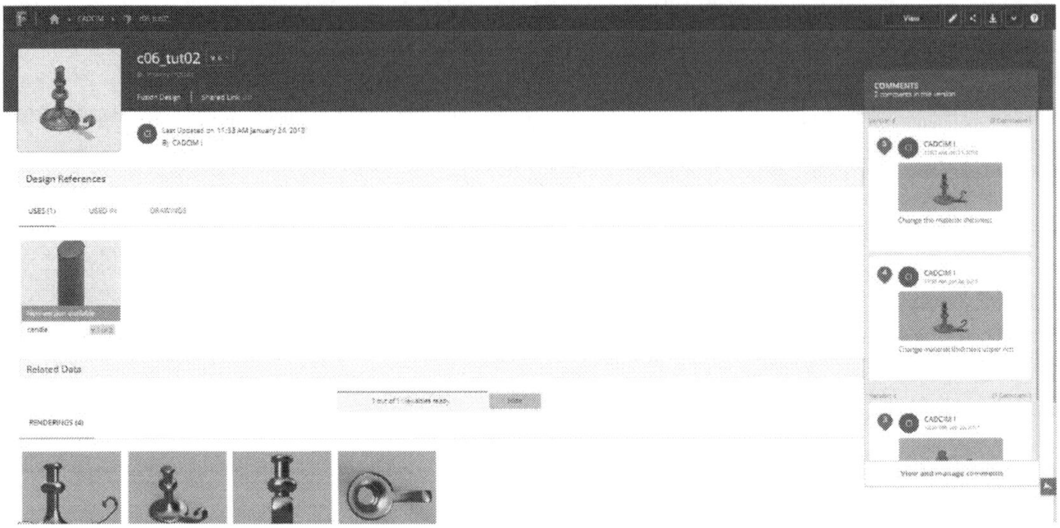

Figure 10-20 *Design reference updated in the* ***USES*** *tab*

Note

When you insert design components to the Fusion 360, all the inserted components will be automatically linked and referenced to their respective source component files. So, in case the original component file is moved to a different folder or is renamed, the design file that references that component will not have any broken links.

10. Move the cursor on the Candle design in the **USES** tab; the **Overview** button is displayed as shown in Figure 10-21.

11. Choose the **Overview** button from the design; the **Design Reference** window is displayed.

12. Choose the **Edit on Desktop** button from the **Edit design** flyout in the **Design Reference** window; the design opens in Fusion 360.

13. Increase the height of the model by 25 mm by using the **Press Pull** tool from the **MODIFY** panel. Next, save the design.

14. Open the c06_tut02 design in Fusion 360. You will notice a triangular yellow icon in the Application bar in front of the **c06_tut02** component node and also in front of the **Candle** node in the **BROWSER** bar. This sign indicates that a component is out of date. Refer to Figure 10-22.

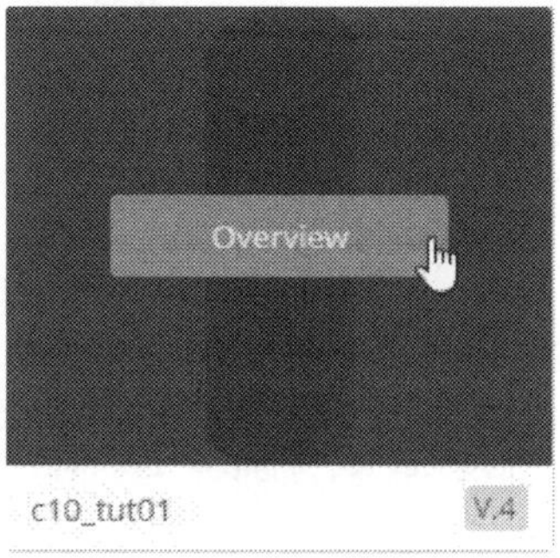

Figure 10-21 The Overview button of referenced design

Figure 10-22 Error symbol for Out-of-date component

When any modifications is made to the referenced design, Fusion 360 will highlight that particular referenced design in your design as the component is out of date and you need to update it.

15. Select the **Out-of-date** icon from the Application bar; the design gets updated.

16. Expand the **COMMENTS** bar in Fusion 360 from the bottom-left of graphics window, as shown in Figure 10-23 and comment in the comment box. Now, choose the **Post** button from **COMMENTS** bar.

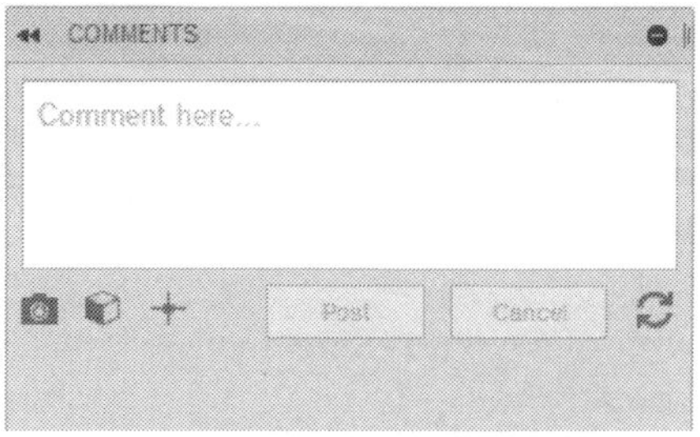

*Figure 10-23 The **COMMENTS** bar*

*The **COMMENTS** bar allows all collaborators to leave comments on shared designs within the group project folder. This fastens the implementation process of suggested changes. For example, in the case of an error in one of the reference designs, an error message will be displayed in the main design file. By following the link of the reference design having error, collaborators can immediately leave a comment to point out the error in the design to the owner of the design. The owner can view the comments made by the collaborators.*

Organizing Live Review Session

Communicating changes in the design with a team of several members can be a giant task, more so if they are physically at different locations. Autodesk Fusion 360 overcomes this problem by offering an integrated live design review feature of its own called the Live Review Session.

1. Choose the **File>Live Review Session** from the Application bar; the **Fusion 360** message box is displayed, as shown in Figure 10-24.

2. Choose the **Enable** button from the message box; the **LIVE REVIEW SESSION** dialog box is displayed, refer to Figure 10-25.

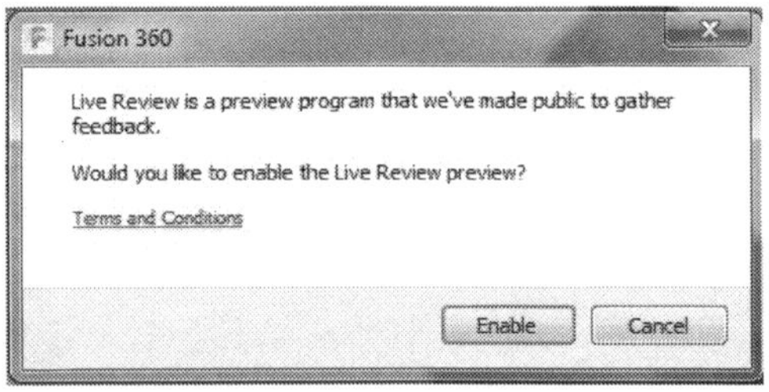

Figure 10-24 *The* ***Fusion 360*** *message box*

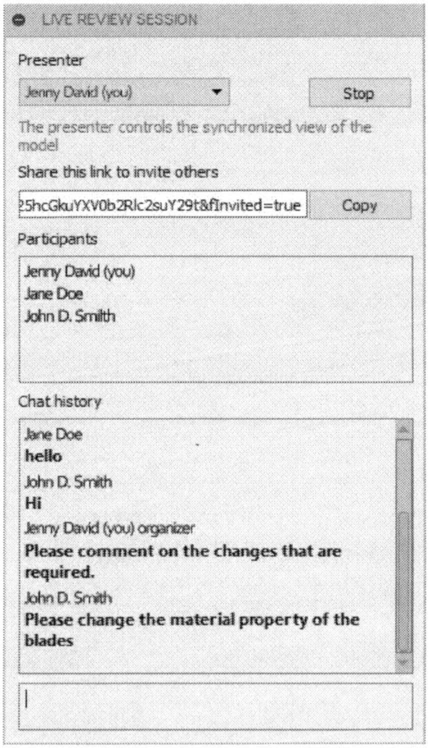

Figure 10-25 *The* ***LIVE REVIEW SESSION*** *dialog box*

3. Invite the participants by copying and sharing the auto-generated link through any media like an e-mail or chat conversation; the name of all the active participants are added to the **Participants** area.

   Once the session is set up, the host can further delegate the task of managing the session either to himself/herself or to any other participant by choosing the name of the participant from the **Presenter** drop-down list in the **LIVE REVIEW SESSION** dialog box.

The role of the Presenter is to control the views of the design being displayed while the host is the only person who can make live changes to the design which will be updated on the screen of the collaborators in real-time. All session participants can make comments and markups on the design, refer to Figure 10-25.

Since these sessions are carried out over web browsers, none of the participants, except for the host, require Fusion 360 installed on their system.

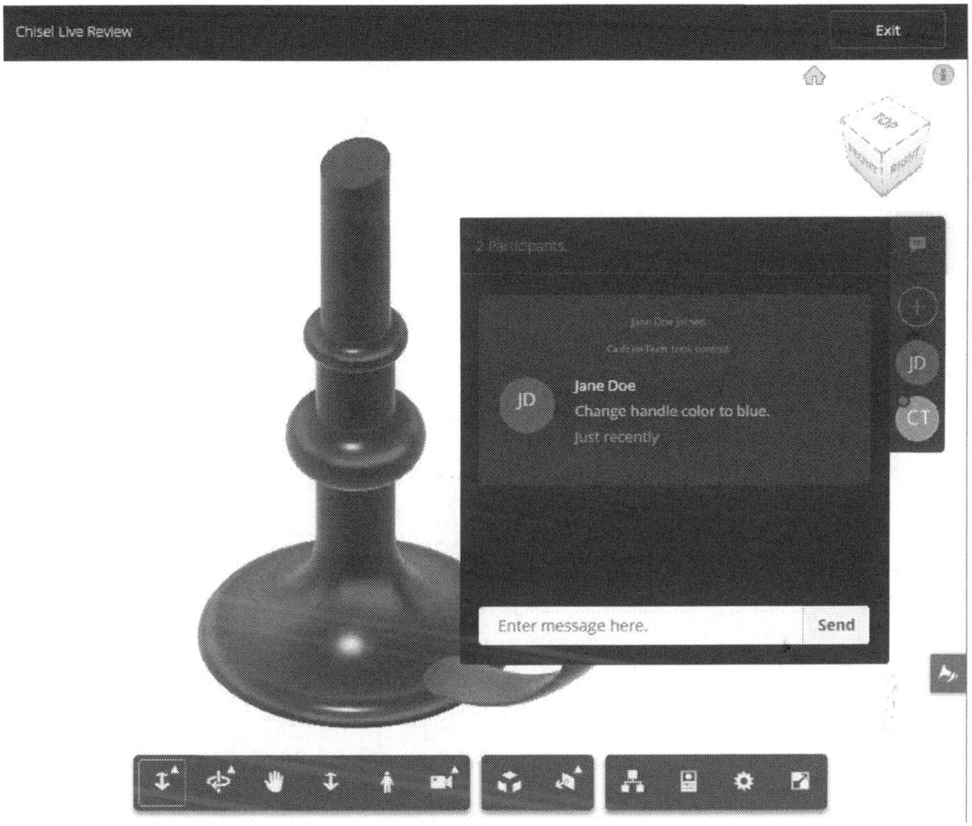

Figure 10-25 *Live review in a web browser*

Note

*If the **Live Review Session** option is not available or is not functioning then select the **Live Review - Collaboration Support** check box from **Preferences** > **Preview** to enable it.*

Self-Evaluation Test

Answer the following questions and then compare them to those given at the end of this chapter:

1. Which of the following privacy settings is enabled by choosing the **Share** button in the menu bar of the A360 workspace?

   (a) Visibility to Collaborators (b) Limited Access to certain Individuals
   (c) Password-Protected Access (d) None of these

2. You can insert designs from other collaborators into a desired design file by directly dragging the file from the __________.

3. Default renderings of a design can be viewed from the __________ page in the A360 Workspace.

4. In a Live Review Session, two participants can take control of the presentation at a time, namely __________ and __________.

5. The __________ tool in the Projects page of A360 enables collaborators to schedule project deadlines and meetings.

6. Fusion 360 allows users to promote older design versions over latest versions. (T/F)

7. A360 supports various file formats for online viewing including 3DS, DWG, MP4, and PDF files. (T/F)

8. The **Activity** panel in A360 also lists the changes made by other collaborators on the designs in the project folder. (T/F)

9. The physical material of an object can be changed using the **Settings** panel in the A360 Viewer. (T/F)

10. A maximum of 12 people can be invited to collaborate on a project. (T/F)

Review Questions

Answer the following questions:

1. Referenced designs are displayed with a __________ symbol beside their names in the **BROWSER** bar.

2. Which of the following is not a component found in the A360 workspace?

   (a) Data Review (b) Overview
   (c) Activity Panel (d) Discussions

3. Associativity of referenced designs can be broken by selecting the design in the **BROWSER** bar and choosing the __________ option from the shortcut menu displayed.

4. To do section analysis of a design in the A360 Viewer, you have to use the __________ panel from the toolbar.

5. The **Live Review Session** feature is enabled by default in Fusion 360. (T/F)

6. Comments made in the A360 Viewer are linked to the exact views in which they were made. (T/F)

7. During a live review session, the presentor must have Fusion 360 installed and running on their system. (T/F)

8. The Design View options in the A360 Viewer includes Design, Simulation, __________, and __________.

9. Using the **Examination** panel, you can measure and explode a design in the A360 Viewer. (T/F)

10. The A360 Viewer can be embedded on an external website using the **Share** button from the A360 Workspace. (T/F)

Answers to Self-Evaluation Test

1. c 2. **Data Panel**, 3. Overview, 4. Host, Presenter, 5. Calendar, 6. T, 7. T, 8. T, 9. F, 10. F

Student Projects

Student Project 1

Create all components of the Wheel Support assembly and then assemble them, as shown in Figure 1. The exploded view of the assembly is shown in Figure 2. The dimension of the components are shown in Figures 3 through 6. **(Expected time: 1hr 45 min)**

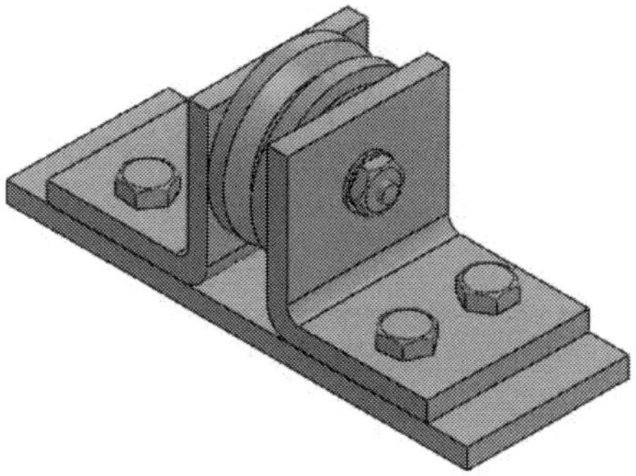

Figure 1 *Wheel Support assembly*

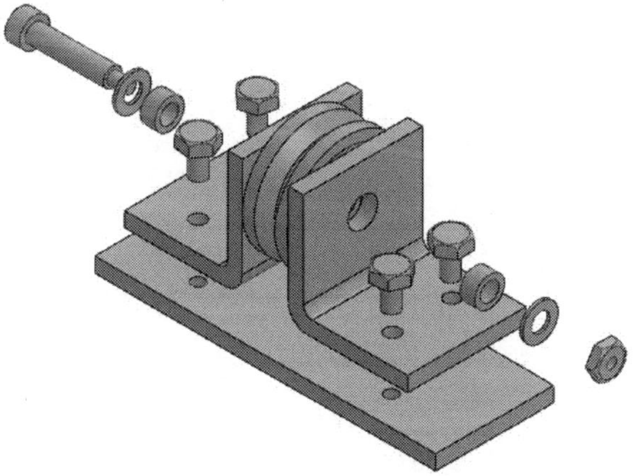

Figure 2 *Exploded view of the Wheel Support assembly*

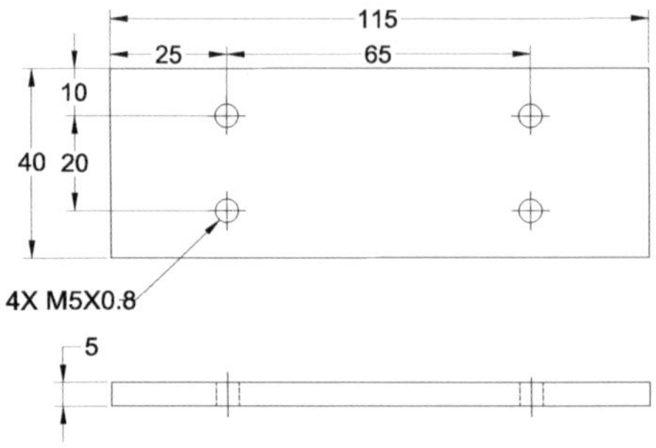

Figure 3 *Front and Top views of the Base*

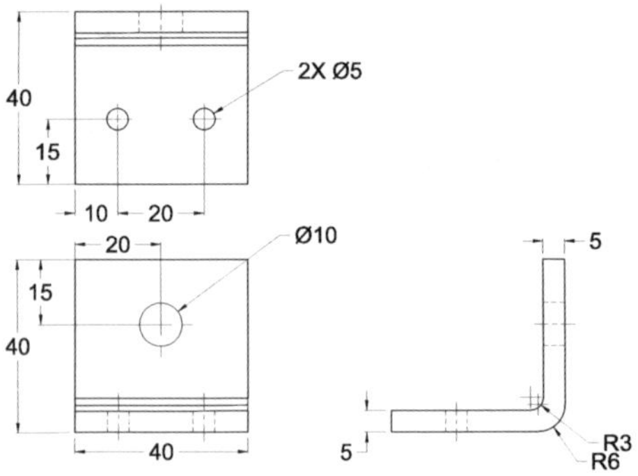

Figure 4 *Top, Front and Right views of the Support*

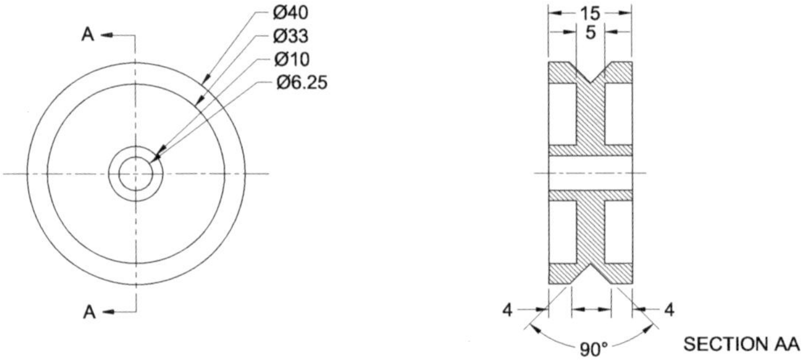

Figure 5 *Front and Section views of the wheel*

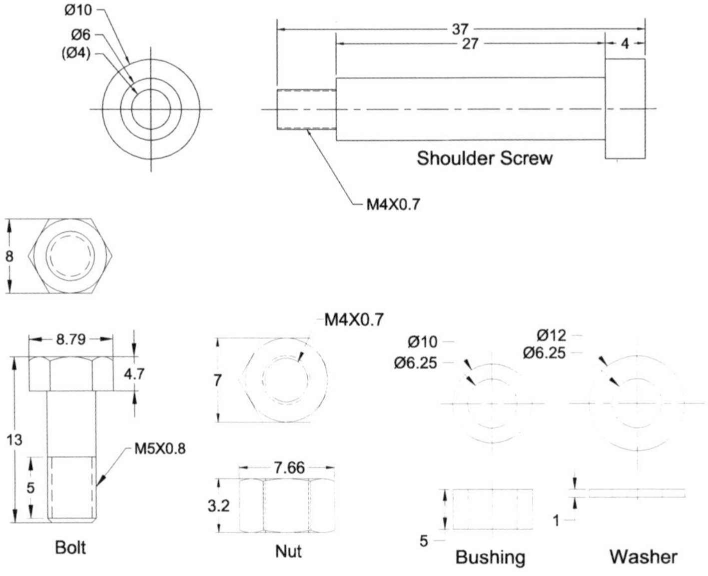

Figure 6 *Dimensions of the Shoulder Screw, Bolt, Nut, Bushing, and Washer*

Student Project 2

Create all components of the Crosshead assembly and then assemble them, as shown in Figure 7. The exploded view of the assembly is shown in Figure 8. The dimension of the components are shown in Figures 9 through 13. **(Expected time: 2hr)**

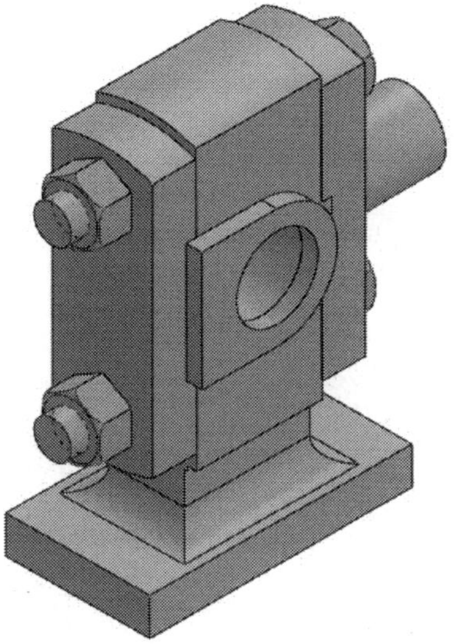

Figure 7 *Crosshead assembly*

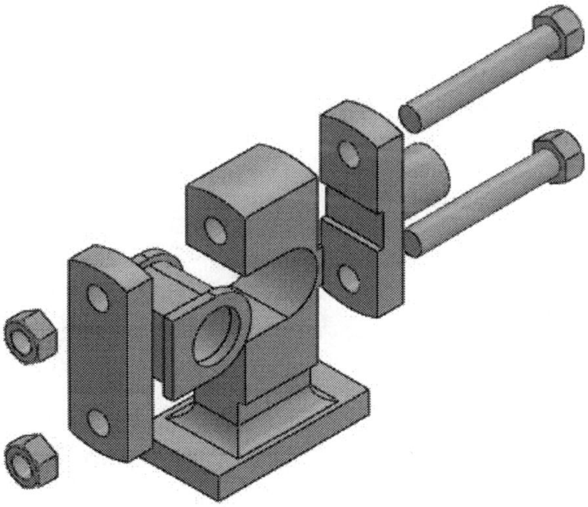

Figure 8 *Exploded view of the Crosshead assembly*

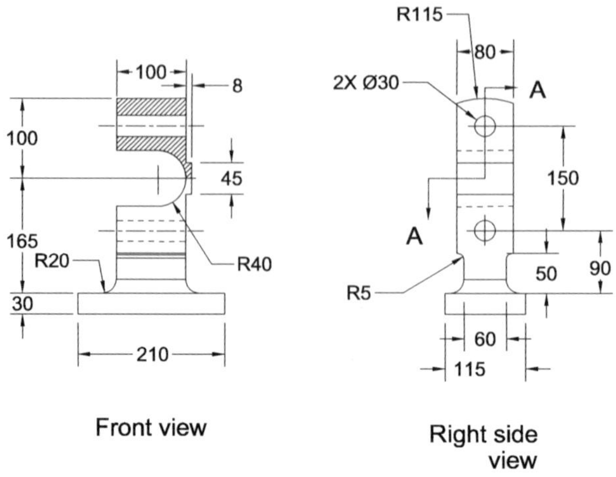

Figure 9 *Front view and right-side views of the Body*

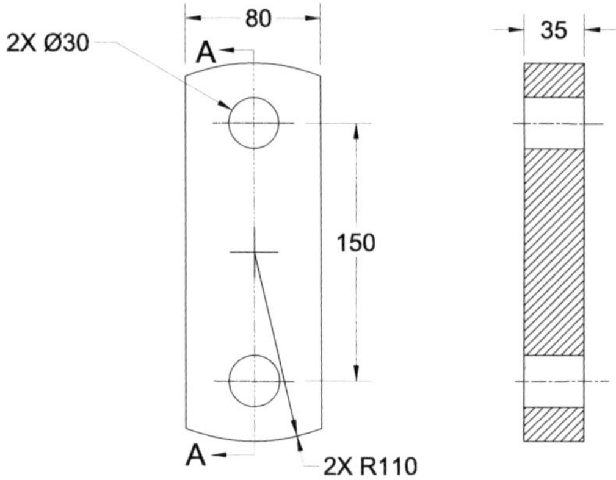

Figure10 *Dimensions of the Keep Plate*

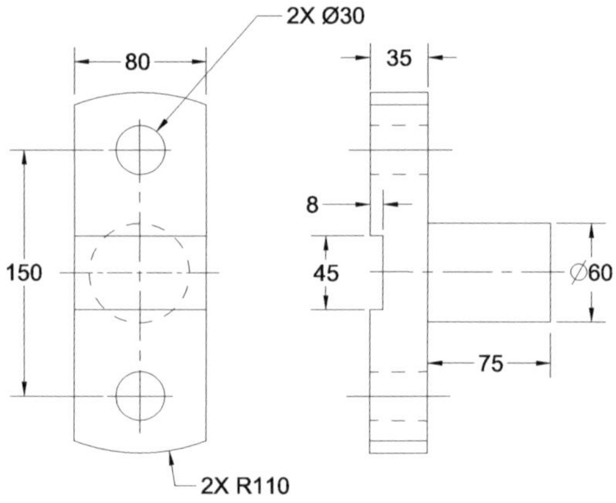

Figure 11 *Dimensions of the Piston Rod*

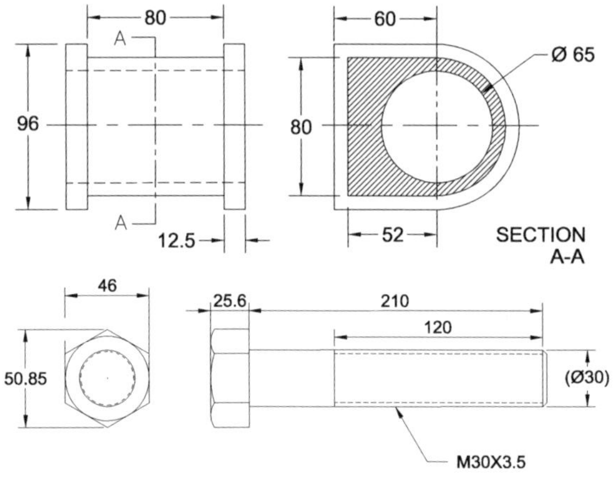

Figure 12 *Dimensions of the Brass and Bolt*

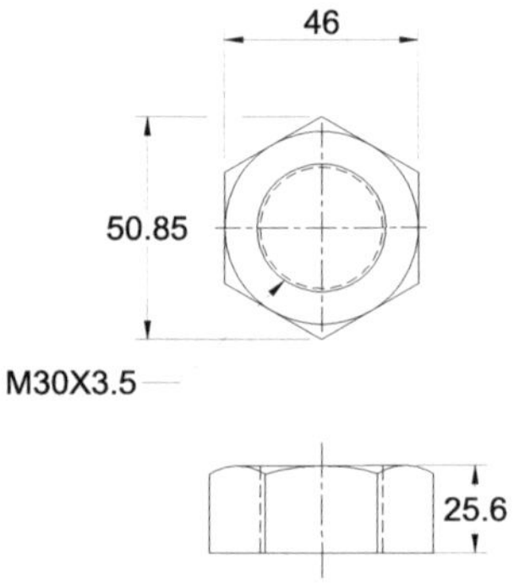

Figure 13 *Dimensions of the Nut*

Student Project 3

Create all components of the Lever assembly and then assemble them, as shown in Figure 14. The exploded view of the assembly is shown in Figure 15. The dimension of the components are shown in Figures 16 through 21. **(Expected time: 2hr)**

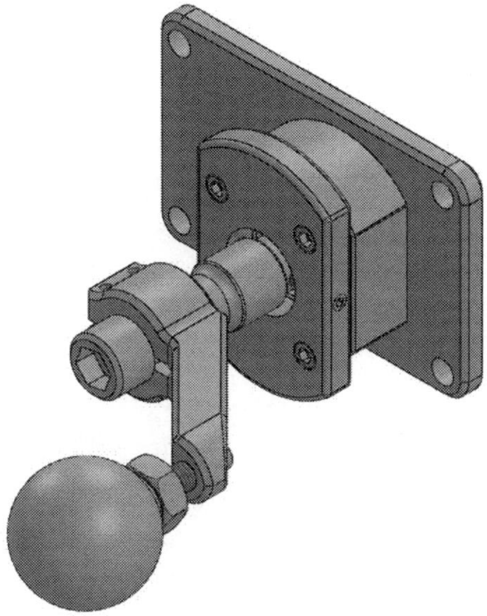

Figure 14 *Lever assembly*

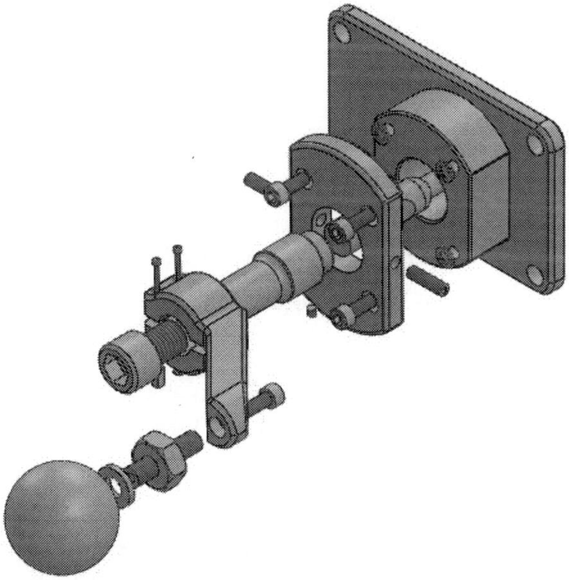

Figure 15 *Exploded view of the Lever assembly*

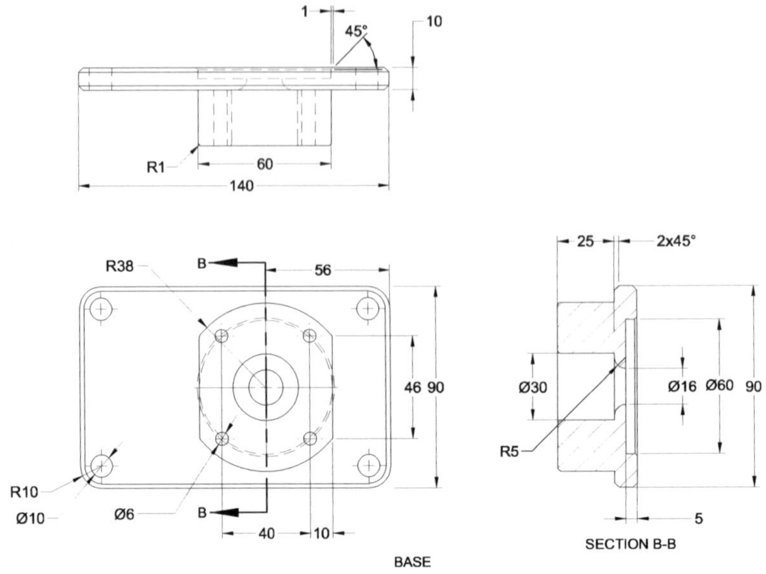

Figure 16 *Dimensions of the Base*

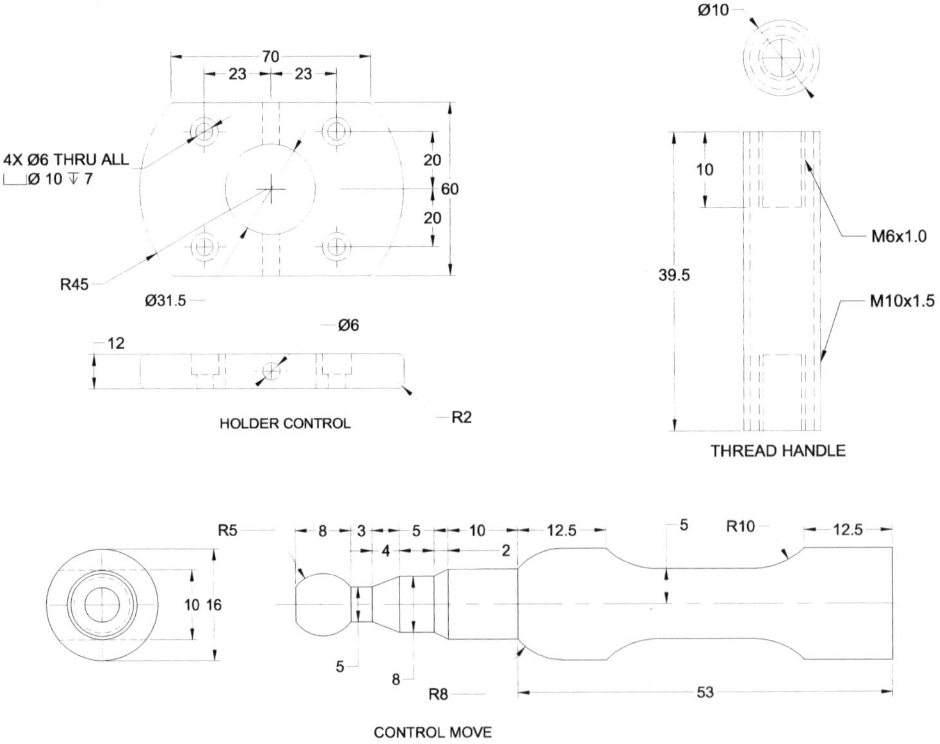

Figure 17 *Dimensions of the Holder Control, Thread Handle, Control Move*

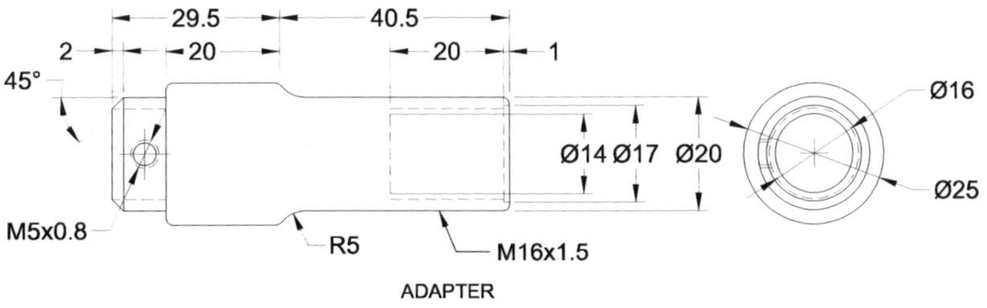

Figure 18 *Dimensions of the Adapter*

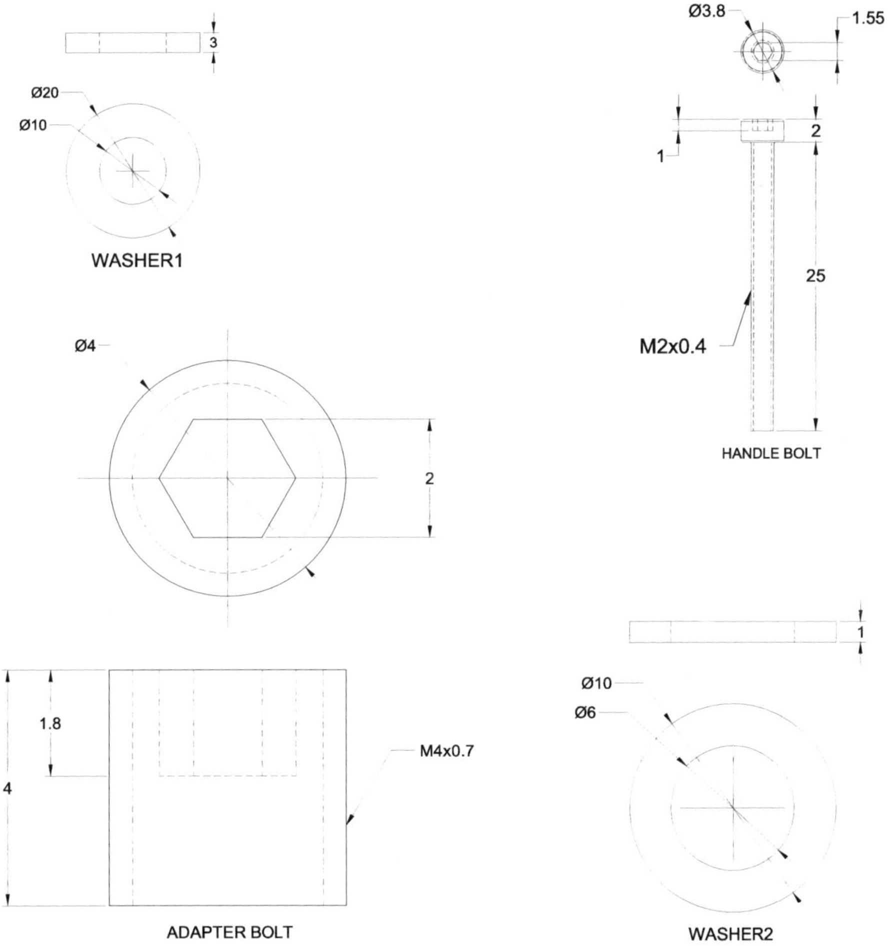

Figure 19 *Dimensions of the Various components*

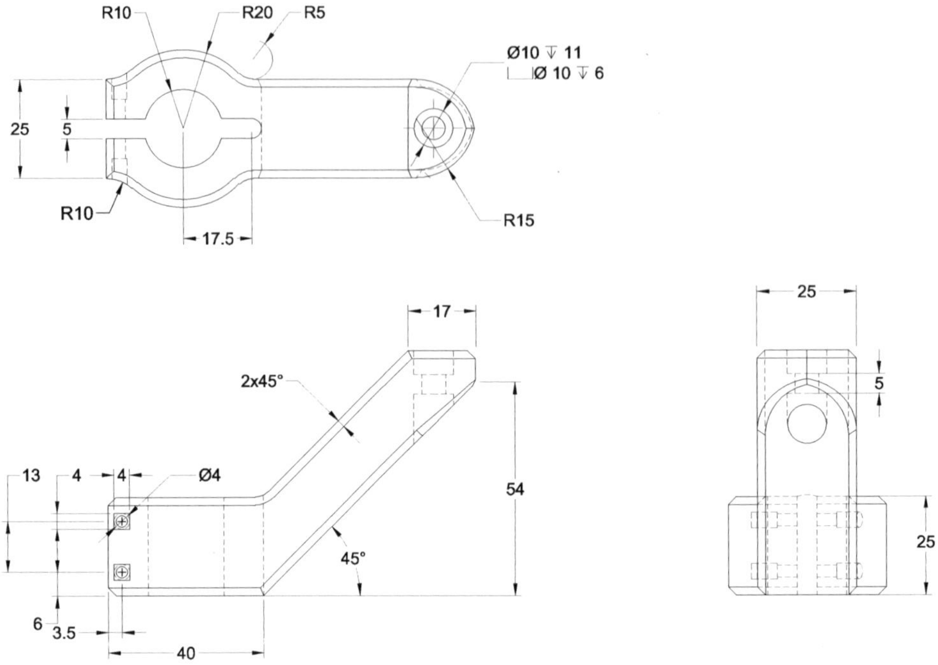

Figure 20 Dimensions of the Handle

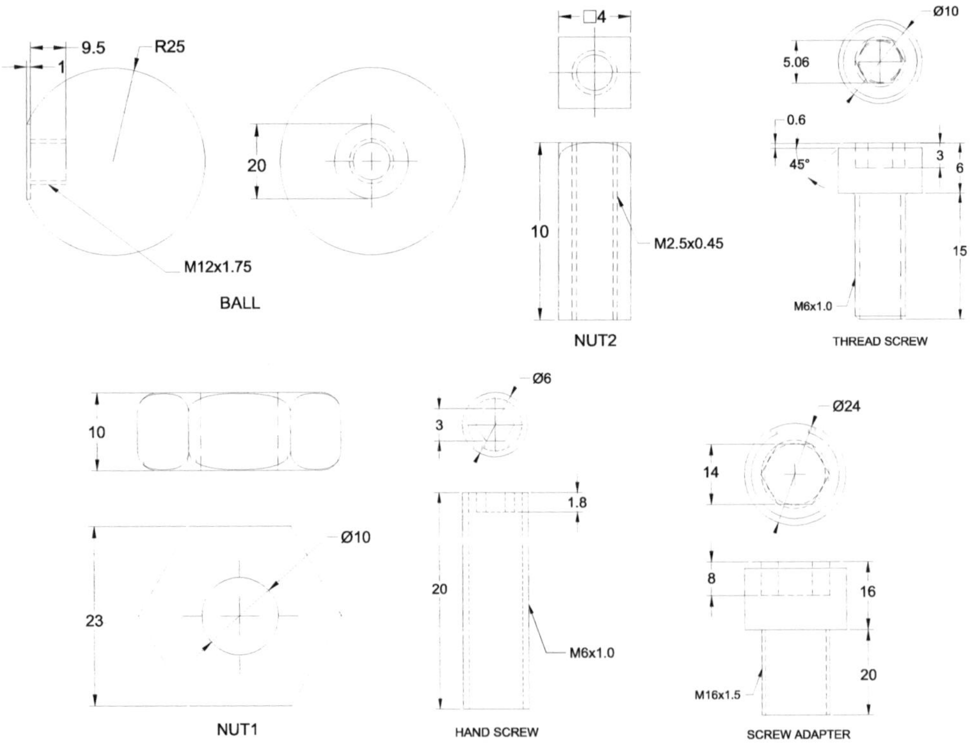

Figure 21 Dimensions of the Various components

Index

W

Z

Other Publications by CADCIM Technologies

The following is the list of some of the publications by CADCIM Technologies. Please visit *www.cadcim.com* for the complete listing.

AutoCAD Textbooks

- AutoCAD 2018: A Problem-Solving Approach, Basic and Intermediate, 24th Edition
- Advanced AutoCAD 2018: A Problem-Solving Approach, 3D and Advanced, 24th Edition
- AutoCAD 2017: A Problem-Solving Approach, Basic and Intermediate, 23rd Edition
- AutoCAD 2017: A Problem-Solving Approach, 3D and Advanced, 23rd Edition

Autodesk Inventor Textbooks

- Autodesk Inventor Professional 2019 for Designers, 19th Edition
- Autodesk Inventor Professional 2018 for Designers, 18th Edition
- Autodesk Inventor Professional 2017 for Designers, 17th Edition
- Autodesk Inventor 2016 for Designers, 16th Edition

AutoCAD MEP Textbooks

- AutoCAD MEP 2018 for Designers, 4th Edition
- AutoCAD MEP 2016 for Designers, 3rd Edition

Solid Edge Textbooks

- Solid Edge ST10 for Designers, 15th Edition
- Solid Edge ST9 for Designers, 14th Edition
- Solid Edge ST8 for Designers, 13th Edition

NX Textbooks

- Siemens NX 12.0 for Designers, 11th Edition
- NX 11.0 for Designers, 10th Edition

NX Nastran Textbook

- NX Nastran 9.0 for Designers

SolidWorks Textbooks

- SOLIDWORKS 2018 for Designers, 16th Edition
- SOLIDWORKS 2017 for Designers, 15th Edition
- SolidWorks 2014: A Tutorial Approach
- SolidWorks 2012: A Tutorial Approach
- Learning SolidWorks 2011: A Project Based Approach

CATIA Textbooks

- CATIA V5-6R2017 for Designers, 15th Edition
- CATIA V5-6R2016 for Designers, 14th Edition
- CATIA V5-6R2015 for Designers, 13th Edition

Creo Parametric and Pro/ENGINEER Textbooks
- Creo Parametric 4.0 for Designers, 4th Edition
- PTC Creo Parametric 3.0 for Designers, 3rd Edition

ANSYS Textbooks
- ANSYS Workbench 14.0: A Tutorial Approach
- ANSYS 11.0 for Designers

Creo Direct Textbook
- Creo Direct 2.0 and Beyond for Designers

Autodesk Alias Textbooks
- Learning Autodesk Alias Design 2016, 5th Edition
- Learning Autodesk Alias Design 2015, 4th Edition

AutoCAD LT Textbooks
- AutoCAD LT 2017 for Designers, 12th Edition
- AutoCAD LT 2016 for Designers, 11th Edition

EdgeCAM Textbooks
- EdgeCAM 11.0 for Manufacturers
- EdgeCAM 10.0 for Manufacturers

AutoCAD Electrical Textbooks
- AutoCAD Electrical 2018 for Electrical Control Designers, 9th Edition
- AutoCAD Electrical 2017 for Electrical Control Designers, 8th Edition

Autodesk Revit Architecture Textbooks
- Exploring Autodesk Revit 2018 for Architecture, 14th Edition
- Exploring Autodesk Revit 2017 for Architecture, 13th Edition

Autodesk Revit Structure Textbooks
- Exploring Autodesk Revit 2018 for Structure, 8th Edition
- Exploring Autodesk Revit 2017 for Structure, 7th Edition

AutoCAD Civil 3D Textbooks
- Exploring AutoCAD Civil 3D 2018, 8th Edition
- Exploring AutoCAD Civil 3D 2017, 7th Edition

AutoCAD Map 3D Textbooks
- Exploring AutoCAD Map 3D 2018, 8th Edition
- Exploring AutoCAD Map 3D 2017, 7th Edition
- Exploring AutoCAD Map 3D 2016, 6th Edition

3ds Max Design Textbooks
- Autodesk 3ds Max Design 2015: A Tutorial Approach, 15th Edition
- Autodesk 3ds Max Design 2014: A Tutorial Approach

3ds Max Textbooks

- Autodesk 3ds Max 2018 for Beginners: A Tutorial Approach, 18th Edition
- Autodesk 3ds Max 2018: A Comprehensive Guide, 18th Edition
- Autodesk 3ds Max 2017 for Beginners: A Tutorial Approach, 17th Edition

Autodesk Maya Textbooks

- Autodesk Maya 2018: A Comprehensive Guide, 10th Edition
- Autodesk Maya 2017: A Comprehensive Guide, 9th Edition
- Character Animation: A Tutorial Approach

ZBrush Textbooks

- Pixologic ZBrush 4R7: A Comprehensive Guide
- Pixologic ZBrush 4R6: A Comprehensive Guide

Fusion Textbooks

- Blackmagic Design Fusion 7 Studio: A Tutorial Approach
- The eyeon Fusion 6.3: A Tutorial Approach

Flash Textbooks

- Adobe Flash Professional CC2015: A Tutorial Approach
- Adobe Flash Professional CC: A Tutorial Approach

Computer Programming Textbooks

- Introduction to C++ programming
- Learning Oracle 12C- APL/SQL Approach
- Learning ASP.NET AJAX
- Introduction to Java Programming
- Learning Java Programming
- Learning Visual Basic.NET 2008

AutoCAD Textbooks Authored by Prof. Sham Tickoo and Published by Autodesk Press

- AutoCAD: A Problem-Solving Approach: 2013 and Beyond
- AutoCAD 2012: A Problem-Solving Approach
- AutoCAD 2011: A Problem-Solving Approach

Textbooks Authored by CADCIM Technologies and Published by Other Publishers

3D Studio MAX and VIZ Textbooks

- Learning 3DS Max: A Tutorial Approach, Release 4
  Goodheart-Wilcox Publishers (USA)
- Learning 3D Studio VIZ: A Tutorial Approach
  Goodheart-Wilcox Publishers (USA)

CADCIM Technologies Textbooks Translated in Other Languages

SolidWorks Textbooks

- SolidWorks 2008 for Designers (Serbian Edition)
  Mikro Knjiga Publishing Company, Serbia
- SolidWorks 2006 for Designers (Russian Edition)
  Piter Publishing Press, Russia

NX Textbooks

- NX 6 for Designers (Korean Edition)
  Onsolutions, South Korea

Pro/ENGINEER Textbooks

- Pro/ENGINEER Wildfire 4.0 for Designers (Korean Edition)
  HongReung Science Publishing Company, South Korea
- Pro/ENGINEER Wildfire 3.0 for Designers (Korean Edition)
  HongReung Science Publishing Company, South Korea

Autodesk 3ds Max Textbook

- 3ds Max 2008: A Comprehensive Guide (Serbian Edition)
  Mikro Knjiga Publishing Company, Serbia

AutoCAD Textbooks

- AutoCAD 2006 (Russian Edition)
  Piter Publishing Press, Russia
- AutoCAD 2005 (Russian Edition)
  Piter Publishing Press, Russia
- AutoCAD 2000 Fondamenti (Italian Edition)

Coming Soon from CADCIM Technologies

- SolidCAM: A Tutorial Approach

Made in United States
Orlando, FL
04 November 2026

90202169R00168